EMERGENCE

Emergence

Towards a New Metaphysics
and Philosophy of Science

MARIUSZ TABACZEK

University of Notre Dame Press

Notre Dame, Indiana

Copyright © 2019 by the University of Notre Dame
University of Notre Dame Press
Notre Dame, Indiana 46556
www.undpress.nd.edu

All Rights Reserved

Published in the United States of America

Library of Congress Cataloging-in-Publication Data

Names: Tabaczek, Mariusz, 1980– author.
Title: Emergence : towards a new metaphysics and philosophy of science / Mariusz Robert Tabaczek.
Description: Notre Dame : University of Notre Dame Press, 2019. | Includes bibliographical references and index. |
Identifiers: LCCN 2019011970 (print) | LCCN 2019013532 (ebook) | ISBN 9780268104993 (pdf) | ISBN 9780268105006 (epub) | ISBN 9780268104979 (hardback : alk. paper) | ISBN 0268104972 (hardback : alk. paper)
Subjects: LCSH: Complexity (Philosophy) | Emergence (Philosophy) | Causation. | Metaphysics. | Science—Philosophy.
Classification: LCC B105.C473 (ebook) | LCC B105.C473 T33 2019 (print) | DDC 110—dc23
LC record available at https://lccn.loc.gov/2019011970

∞ *This book is printed on acid-free paper.*

*To Michael J. Dodds, O.P.
for all he has taught me*

See skulking Truth *to her old Cavern fled,*
Mountains of Casuistry heap'd o'er her head!
Philosophy, *that lean'd on Heav'n before,*
Shrinks to her second cause, and is no more.
Physic *of* Metaphysic *begs defence,*
And Metaphysic *calls for aid on* Sense*!*
See Mystery *to* Mathematics *fly!*
In vain! they gaze, turn giddy, rave, and die.
—Alexander Pope, *The Dunciad* (1743)

Hope springs eternal in the human breast;
Man never Is, but always To be blest.
The soul, uneasy, and confin'd from home,
Rests and expatiates in a life to come.
—Alexander Pope, *Essay on Man* (1733–34)

CONTENTS

List of Figures	xi
List of Tables	xiii
Preface	xv
Acknowledgments	xix
Abbreviations	xxi
Introduction: Causation in Philosophy and Scientific Explanation	1

PART 1   Metaphysical Aspects of Emergence and Downward Causation — 41

CHAPTER 1   The Central Dogma of Emergentism — 45

CHAPTER 2   Metaphysical Challenges for Emergence and Downward Causation — 63

CHAPTER 3   Dynamical Depth and Causal Nonreductionism — 99

PART 2   Dispositions/Powers Metaphysics and Emergence — 135

CHAPTER 4   Theories of Causation in Analytic Metaphysics — 139

CHAPTER 5   Dispositional Metaphysics and the Corresponding View of Causation — 181

CHAPTER 6	From Powers to Forms and Teleology: New Aristotelianism	201
CHAPTER 7	Dispositional Metaphysics, Downward Causation, and Dynamical Depth	247
	Conclusion	271
APPENDIX 1	Potency and Act	275
APPENDIX 2	Process Metaphysics of Emergence	279
	Notes	285
	Bibliography	349
	Index	369

FIGURES

FIGURE 2.1.	An example of the collapse of DC into physical causation, according to Jaegwon Kim	80
FIGURE 2.2.	Examples of feedback causal loops according to Scott	82
FIGURE 2.3.	O'Connor and Wong's dynamical EM model of the evolution of system S over time	84
FIGURE 2.4.	The final step of Kim's causal exclusion argument	85
FIGURE 2.5.	Macdonald and Macdonald's refutation of Kim's causal exclusion argument	86
FIGURE 3.1.	Two general classes of autocells depicted as geometric constructions	110
FIGURE 4.1.	A neuron-firing network depicting Menzies's theoretical example of causes decreasing probability	153
FIGURE 4.2.	Lac operon in *Escherichia coli* as an example of double prevention	166
FIGURE 4.3.	Interactive fork	170
FIGURE 4.4.	Conjunctive fork	171
FIGURE 5.1.	Classification of the major positions among dispositionalists concerning the relation between dispositional and categorical properties	186
FIGURE 5.2.	Neuron causal diagrams	194
FIGURE 5.3.	Vector modeling of causes	196

FIGURE 5.4.	Composition of causes and vectors addition	197
FIGURE 5.5.	Multidimensional causal vector diagrams	198
FIGURE 6.1.	The process of NaCl dissolving in water	212
FIGURE 7.1.	Vector model of the reciprocity and simultaneity of causes in autogenesis	262
FIGURE 7.2.	Vector model depicting "causality of absences"	267
FIGURE A.1.	Various kinds of potency	276
FIGURE A.2.	Various kinds of act	277

TABLES

TABLE 2.1.	Various interpretations of the causal factor in DC grouped in four categories	74
TABLE 2.2.	Various interpretations of the object of DC grouped in four categories	75
TABLE 4.1.	Strengths and weaknesses of the six main views of causation in analytic metaphysics	176

PREFACE

Emergence is an extended and reconsidered version of the two central parts of a large interdisciplinary research project I pursued during the five years of my doctoral studies at the Graduate Theological Union in Berkeley, California (2011–16). The original project places the metaphysics of emergence and downward causation within a more elaborate and thorough account of the history of causal explanation in philosophy of nature and in natural science, which is only briefly summarized in the introduction. It also explores the panentheistic theory of divine action through emergent processes in nature, offering an alternative theological interpretation of emergence in terms of the new Aristotelianism and the Thomistic concept of the concurrent action of God in the world. This new analysis of the theological implications of emergence is a subject of a separate and upcoming book project of mine.

The theory of emergence in its contemporary version and the concept of downward causation that is related to it were both developed in a very specific context of the recent debate on nonreductionism in metaphysics and philosophy of science. This debate, in turn, is deeply related to the origins of the systems approach to biology, ecology, climatology, psychology, medicine, economy, engineering, and other branches of natural science. Moreover, the theory of downward causation is even more specifically related to the research on the character and typology of cause-effect dependencies in nature. This topic has recently become an object of a hot debate, leading to a number of important publications. Some of them take the interdisciplinary approach, trying to bring metaphysics and philosophy of nature into a conversation with the methodology of natural science and metascientific reflections on the outcomes of empirical research.

As the new approach to the analysis of dynamical systems progressed, it became more and more apparent that the modern general conceptual commitment to explaining causation in terms of mechanistically understood efficient causes alone is inherently limited and insufficient. What is more, Hume's skepticism about the reality of ontological relations and dependency between causal relata—together with his definition of causation in terms of atomic events conjoined in the mind of the observer—became even more irrelevant in the context of the deep conviction of many contemporary scientists about the intrinsic and multilevel interrelatedness of parts building dynamical nonliving and living wholes. In addition to the analysis of the emergence of those wholes, contemporary scientists are also investigating ways in which the participant mereological substrates of wholes are changed due to their participation in larger dynamical systems.

The acknowledgment of the need to extend our notion of causation leaves contemporary scientists and philosophers of science with the question concerning the most suitable way of proceeding in such revision. Many of them think that what we need is a plurality of ways and approaches that will jointly contribute to a reductionist account of ontologically real situations of multifaceted causal dependencies in nature. In the context of numerous difficulties challenging all causal theories proposed within this project—pursued in analytic metaphysics—it has been suggested that we should return to Aristotle's fourfold division of causes as the most fitting metaphysical response and background of the new science of complex systems. This is the argument brought up by Claus Emmeche, Simo Køppe, and Frederic Stjernfelt; Michael Silberstein; Charbel Niño El-Hani and Antonio Marcos Pereira; Alvaro Moreno and Jon Umerez; and others, which we will discuss in the main text. As we shall see, their suggestion finds a response in analytic metaphysics, as one of its branches of research rediscovers the importance of the distinction between potentiality and actuality, expressed in the realization of dispositions (powers) of entities (causal relata) to act and react in a specific way. The main goal of this book is to proceed from the preliminary suggestion of the thinkers mentioned here and offer a constructive proposal of the reinterpretation of emergence in terms of both the old and the new (analytic) versions of Aristotelianism.

Having specified the goal of *Emergence*, I must briefly describe some aspects of its content and structure. The story about the origin of Aristotle's metaphysics and the history of its reception, and the history of his theory of four causes in particular, are both long and complicated. I cannot offer a complete and comprehensive critical account of their development here. At the same time, my analysis of the metaphysical aspects of emergentism and of the reinterpretation of emergentism in terms of the old and the new Aristotelianism would be lacking if it failed to refer to, or to try to place them within, the context of the whole history of causation in philosophy and scientific explanation. In an attempt to forge a middle way, I will offer in the extended introduction a short overview and evaluation of some major turning points in this history. I will naturally spend more time presenting the origin and the definition of Aristotle's theory of four causes and its further qualification in the teaching of Aquinas. I will describe the causal reductionism in modern science and philosophy rather briefly, but still informatively. Moving to causal nonreductionism in the most recent philosophy of science, I will spend some more time analyzing and evaluating the new mechanical approach in philosophy of biology, which seems to question the plausibility of emergentism.

Hence, even if not exhaustive, the reflection offered in the opening of the book should help readers contextualize the research presented in it and better understand both the objectives and the plan of the project, which I outline in the closing section of the introduction. At the same time, the opening overview of the crucial historical aspects of the causality debate will find its logical continuation in chapter 4, which begins the second part of the project. In this chapter, I invite readers to accompany me in a more detailed investigation of the six main theories of causation developed in analytic metaphysics. I present this analysis as a necessary introduction to the new Aristotelianism of dispositional metaphysics and its view of causation, as well as to my constructive proposal of their application in emergence theory.

Finally, the book analyzes metaphysical puzzles of both the classical version of the contemporary account of emergence—defined in terms of downward causation—and its more recent version, the theory of dynamical depth, proposed by Terrence Deacon from the University of California in Berkeley. As I have said already, my objective goal is to propose a new metaphysical ground for both accounts of emergentism, which I find in the classical and the new Aristotelianism. I leave it to readers to decide whether my endeavor succeeds. At the same time, I acknowledge that the debate on metaphysics of emergence is open. I hope to continue my participation in it in the future.

ACKNOWLEDGMENTS

I would like to thank Fr. Michael Dodds, O.P., my teacher and mentor, for sharing his academic resources and expertise and for all his encouragement, support, sense of humor, patience, and fraternal friendship. Many thanks to Terrence Deacon, whose dynamical depth model of emergence is one of the central topics of this book. My presentation of Terrence's project would not be as accurate without our long conversations, his invitation to become a friend of the "Pirates," and his giving me the opportunity to study with him and his fellow researchers at the University of California in Berkeley. Many thanks to Robert Russell, the director of the CTNS in Berkeley, for our helpful discussions and for his insights, encouragement, and friendship. Thanks to Marga Vega from the Dominican School of Philosophy and Theology in Berkeley for her expertise in contemporary analytic metaphysics, and to Paul Skokowski from Stanford University for triggering my interest in philosophy of science. I would also like to thank the Dominican Community at the Priory of St. Albert the Great in Oakland, California, for their hospitality during the five years of my research, studies, and writing. Many thanks to all whom I was honored to serve as a priest, all those who supported me spiritually and financially during this time, and to my family for their always faithful love and presence. Thanks to my former and present Provincials, Fr. Krzysztof Popławski, O.P., and Fr. Paweł Kozacki, O.P., and to the former regent of studies of the Polish Dominican Province of St. Hyacinth, Fr. Michał Paluch, O.P., for their support in allowing me to study at the GTU in Berkeley. Thanks to the organizers and the participants of the 2016 Classical Theism Workshop founded by the Templeton Foundation at the University of St. Thomas in St. Paul, Minnesota, as well as the directors, seminar guest professors, and 2016/2017 fellows of the Notre Dame Institute for Advanced Study, who discussed with me some parts of this project at the final stage of its realization. I also received many valuable comments and insights from the reviewers Robert Koons and Ignacio Silva, as well as other fellow philosophers and friends, including Therese and David Cory, Sean Costello, Gloria Frost, David Bentley Hart, Christopher Shields, and Carol Trabing. I greatly benefited from reading the works of Sophie R. Allen; Helen Beebee, Christopher Hitchcock, and Peter Charles Menzies; Edward Feser; John Losee; Michael J. Loux; Robert C. Koons and Timothy H. Pickavance; Alyssa Nay; David S. Oderberg; Paul Humphreys; William Wallace, O.P.; and many others. I am grateful to all of them. Thanks to Fr. Reginald Martin, O.P., Emily Zion, and Bob Banning for helping me to polish my English. Thanks to NDIAS for sponsoring a professional indexer. And last, but not least, thanks to Stephen Little, Matthew Dowd, and their coworkers from the University of Notre Dame Press for their patient work on my book.

ABBREVIATIONS

ABBREVIATIONS FOR THE WORKS OF ARISTOTLE

De an.	*De anima* (*On the Soul*)
De cae.	*De caelo* (*On the Heavens*)
De gen. an.	*De generatione animalium* (*On the Generation of Animals*)
De gen. et corr.	*De generatione et corruptione* (*On Generation and Corruption*)
De part. an.	*De partibus animalium* (*On the Parts of Animals*)
Meta.	*Metaphysica* (*Metaphysics*)
Meteo.	*Meteorologica* (*Meteorology*)
Phys.	*Physica* (*Physics*)

ABBREVIATIONS FOR THE WORKS OF ST. THOMAS AQUINAS

De ente	*De ente et essentia*
De mixt. elem.	*De mixtione elementorum ad magistrum Philippum de Castro*
De prin. nat.	*De principiis naturae*
In De an.	*In Aristotelis librum De anima commentarium*
In Meta.	*In Metaphysicam Aristotelis commentaria*
In Phys.	*In octo libros Physicorum Aristotelis expositio*
In II Sent.	*Scriptum super libros Sententiarum Magistri Petri Lombardi Episcopi Parisiensis*
Q. de ver.	*Quaestiones disputatae de veritate*
SCG	*Summa contra gentiles*
ST	*Summa theologiae*

OTHER ABBREVIATIONS

ArelVC	action-related view of causation
AVC	agency view of causation
CVC	counterfactual view of causation
DC	downward causation
DVC	dispositional view of causation
EM	emergence
InfVC	inferability view of causation
IntVC	interventionist view of causation
INUSConVC	INUS-conditions view of causation
M-basedVsC	manipulability-based views of causation
MDC	medium downward causation
MVC	manipulation view of causation
NP	nonreductionist physicalism
ProbVC	probability view of causation
ProcVC	process view of causation
QFT	quantum field theory
RVC	regulatory view of causation
SDC	strong downward causation
SUP	supervenience
SVC	singularist view of causation
UAC	unity by *alteratio componentium* (through alteration of components)
USCC	unity *secundum contactum et colligationem* (through contact and bond)
USO	unity *secundum ordinem* (through arrangement of parts)
WDC	weak downward causation

Introduction

Causation in Philosophy and Scientific Explanation

> *Nature is a source or cause of being moved and of being at rest in that to which it belongs primarily, in virtue of itself and not in virtue of a concomitant attribute.*
>
> —Aristotle, *Physics* 2.1.192b22–23

The theory of emergence (EM) and downward causation (DC) has become over the last few decades probably the most popular conceptual tool in scientific and philosophical attempts at explaining the nature and character of global organization observed in various biological phenomena, ranging from individual cell organization, to neural networks, to physiological cycles, to stability of an organism, to collective behavior, to ecological systems. However, a closer analysis of emergentism shows that it faces a number of philosophical problems. The main goal of the research presented here is to analyze and answer these problems, offering a constructive proposal of a new metaphysical foundation for the theory of EM. First, however, we need to realize that the concept of DC, which is both the source of the ontological weight of emergentism and the source of its insurmountable difficulties, is necessarily related to the broader topic of causal explanation in philosophy and natural sciences. Thus, it seems most appropriate to begin our journey with a short introductory overview and critical evaluation of some major turning points of this fascinating interdisciplinary inquiry, which shapes the foundations of Western philosophy, empirical science, and technology.[1]

1. CAUSATION AND THE ORIGINS OF ANCIENT PHILOSOPHY

When reflecting on the most basic human experiences, we naturally and spontaneously use verbs such as "to do," "to act," "to make," "to receive," "to undergo," and so on. Moreover, the most basic questions we ask about various circumstances and changes taking place around us are the questions of how and why things happen. Hence, it is not surprising that these questions stand at the very beginning of philosophy, which enabled man to raise his reflection to the metalevel of description, asking about causes of things and dynamical systems. This ascent of human reflection is mirrored in the basic terminology characteristic of the philosophy of causation. Although in classical Greek the first word for cause, occurring as early as Homer, is ἀρχή (*archē*), which translates as "beginning" or "point of departure," the more popular term that comes into use in the fifth century BC is αἰτία (*aitia*), which has its origin in human action and moral responsibility and is related to terms such as "blame" or "accusation." Thus, we can see that the philosophical meaning of cause was transferred from the experience of human action, and for this reason it still finds in experience an important analogue and example.[2]

1.1. Early Milesians

Greek reflection on causality began in the philosophy of the early Milesians, who presented three positions on the problem of change in nature.[3] According to the first and predominant option, some things change (totally or partially), while other things do not. Led to this conclusion by their observations, some philosophers looked for a fundamental principle which does not change and remains constant (a principle persisting through change). Thus, we can say they laid the foundation for reflection on material cause.[4] They understood it in several different ways. Thales (620–550 BC) claims that the fundamental constituent that remains unchanged is water. It can exist as solid (e.g., nutriment of all things is moist), liquid, or vapor (e.g., heat is generated from and kept alive by the moist). Hence, all things are water in various states.[5] Anaximenes (570–500 BC) takes air for the basic stuff and says everything is really air in different forms of rarity or density.⁶ Anaximander (610–525 BC) takes an important step by saying that none of the basic elements (earth, air, fire, water) can be the first principle, for it would destroy other elements. He enters a metalevel of philosophical reflection, arguing that the undetermined principle is most fundamental and may be called ἄπειρον (*apeiron*, the infinite). It is by the joining and separating of opposite qualities present in "the infinite" that all things come into being and change.⁷

The second view on change among early Greek philosophers is represented by Heraclitus (fl. 500 BC), who assumes all things are constantly changing. For him, things are simply ephemeral patterns of continuity in the perpetual flux of the world, which is made of changing fire.⁸ The third view, represented by Parmenides (fl. 500 BC), stands opposed to that of Heraclitus, since it assumes that nothing really changes in nature. Things either "are" or "are not"; there is no other option, and thus nothing can change. For a new thing could come into being only from nonbeing (which is impossible) or from being (which would mean that it existed before). Therefore, change is an illusion.⁹

The conceptual level of Anaximander's reflection (as distinct from the more empirically grounded philosophy of other Milesians) continues in Anaxagoras (ca. 510–428 BC), who declares cosmic intelligence (mind, νοῦς, *nous*) to be what causes everything. But since intelligence seeks what it values, this kind of explanation exceeds the interest in the material cause predominant among Anaxagoras's predecessors and enters a new level of explanation, teleological in style. Hence, we can find here the first primitive notion of final causation (causal explanation of directionality toward an end/goal).¹⁰ However, Plato (Socrates in *Phaedo*) claims that the details of Anaxagoras's reflection show he was not aware of this commitment, as he concentrates mainly on materialistic factors such as air, water, and *vortex* (δίνη, *dinē*).¹¹

Empedocles (ca. 495–435 BC) further defines causality by presenting the first reflection on efficient cause. He claims that in addition to the elements of earth, air, fire, and water, two further elements, "love" and "strife," are needed to combine or keep apart the basic elements.¹² Later, Democritus (ca. 460–ca. 370 BC) and the atomists develop the idea of efficient cause, remaining opposed to the idea of a final cause of the universe and arguing in favor of the purely mechanical view of causation (movement and collisions of atoms).¹³

1.2. The School of Pythagoras and Plato

Among the earliest philosophical schools, we find one more, which remains at odds with those described so far. It is the school of Pythagoras (ca. 570–ca. 490 BC), who was the first to try to understand the cosmos and its phenomena in terms of number. His reflection was rooted in Egyptian geometry and Babylonian arithmetic. Pythagoras used these to develop a philosophy of nature based on mathematics. He suggested a parallelism between the idealizations of geometry and the physical patterns encountered in the universe. He argued that number underlies all physical objects. Its study reveals, therefore, a deeper level of reality than appears on the surface.[14] Aristotle would later say that Pythagoreans considered number to be the principle of both matter (even or "unlimited" elements of number) and form (uneven or "limited" elements of number).[15] Thus, every physical entity is explainable in terms of mathematics. From this basic point the Pythagoreans went on to a detailed analysis of mathematical proportions, harmonic relationships, and irrational numbers, which led to a study of continuum and infinite divisibility.

Pythagorean ideas, combined with Democritus's atomism, inspired Plato (b. 428/427 or 424/423; d. 348/347 BC) to develop his geometric theory of matter. The originality of his theory is based on his regarding matter ("the mother of all becoming") as a stable and eternal receptacle for Ideas or Forms, and this concept of matter introduces for the first time the notion of formal causation. He is also probably the first to state the principle of causality explicitly, as he does in *Timaeus*: "Again, everything which becomes must of necessity (ἐξ ἀνάγκης [*ex anangkēs*]) become owing to some cause (ὑπ' αἰτίου [*hup' aitiou*]); for without a cause it is impossible for anything to attain becoming."[16] For Plato, the world accessible to our sensual experience is changing and made of appearances. It is merely a shadow of the true reality, which is the world of autonomous and immaterial Forms. If formless matter is "the mother," eternal form is "the father," and the transitory phenomena available to our senses are their offspring. Plato construed the elements of the physical world on the model of geometrical patterns: fire takes the shape of a tetrahedron, air of an octahedron, water of an icosahedron, and earth of a cube. Moreover, all these shapes are made of multiplied triangles. Therefore, the triangle can be regarded as an atomic element in the Democritean sense.[17]

Wallace emphasizes that what remains crucial for our reflection is that neither Pythagoras nor Plato regarded mathematics (as did Aristotle) as a formal and abstract discipline that, applied to reality, enables us to construe a subalternated type of scientific knowing, abstracted from empirical and sensual experience.[18] They saw it, rather, as a way of arriving at the ultimate and basic reality itself, which in sensual experience is given only in ephemeral appearances.[19]

2. ARISTOTLE AND THE FOURFOLD NOTION OF CAUSATION

Of all philosophers in ancient Greece, Aristotle (384–322 BC) presents the most developed theory of causation. Regarded by many as the first scientist, he not only points out the connection between causation and scientific explanation but also systematizes and further develops the teaching of his predecessors in philosophy on the nature of causation. He shows that the pioneers of the theory of causality failed to distinguish more precisely between the two basic levels of inquiry: the empirical (practical), where causes are perceivable in sensual observation and experience, and the philosophical (conceptual), where the questions concerning causal descriptions have metaphysical and epistemological character. Having this distinction in mind, Aristotle sees causes as ways of rationally explaining stability and change in nature. He proposes a clear division of four types of causality.

2.1. Material and Formal Causes

Criticizing Plato for emphasizing the importance of form and its transcendent character, which not only results in a certain kind of dualism and devaluation of material reality but also makes it difficult for his philosophy to explain change and stability (becoming requires participating in various forms), Aristotle proposes a doctrine of matter and form understood as two types of causality closely related to each other. He thus defines them, respectively, in the *Physics* and the *Metaphysics*:

In one sense, then, (1) that out of which a thing comes to be and which persists, is called "cause," e.g. the bronze of the statue, the silver of the bowl, and the genera of which the bronze and the silver are species. In another sense (2) the form or the archetype, i.e. the statement of the essence, and its genera, are called "causes" (e.g. of the octave the relation of 2:1, and generally number), and the parts in the definition.[20]

"Cause" means (1) that from which, as immanent material, a thing comes into being, e.g. the bronze is the cause of the statue and the silver of the saucer, and so are the classes which include these. (2) The form or pattern, i.e. the definition of the essence, and the classes which include this (e.g. the ratio 2:1 and number in general are causes of the octave), and the parts included in the definition.[21]

What is crucial regarding the material cause is its not being reducible to basic chunks of stuff out of which things are made.[22] Although one may find it difficult to grasp in the oft-cited quotations from the *Physics* and the *Metaphysics*,[23] for Aristotle matter is the most basic principle of potentiality—that is, primary matter (πρώτη ὕλη, *prōtē hylē*), underlying nature (ὑποκείμενον φύσις, *hypokeimenon physis*), or primary substratum (πρῶτον ὑποκείμενον, *prōton hypokeimenon*)—which persists through all changes that a given substance can be exposed to. As something that constitutes the very possibility of being a substance, it is called primary matter and should be distinguished from secondary (proximate) matter, which is perceptible to our senses and quantifiable. This becomes clear in the following passages from the *Physics* and the *Metaphysics*:

> The underlying nature [ὑποκείμενον φύσις, *hypokeimenon physis*] is an object of scientific knowledge, by an analogy. For as the bronze is to the statue, the wood to the bed, or the matter and the formless before receiving form to any thing which has form, so is the underlying nature to substance, i.e. the "this" or existent.[24]

> The matter comes to be and ceases to be in one sense, while in another it does not. As that which contains the privation, it ceases to be in its own nature, for what ceases to be—the privation—is

contained within it. But as potentiality it does not cease to be in its own nature, but is necessarily outside the sphere of becoming and ceasing to be. . . . For my definition of matter is just this—the primary substratum [πρῶτον ὑποκείμενον] of each thing, from which it comes to be without qualification, and which persists in the result.[25]

By matter I mean that which in itself is neither a particular thing nor of a certain quantity nor assigned to any other of the categories by which being is determined. . . . The ultimate substratum is of itself neither a particular thing nor of a particular quantity nor otherwise positively characterized; nor yet is it the negations of these, for negations also will belong to it only by accident.[26]

And if there is a first thing, which is no longer, in reference to something else, called "thaten," this is prime matter.[27]

Furthermore, in the second book of his *De generatione et corruptione*, after analyzing the reciprocal transformation of the elements (fire, air, water, and earth), Aristotle emphasizes that there is no first element that would underlie all other elements. Consequently, in case of any two (or more) elements taken together, there will be "something else which is the same; i.e. there will be some 'matter,' other than either, common to both."[28] In reference to contrarieties (contrary qualities) characterizing elements and their transformations, Aristotle notes: "Assuming, then, that the contrariety, in respect to which they are transformed, is one, the 'elements' will inevitably be two: for it is 'matter' that is the 'mean' between the two contraries, and matter is imperceptible and inseparable from them."[29] The latter assertion suggests that primary matter is real, that it grounds all four elements, and that it is crucial for their transformations.

It is not easy to grasp the idea of primary matter, since existence is not properly predicated directly of primary matter, but of substance (the compound of primary matter and substantial form).[30] And yet primary matter does exist, even if not with its own independent act of existence, but with existence of the substance. We can say primary matter underlies each and every substance, remaining a principle of continuity in the process in which one substance (S_1) becomes another substance (S_2). Even if all physical aspects of S_1 change on the way to its becoming S_2, we are not dealing with a total annihilation of S_1 and coming to be out of nothing of S_2. Rather, due to primary matter as principle of potentiality underlying all existing substances, we observe the continuity of the process of S_1 changing into S_2. Moreover, it is due to primary matter that both S_1 and S_2 are characterized by the persistent passive potentiality for change, which is actualized by substantial form.[31]

Concerning formal cause, Aristotle radically opposes the transcendental character of Ideas in Plato. For Aristotle, forms must be in things, determining their actuality. This becomes clear from the quotations from the *Physics* and the *Metaphysics* cited above. In both passages, Aristotle, speaking of formal causality, uses the term ὁ λόγος τοῦ τί ἦν εἶναι (*ho logos tou ti ēn einai*), which Hardie and Gaye translate as "the statement of the essence," and Ross as "the definition of the essence."[32] In reference to the meaning of λόγος (*logos*), understood in this context as "ground," we can tell that form is for Aristotle a principle of each existing substance that grounds it—that is, makes it to be the particular kind of thing it is—a principle actualizing (determining) a pure possibility-of-being (primary matter) to be a concrete substance. As such, similar to primary matter, form is a simple metaphysical principle (not a thing) which does not have the property of quantity or extension. For this reason, says Dodds, "we cannot make an imaginative picture of a substantial form. It is not imaginable, but it is intelligible."[33] Form cannot increase or decrease. It is "educed" from the potentiality of primary matter and remains present in the entire substance and its parts as a fundamental principle of operation. It is expressed in essential qualities of a given substance, which classifies Aristotelian ontology as essentialist. In other words, form can be taken as an essence—that is, "the property of x that makes x the same subject in different predications—the property that x must retain to remain in existence."[34] It embodies the criterion of identity appropriate to a given entity and can be characterized as the ratio or proportion between the different elements that go through a substantial change in the process of that entity's formation. As such, it becomes an inner cause of a given entity's natural behavior.

However, in his definition of formal cause Aristotle uses at least three other terms: παράδειγμα (*paradeigma*), μορφή (*morphē*), and εἶδος (*eidos*). The first one is usually translated as "archetype" (Hardie and

Gaye), "pattern" (Ross, Tredennick, and Kirwan), "model" (Charlton), or "characteristics of the type" (Wicksteed and Cornford), which seems to be in line with the understanding of form as the metaphysical principle of actualization. The second term typically translates as "shape," while the third—in most cases interpreted simply as "form"—has its primary nonphilosophical meaning defined as "outward appearance" or "visible form."[35] Defining form as "shape" or "outward appearance" may bring confusion and reduction of this metaphysically robust principle to a geometrical shape (available for sensory perception and mathematical description), which flattens out Aristotle's original idea.[36] Trying to avoid this error, Irwin rightly notes that "if the form of the statue is essential to it, then other features besides shape must constitute the form, and the reference to shape can at most give us a very rough first conception of form. If we turn from artifacts to organisms, it is even clearer that form cannot be just the same as shape."[37]

In addition to all these terms, we find Aristotle describing form, on several occasions, in terms of ἐντελέχεια (*entelecheia*), which relates formal to final causation (anticipating our analysis of it) and denotes form as actualized in the final state of a being.[38] Commenting on this term, O'Rourke says, "It is form (μορφή), therefore, which is nature (φύσις [*physis*]). It is form as ἐντελέχεια which is the τέλος [*telos*] of γένεσις [*genesis*], that is, of the coming-to-be of φύσις. In its state of completion, φύσις is synonymous with ἐντελέχεια, the fulfillment of εἶδος."[39]

Contrary to primary matter, which is a principle of continuity and a passive principle of change (as pure potentiality), form is a principle of novelty and an active principle of change in causal processes. Hence, even if in a process of change from S_1 to S_2 primary matter does not change, we distinguish S_1 and S_2 as separate substances due to different forms that in-form primary matter in them and are educed from its potentiality. But what if S_1 changes in a way that makes it different but does not lead to its transformation into a completely new substance S_2 (e.g., a puppy growing up and becoming a mature dog)? Here, Aristotle introduces an important distinction between what was later on classified as substantial form and accidental form. The distinction between these two types of form is best explained in his example of the two types of changes in nature. In *On Generation and Corruption* Aristotle first mentions "alteration" (an accidental change), which occurs when a thing or being changes in its properties while remaining the same substance: "The body, e.g., although persisting as the same body, is now healthy and now ill; and the bronze is now spherical and at another time angular, and yet remains the same bronze." He then contrasts it with a situation "when nothing perceptible persists in its identity as a substratum, and the thing changes as a whole." He calls the latter "a coming-to-be of one substance and a passing-away of the other" (a substantial change)—for example, a wooden plank burning into ashes.[40] Thus, in our example of a puppy growing up and becoming a mature dog, what really changes is a number of accidental formal features of its organism (the size of its bones and muscles, secretion of hormones, its vocal chords, etc.), while its substantial form of a particular dog remains the same. It is only at the moment of its death that—due to substantial change—it will cease to be a dog (S_1) and will turn into a carcass (S_2).

2.2. Efficient and Final Causes

The complete explanation of the fact that things are what they are, keep their identity while going through accidental changes, and change into something else on entering into a process of substantial change requires an introduction of two other types of causality. Aristotle says that we need to know (a) what the source of change and rest is and (b) what is the end of this activity. He defines efficient and final causality in the *Physics* thus:

> Again (3) the primary source of the change or coming to rest; e.g. the man who gave advice is a cause, the father is cause of the child, and generally what makes of what is made and what causes change of what is changed.
>
> Again (4) in the sense of end or "that for the sake of which" a thing is done, e.g. health is the cause of walking about. ("Why is he walking about?" we say. "To be healthy," and, having said that, we think we have assigned the cause.) The same is true also of all the intermediate steps which are brought about through the action of something else as means towards the end, e.g. reduction of flesh, purging, drugs, or surgical instruments are means towards health. All these things are "for the sake of" the end, though they differ from one another in that some are activities, others instruments.[41]

Efficient cause, defined simply as an activity of an agent that is a source of change or coming to rest, seems to be the least problematic from the point of view of contemporary science. But it should not escape our attention that, first, Aristotle sees it always as "the fulfillment of . . . potentiality . . . by the action of that which has the power of causing motion."[42] As such, efficient cause is necessarily related to intrinsic features (natures) of entities that enter into a given causal relation. In other words, the possibility of an agent to act, and of a patient to be acted upon, is in each one of them a property dependent on substantial form. This fact helps us understand that the contemporary processual descriptions of efficient causation in terms of causal events remain incomplete without a reference to natures of its participants, analyzed in terms of actuality and potentiality.

Second, it is important to realize that Aristotle develops the theory of action and reaction in the context of efficient (physical) causation. We find him saying in *On the Generation of Animals* 4.3.768b16–20 that "that which cuts is blunted by that which is cut by it, that which heats is cooled by that which is heated by it, and in general the moving cause (except in the case of the first cause of all) does itself receive some motion in return; e.g. what pushes is itself in a way pushed again and what crushes is itself crushed again." Obviously, then, the action is not all on the side of one causal relatum and the passion on the side of the other. Rather, efficient causation is an interplay of efficient causality whereby the superior and more actively potential agent itself suffers from the resistant action or reaction of the patient. The only mover free from such re-agency or reciprocal action is the first mover. In the cases of all contingent entities which have "the same matter"—that is, are paired categorically—efficient action is accompanied by reaction. But in order for this to happen, potential causal relata need to meet some essential conditions or prerequisite and concomitant circumstances, such as (1) being in contact (immediate or mediate, i.e., virtual), (2) proximate disposition of potential causal relata to act or to be acted upon in a specific way (e.g., earth is not potentially a human being, while an ovum and a sperm are), and (3) absence of impediments.

Third, it should not escape our attention that in his account of efficient causality Aristotle makes an important distinction, which would be later on classified as the one between primary and secondary causation.

He alludes to it briefly in *Phys.* 2.2.194b13, saying that "man is begotten by man and by the sun as well." What we learn from this short assertion is that one efficient cause (sun) can exercise its action through another efficient cause (man). This observation will be further developed by Aquinas.

Finally, we should remember that for Aristotle efficient causation need not always be predicated in terms of some physical interaction and an exchange of energy. We find him saying that giving advice, which can hardly be associated with the realm of masses, forces, and physical action, is also an example of efficient causation.

Similar—that is, irreducible to the physical push-and-pull dynamics—is the character of the fourth, final cause, which Aristotle defines as "that for the sake of which" a thing is done, or a good that can be attained and that is proper for a being. It takes its other name, "teleology," from the Greek τέλος, which translates as "end" or "goal." Although he invokes necessity as an explanation of the availability of suitable matter, Aristotle acknowledges the need of an explanation in terms of purpose as a function of nature, to explain why given matter acquires the particular shape and structure it does.[43] It can be clearly seen in the case of animals, whose natures transmitted in the processes of generation function as causes, by way of being goals toward which animals develop, and which are their proper good.[44] This observation inspires Aristotle to suggest a general proposition stating that "generation is a process from something to something, from a principle [ἀρχή, *archē*] to a principle—from the primary efficient cause, which is something already endowed with a certain nature, to some definite form or similar end [τέλος]."[45]

This claim helps us realize that Aristotle extends teleology (goal directedness)—which is usually associated with conscious human decisions—to other living and nonliving entities. Indeed, as notes Bostock, in *Meteorology* 4.12.389b25–390a21, "Aristotle does explicitly say that the elements, and the inorganic compounds that are formed from them, are 'for the sake of something,' equating this with the view that they have a 'function' (ἔργον [*ergon*]) which in turn is a power (δύναμις [*dynamis*]) to act or be acted upon."[46] Most importantly, when predicated about inanimate and animate but unconscious nature, teleology should not be understood as a mysterious—quasi-efficient—cause, directing things according to a preestablished harmony. Quite the contrary, it should be seen as a natural tendency of things to realize what is proper to their nature (e.g., a tree blossoming and bearing fruit)—a tendency that does not have to be known or intended by a conscious agent. That is why Aristotle delineates in *Phys.* 2.8.199b26–27 that "it is absurd to suppose that purpose is not present because we do not observe the agent deliberating."[47] Obviously, an inorganic compound does not have a soul. But it does have a substantial form, which might fulfill the same task of holding elements and transforming them into a composite entity. Hence, by analogy, just as the soul of a living organism can be identified with its goal and actions that realize it, so the form of an inorganic compound has a similar task to perform.

Moreover, in this context, it becomes apparent that teleology has a normative import not only at the level of living organisms but also in the case of inanimate entities (objects). In other words, teleology is—as notes Mark Bedau—"value-centered," and this fact cannot be neutralized by either (1) relating it, in a Humean way, to minds and their designs (the mental approach) or (2) explaining it in terms of goal-directedness of natural systems (the systems approach) or (3) rationalizing it in reference to causal histories of teleological phenomena (the etiological approach). Bedau successfully shows that all three ways of striving to eliminate teleology's reference to value end up reorienting the conversation back to the value approach.[48] Naturally, the evaluative element of teleology thus understood—that is, understood as the proper good of a given entity—need not consist in any moral good. Neither does it have to confer an important or the best good. Quite the opposite, it is enough for it to refer to some good: "moral or non-moral or even immoral, important or insignificant, and intrinsic or merely instrumental."[49] The good in question is always related to what kind of thing a given entity is (its formal principle). Thus, even if properly classified and named in philosophical investigation performed by conscious beings, it is, nonetheless, real and intrinsic to the entity in question, independently of the operation of any human mind.

To better understand Aristotle's notion of teleology, Wallace distinguishes three different meanings of the word "end":

1. A terminus—that is, the moment when an action stops (e.g., a terminus of a natural fall of an object, or a chemical reaction reaching equilibrium).
2. A higher level of perfection of a nonliving or living entity reached in a natural process (e.g., a certain configuration of a chemical compound which makes it fitting for becoming a building block of an organic substance, or a higher survival rate of an organism, reached through the process of chance mutation and natural selection).
3. An intention or an aim (proper for cognitive agents: animals and humans).[50]

Note that only the third meaning of the word "end" is restricted to conscious beings, capable of articulating and realizing particular goods (including moral goods). The other two meanings refer to innumerable cases of inanimate entities and human artifacts and to the proper goods intrinsic to their natures. This shows once again the scope of Aristotle's understanding of teleology and his ability to recognize, describe, and defend the reality of teleological phenomena on each and every level of complexity of matter. Contemporary analyses of Aristotle's final cause which do not pay attention to this fact may lead to somewhat confusing interpretations, which are sometimes proposed in scientific or philosophical circles.[51]

2.3. Interrelatedness of Causes

Finally, after defining all four causes, Aristotle notes in *Phys.* 2.7.198a25–27: "The last three [the form, the mover, that for the sake of which] often coincide; for the 'what' and 'that for the sake of which' are one, while the primary source of motion is the same in species as these." To this he adds in *On the Parts of Animals* 1.1.639b14–16: "Plainly, however, that cause is the first which we call that for the sake of which. For this is the account of the thing, and the account forms the starting-point, alike in the works of art and in works of nature."

This notion of the interrelatedness of causes listed by Aristotle is crucial for a proper understanding of the concrete causal occurrences. As we have seen in the motto opening our reflection, Aristotle speaks about the "nature" of each concrete entity being "a source or cause of being moved and of being at rest in that to which it belongs primarily."[52] In the context of his teaching it becomes obvious that each particular "nature" is, first of all, a function of material and formal causes, intertwined such that their specification and distinction become possible only at the level of mental analysis. What we encounter in the world is always primary matter in-formed by different substantial forms, or—in other words—substantial forms actualizing different potentialities of primary matter. Hence, it is the principle of hylomorphism (i.e., of ὕλη [*hylē*] intrinsically related to μορφή [*morphē*]) that lies at the very foundation of Aristotle's ontology. Such understood "nature" becomes a source of activity and reactivity, which links it to the other two causes listed by Aristotle. For he always sees efficient cause as exercised "by" or "over" concrete entities characterized by concrete "natures." And it is only with reference to those entities that we can speak about natural teleology—that is, about beings realizing their natural possibilities and propensities.[53]

Awareness of the complex and multilevel character of causal dependencies proves to be crucial for the proper understanding of the methodology of natural science. It helps us realize that our use of scientific methods enables us to "identify the basic persistent subjects by reference to the properties that provide efficient causal explanations of change and stability. Some of these properties are material [i.e., defined in terms of secondary or proximate matter], but some are formal [i.e., defined in terms of substantial form and primary matter] and final [teleological]."[54] An analysis acknowledging all types of causation needs a reference to, and becomes an argument in defense of, metaphysics, which

> inquires into the presuppositions of empirical science; for an empirical science assumes that it deals with an objective world, and with substances and their essential and coincidental properties. First philosophy [metaphysics] shows why we should accept these presuppositions, and what happens if we attempt to give them up. . . . While it would be quite wrong to claim that arguments in first philosophy are wholly non-empirical, it is still true that . . . they are prior to empirical inquiry, in so far as they defend the assumptions taken for granted in empirical inquiry.[55]

2.4. Chance and Necessity

Before we move forward, we need to refer to Aristotle's analysis of the phenomenon of chance occurences in nature, which follows his account of four causes in the *Physics*. To explain chance—which can refer to events happening either with or without a deliberate intention—Aristotle uses the distinction between *per se* (καθ' αὐτὸ αἴτιον, *kath' hauto aition*) and incidental (κατὰ συμβεβηκὸς, *kata symbebēkos*), or *per accidens*, causes. *Per se* causes are fundamental and essential efficient causes that come from nature (φύσις) or intellect (νοῦς). They are naturally related to the formal and final causality of an agent. In other words, an efficient cause is acting *per se* when its act is a function of an agent in accord with its substantial form to produce its proper effect.[56] Incidental causes, on the other hand, have a peculiar character, which we may try to explain by reference to Aristotle's metaphysics of substance. Just as an accident (accidental formal feature) has no existence of its own but is a function of a substance (having substantial formal features), similarly an accidental cause must be related to a *per se* cause.[57] To give an example taken from Aristotle, the essential efficient cause of a statue is the sculptor. If he happens to be fair-skinned and musical as well, it seems just to say that a musician or a fair-skinned man made a statue. But his musical skills and the fact that he is fair-skinned are only incidental (coincidental, *per accidens*) causes related to the *per se* cause of his being a sculptor.[58]

With this distinction in mind, Aristotle defines chance as an unusual incidental cause, which, as such, is inherently unpredictable, although it still falls in the category of events that "happen for the sake of something" (since it refers and is related to such occurrences). He emphasizes that chance events are due to nothing in the substance or *per se* cause, which happens to concur with these unexpected occurrences. Thus, he states that "chance is an incidental cause. But strictly it is not the cause—without qualification—of anything."[59] As an incidental cause, chance occurs always in reference to *per se* causes. Therefore, chance for Aristotle is posterior. It can be distinguished as a unique type of occurrence which is not primary, and yet is inherently related to nature (φύσις) and intellect (νοῦς). Hence, it needs to be defined in reference to *per se* formal and final causality rather than blind material necessity.[60]

At the same time, however, our reference to *per se* causes in chance occurrences does not make the latter epistemological in their nature. They cannot be described merely as an unexpectedness due to the limitations of human understanding. For Aristotle chance has primarily an ontological character. It is a kind of event that (ontologically) demands the agency/intentionality of nature or human will, since it appears to "happen for an end," but no such agency is involved in its occurrence.

Therefore, the need of reference to *per se* causes in the case of chance occurrences protects Aristotelian metaphysics not only from blind material necessity, defined as "tychism" (from the Greek τύχη, *tychē*)[61]—that is, attributing everything to chance—but also from absolute determinism, which sees chance simply as lack of human knowledge of causes.[62] This fact has a significant influence on Aristotle's philosophy of nature and his understanding of necessary occurrences in the physical world. Necessity as such is for him never absolute, but always suppositional. Things happen in accordance with causal patterns, but on the supposition that nothing interferes with given causal occurrences. In other words, what we observe in nature is a nomological necessity governing relations between metaphysically contingent entities and dynamical systems they enter.

3. CAUSALITY IN THE MIDDLE AGES

Apart from the school of Oxford (Robert Grosseteste [ca. 1168–1253], Roger Bacon [ca. 1214–94], and the group of Mertonians), which fostered the more mathematical component of philosophy of nature along the Platonic tradition, other medieval centers of education (especially the one in Paris) followed the methodology and the theory of causation offered by Aristotle. It is true that they did not resist the influence of the Neoplatonic overtones introduced by Arabic commentators on Aristotle, who kept and transmitted his heritage to the Middle Ages. But it seems that it was precisely the combination of these two great ancient schools of thought that moved forward our understanding of causation. It can certainly be seen in the philosophical works of Thomas Aquinas (1225?–74).[63]

Commenting on material cause, Aquinas dispels uncertainty concerning primary matter, emphasizing that it is a principle and source of potentiality and not the physical stuff building up concrete entities: "Bronze itself is a composite of matter and form. Accordingly, since it possesses matter, bronze cannot be called prime matter. Only that matter which is understood without any form or privation, but which is subject to form and privation, is called prime matter, inasmuch as there is no other matter prior to it. It is also called 'hyle.' . . . We know prime matter as that which is related to all forms and privations, as bronze is related to the form of a statue and to the privation of some shape. It is called *primary* without qualification. . . . We should note also that prime matter is said to be numerically one in all things."[64]

He defines formal cause as that which makes a thing to be what it is, saying that—as such—form "causes the quiddity of the thing."[65] He distinguishes between substantial form, which gives being to matter in an absolute way, and accidental form, which does so merely "in a qualified sense."[66] Dismissing another ontological uncertainty, Thomas distinguishes among form as (1) arrangement of parts, (2) union by contact and bond, and (3) union effecting an alteration of the component parts. Only the last refers to substantial form, which is thus not a mere aggregation of building blocks, but a source of the quiddity of an entity.[67] Finally, in reference to the Platonic idea of participation, Aquinas introduces one more crucial ontological composition of *esse* and *essentia*. He sees primary matter actualized by substantial form as essence, which is still in potency to exist. It receives its being through participation in God, who is pure act and the source of all being—a principle strategic for Aquinas's view of divine action.[68]

Discussing efficient causation, Thomas distinguishes its four types: (1) perfective—that is, bringing entities to perfection; (2) dispositive—that is, disposing matter to receive form; (3) auxiliary—that is, acting for an end which is not the agent's own end; and (4) advisory—that is, specifying for someone else an end and the form of activity to achieve it.[69] He also introduces another three important distinctions, crucial for his metaphysics:

1. The distinction between primary and secondary efficient causes. For example, we can see a conductor as a primary cause, and

members of an orchestra as secondary causes, of a symphony (the conductor acts "through" members of an orchestra).
2. The distinction between principal and instrumental efficient causes. For example, when a writer uses a pen, the action of the pen—unlike causality of the members of an orchestra using natural skills to play their instruments—exceeds its own capacity for action (a pen cannot write by itself). Thus, a pen in the hand of a writer becomes a special kind of secondary agent, an instrumental agent, operated by a principal agent (the writer).
3. The distinction between univocal and equivocal efficient causes, where the *causa univoca* has the same essence as the effect (or belongs to the same realm of being), while the *causa equivoca* differs with its effect in quiddity (or belongs to a different realm of being).[70]

Aquinas's teaching on final causation and the interrelatedness of causes follows—for the most part—the position of Aristotle.[71] Similar is his reflection on the nature of chance. He agrees with Aristotle, arguing in favor of contingency and against absolute determinism. Although natural teleology is inherently present in all beings, it does not produce absolutely necessary results. The world is not a machine in which one cog necessarily moves another. Accidents and chance are inherent in nature. Thus, we may say that Aquinas stands, together with Aristotle, in opposition to the Presocratics, who—as we can tell based on our analysis presented above—accepted, for the most part, a determinist view of the physical world.

Finally, commenting on Aristotle's *De gen. et corr.* 1.10.327b24–32, Aquinas develops a theory traditionally referred to as the doctrine of the virtual (*virtute*) presence of elements in mixed substances. Trying to answer the question of what happens with basic elements and their causal activities as they go through substantial changes that issue in the emergence of complex substances, Aquinas states: "The powers of the substantial forms of simple bodies are preserved in mixed bodies. The forms of the elements, therefore, are in mixed bodies; not indeed actually, but virtually (by their power). And this is what the Philosopher says in book one of *On Generation*: 'Elements, therefore, do not remain in a mixed body actually, like a body and its whiteness. Nor are they corrupted, neither both nor either. For, what is preserved is their power.'"[72]

Christopher Decaen rightly notes that since in modern English the word "virtually" means "more or less," "practically," or "pretty much but not quite," and may suggest a modality of existence that is in some sense "between" potency and actuality, it is better to translate *virtute* as "by power." He claims that according to Aquinas's theory the active and passive qualities (powers) that differentiate the elements and allow them to act upon each other are being preserved in mixed substances, yet not in their actuality, nor to their full "excellence."[73] We will say more about the theory of virtual presence in chapter 6.[74]

4. CAUSAL REDUCTIONISM OF THE MODERN ERA

The classical period of modern science and philosophy (from the seventeenth to the nineteenth centuries) brought a radical change and revision of Aristotelian theory of causation. Although not suddenly nor definitively, the course of both scientific and philosophical reflection on causality—influenced by the new appreciation of mathematics and its application in analysis and description of nature—shifted gradually toward dismissing teleology and formal causes.[75] Not amenable to empirical investigation and mathematical description, final and formal causes seemed to be more and more obscure for those who followed the mainstream of the development of modern academia.

4.1. Causation and Modern Science

Those working in various fields of natural science wanted to replace the causes described by scholastics—which they found rather occult—with their own explanation in terms of "true causes." Even if they were moving slowly toward defining the latter in terms of physical (efficient) interactions, some of them would still follow Aristotelian typology of causes, trying to explain, for instance, magnetic and electric phenomena with reference to substantial and accidental forms (William Gilbert [1544–1603]), or the function of the heart in terms of all four causes (William Harvey [1578–1657]). Others, like Newton, strove to approach the "true causes" of phenomena through formulation of general laws of nature (e.g., gravity, composition of light), at the price of remaining skeptical about these phenomena's ultimate explanation, which they thought was beyond the scope of their investigation. Finally, we find those who—like Francis Bacon—fascinated with inductive and experimental method, openly reduced formal causes to "laws and determinations of absolute actuality" and dismissed teleology, saying that "the inquisition of final causes is barren, and like a virgin consecrated to God produces nothing."[76]

4.2. Causation and Modern Philosophy

The course of modern philosophy went even further in the process of reducing the character and nature of causal explanation. The father of ontological dualism and modern rationalism, René Descartes (1596–1650), became one of the first philosophers who openly questioned Aristotelian theory of causality. In the spirit of reemerging atomism, he strove to replace scholastic substantial forms with the idea of bodies composed of parts in motion. He also rejected final causes as unknowable, unscientific, and spurious, and restrained the concept of matter to basic physical constituents. Even if he still spoke of God as a special kind of total efficient cause of being and existence—in opposition to efficient causes producing an effect by contact action in different phenomena and occurrences in the physical world—the truth is that Descartes found himself left with this one type of causation only. Moreover, it seems that it was precisely his combining causal reductionism and faith in God that gave his philosophy a strong determinist and occasionalist flavor, which became a point of departure for modern schools of materialism and occasionalism.[77]

Those supporting the materialist position followed Thomas Hobbes (1588–1679) and saw only one universal cause, consisting in "certain accidents both in the agents [efficient cause] and in patients [material cause]; which when they are all present, the effect is produced; but if any of them be wanting, it is not produced."[78] They found appealing Hobbes's reduction of formal and final causes—of which he said that they "are nevertheless efficient causes"[79]—as well as his acceptance of strict necessity of events in nature, assuming that chance, contingency, and possibility are just results of our lack of knowledge of the necessary cause.

Followers of modern occasionalism sided with Nicolas Malebranche (1638–1715), who defined a "true cause" as the one "such that the mind perceives a necessary connection between it and its effect," concluding that only one such cause exists—namely, the will of all-powerful God.[80] This necessitarian and determinist strain of modern philosophy finds its logical conclusion in the rationalist system of Baruch Spinoza (1632–77), who identifies God with Nature (*Deus sive Natura*) and sees all individual things as modes of the one Substance. His ontology leaves no space for free will and contingency of cause-effect relations, rejecting also final causes as anthropomorphic fictions.

Although earlier in his career he was seduced by the agenda of mechanical philosophy, Gottfried Wilhelm Leibniz (1646–1716) proposed in his mature philosophical thought a new solution to the question of the causes of motion. He introduced the idea of force, which he analyzed in Aristotelian terms, distinguishing between its active and passive aspects—the former defined in reference to entelechy, soul, or substantial form, the latter in terms resembling the idea of primary matter. His response to the rationalist reduction of four causes went even further when he argued in favor of final causality, which he found essential for the explanation of matter and natural mechanisms. This retrieval of teleology and formal causation found a new expression in Leibniz's *Monadology*. The internal forces of monads can be identified with substantial form. When conceived as appetites, they seem to have a teleological aspect to them as well. Whether Leibniz managed to avoid occasionalism remains controversial. Although he argued that God's eternal law is carried out by the activity of creatures, his idea of preestablished harmony brings him dangerously close to the position of Malebranche.

Leibniz's defense of formal and final causes was significant. But it did not stop the ongoing process of causal reductionism in philosophy, the next stage of which was dominated by the empiricism of Locke, Berkeley, and Hume. Despite approving the reality of causal activity, John Locke (1632–1704) saw no other way of explaining it but in terms of the mechanics of motion, which places him alongside scientists of his time and their approach to cause-effect relationships. Moreover, anticipating David Hume (1711–76), Locke questioned the necessary connection between causes and effects, assumed by rationalists and stressed by Hobbes. His giving up metaphysical reflection on the nature of causality was followed by George Berkeley (1685–1753), who saw no space for causal explanation in physical science either. He argued that "all forces attributed to bodies are mathematical hypotheses,"[81] with no grounding in the nature of things. He saw them as dependent on the notion of the one who speaks about them.

This preliminary development of empiricism, fostered by Locke and Berkeley, became even more pervasive in the thought of Hume, who remained critical about all theories holding that causes somehow "produce" their effects. His disapproval of all known views of causal relationships made Hume dismiss the ontological dimension of the causation debate altogether. Questioning the contiguity in space and time and the temporal priority of a cause to its effect—as affording a complete idea of causation—as well as the necessity of the connection between cause and effect, Hume saw causation as an outcome of the constant sequential conjunction of events which produces an association of ideas in our mind. Because we do not have any sensory impression of necessary connections between events, the tie of necessary connection "lies in ourselves, and is nothing but the determination of the mind, which is acquired by custom, and causes us to make a transition from an object to its usual attendant, and from the impression of one to the lively idea of the other."[82] What saves us from the paralysis of not being able to have genuine causal knowledge is our human nature, which—despite epistemological causal skepticism—operates by reliance on expectation and custom.

Based on this foundation, Hume offered two important definitions of causation that heavily influenced philosophical debate on this topic. The first—classified today as the regularity view of causation—states that cause is "an object precedent and contiguous to another, and where all the objects resembling the former are placed in like relations of precedency and contiguity to those objects, that resemble the latter."[83] The second—the counterfactual approach to causation—infers it from a rule of dependency, saying, "*If the first object had not been, the second never had existed.*"[84] Even if both definitions seem to interpret causal relata as objects, their context leaves no doubt that Hume's objects are in fact events. This ontological shift proves to have an important influence on the late modern and the contemporary causal debate.

Hume's attack on the meaningfulness of all theories of causation was fierce and thorough. The last attempt at defending the ontological dimension of causal claims, as well as the legacy and need of causal explanation in science and philosophy, came from Immanuel Kant (1724–1804), who strove to find the middle way between the rationalism of Leibniz and the empiricism of Hume—the one which would implement both a priori elements of human knowledge (stressed by rationalists) and the necessity of synthetic judgments based on experience (valued by empiricists). Hence his proposition of a new type of judgment that is both synthetic and a priori.

Faithful to this new methodological strategy, Kant grounded the principle of causality in the structure of reason, as one of the twelve categories of understanding, which organize spatially and temporally atomistic and unstructured sensations produced by "things-in-themselves" that are external to us. Kant defined causation as superimposing on these sensations "the principle of succession in time in accordance with the law of the connection of cause and effect."[85] He classified it as a principle explaining the synthetic a priori judgments of science. An event A is the cause of an event B iff there is a universal law which says that events of type A are necessarily followed by events of type B. But because neither the necessity nor the universality of the causal relation can be established empirically, such a relation needs to be grounded in the a priori conditions of judgment of a possible experience. It is true that this way of reasoning does not give us access to a knowledge of substance and substantial change in themselves, but to claim that such knowledge is possible at all—says Kant—is to fall victim to "transcendental illusion."[86]

4.3. Decline of Causal Explanation

Despite Kant's defense of the importance of causality, causal reductionism and an antimetaphysical attitude in both science and philosophy grew even stronger in the nineteenth and the first half of the twentieth centuries. The Humean dismissal of the ontological aspect of causation inspired Auguste Comte (1798–1859) to propose his positivist methodology of science. Even if he was in favor of including a "positive philosophy" in his program, he saw the role of such a philosophy as merely to pursue an accurate discovery of natural laws, without any attempt to explain them. The founder of the Vienna Circle, Moritz Schlick (1882–1936), went even further, stating that "the words 'cause' and 'effect' do not occur at all in the laws of nature," which describe "the interconnection of events expressed by mathematical functions" and "are expressed in the form of differential equations."[87]

From this attitude the three main approaches to scientific methodology in the twentieth century originated: (1) the phenomenalism of Ernst Mach (1838–1916), who saw the task of science as analyzing human sense impressions and organizing them into a type of synthesis, enabling man to adapt himself to natural conditions; (2) the conventionalism of Henri Poincaré (1854–1912), who claimed that scientific theories were just conventions, having their value in utility rather than truth; and (3) the operationalism of Percy Bridgman (1882–1961), who identified all physical concepts with operations used to define and measure them. Operationalism was applied in the Copenhagen interpretation of quantum mechanics and Niels Bohr's (1885–1962) "principle of complementarity," which acknowledged "the impossibility of any sharp separation between the behavior of atomic objects and the interaction with the measuring instruments which serve to define the conditions under which the phenomena appear."[88] Another contributor to the "orthodox" interpretation of quantum mechanics, Max Born (1882–1970), went even further in his philosophical conclusions. Recognizing the importance of the search for causes in natural science, he ascribed to chance a more fundamental role than to causality, arguing for "a complete turning away from the predominance of cause (in the traditional sense, meaning essentially determinism) to the predominance of chance."[89]

The whole process of constraining and reducing the notion and role of causation in scientific explanation culminated in a bold claim made by an Austrian mathematician, physicist, and philosopher, Friedrich Waismann (1896–1959), who in 1958 at Oxford gave a lecture titled "The Decline and Fall of Causality."[90] He thought that because determinism and predictability were replaced in contemporary scientific theories by chance and probability, causality could no longer be regarded as an ultimate category of explanation.

5. NONREDUCTIONISM AND EMERGENTISM

And yet the influence of the positivist and empiricist approach to the methodology of natural science, predominant in the first half of the last century, has been seriously challenged and questioned with the advance of the most recent scientific research and with the rise of philosophy of science as a separate discipline. The latter has been accompanied by the revival of metaphysics in the analytic tradition, which has offered a number of theories of causation. These new concepts of causal dependencies range from the neo-Humean regulatory and counterfactual, to the probabilistic, to the singularist, to the manipulation-based, to process views of causation, to the dispositional concept of cause-effect relationships—developed in the context of dispositional metaphysics. I will discuss all of these theories in the second part of this book. For now, it suffices to say that, even if they share a common attempt at defending the realist account and the ontological character of causal dependencies, all of these theories, except for the dispositional one, concentrate mainly on the physical—that is, efficient causation.

At the same time, the most recent philosophy of science seems to suggest that the language of physics and mathematics is unable to give a proper and exhaustive account of the robust complexity of natural phenomena. Our metareflection on the nature of scientific progress and some important philosophical aspects of the new scientific discoveries, theories, and hypotheses helped us realize that the monolithic, cleansed, and tidy reductionist picture—which attracted many scholars and thinkers in the past—does not stand anymore. This realization came with new models questioning the Copenhagen interpretation of quantum mechanics and addressing philosophical aspects of the theories of chaos, Big Bang, or anthropic coincidences. It can also be noticed with a particular strength in the field of biology, where the agenda of reducing biological sciences to chemistry and chemistry to physics, as well as limiting all types of causation inherited from ancient and medieval science and philosophy to efficient (physical) causation, failed to prove to be the only viable and truly scientific method of research. Contrary to the expectation of scientific reductionism, our ability to enter the molecular level of organisms and biochemical processes has opened us to the incredible complexity of the structures, processes, and patterns of living organisms, which seem to be irreducible. The intrinsic interrelatedness of different components of natural processes, such as metabolic or cell signaling networks, and their influence on the behavior of organisms, have led many bioscientists not only to distinguish between various levels of organization of matter in biology but also to propose a more holistic approach and methodology in biological sciences, which is known as "systems biology."

Introduction   27

This approach inspired, in turn, a recent revival of the theory of EM, which speaks about the novelty of properties, entities, and dynamical systems at higher levels of complexity of matter. It not only strives to develop an ontology of these levels of complexity and characterize the laws of nature which are proper for them but also speaks about nondeducibility, nonpredictability, and irreducibility of emergents. Most importantly, trying to defend the novelty of emergent properties and phenomena in ontological terms (strong EM), the followers of emergentism introduce the category of DC—that is, a new type of causal power characteristic of complex systems, influencing in a top-down manner their basal constituents, and irreducible to the causal operations proper for the lower levels of complexity (e.g., mind exercising causal influence on the brain, which underlies it).

6. EMERGENTISM AND THE NEW MECHANICAL PHILOSOPHY

Even if explanatorily promising and intriguing, the theory of EM and DC raises a number of metaphysical questions, which become an inspiration for the research presented in this book. But before I outline these questions, we need to realize that the theory of EM and DC becomes a challenge to philosophical reflection on the methodology of life sciences and systems biology, which triggered the very revival of emergentism.

6.1. Organization of Mechanisms

Trying to specify the framework of the philosophical assumptions underlying areas of science such as biology, neuroscience, or psychology—with reference to the topics of causation, levels of complexity, explanation, laws of nature, reduction, and discovery—a number of philosophers raised in the postlogical empiricist milieu argue that natural science is driven by the search for mechanisms.[91] Although they inherit the term "mechanism" from their modern predecessors, the followers of the new mechanical philosophy want to distance themselves from the idea that mechanisms are machines and that a real change involves only fundamental physical forces. These new mechanists argue that "biological mechanisms do things. They move things. They change things.

They synthesize things. They transmit things. They may even hold things steady."[92] Thus these philosophers want to differentiate mechanisms from simple aggregation.

For, unlike an aggregate, whose parts can be rearranged or substituted with no change of the operation of the whole, a biological mechanism is more than just a sum of its parts.[93] It shows a unique spatial (location, size, shape, position, orientation) and temporal (order, rate, duration of components) organization. Hence, even if it is characterized by modularity—which makes possible a physical intervention on its putative causal variable without the disruption of the functional relationships among the other variables—each biological mechanism is in some way irreducible. Its components show unique "meshing properties" and "jointness," which lead the followers of the new philosophical mechanicism to state:

> A mechanism for a behavior is a complex system that produces that behavior by the interaction of a number of parts, where the interaction between parts can be characterized by direct, invariant, change-relating generalizations.[94]

> A mechanism is a structure performing a function in virtue of its component parts, component operations, and their organization. The orchestrated functioning of the mechanism is responsible for one or more phenomena.[95]

> Mechanisms are entities and activities organized such that they are productive of regular changes from start or set-up to finish or termination conditions.[96]

The use of the language of complex systems consisting of component parts and operations, organized and characterized by orchestral functioning, clearly alludes to the irreducible nature of higher features of biological mechanisms.

6.2. From Methodology to Metaphysics

The new mechanistic philosophy—based on a thorough analysis of discoveries in the history of life sciences and concrete examples taken from contemporary molecular biology and biochemistry—may be accurate in its description of an actual modus operandi of practitioners of natural science working in their laboratories. They may be truly driven by a search for mechanisms, explaining various phenomena in living organisms. But the metaphysical foundations of the new mechanicism in philosophy of life sciences is rather sketchy and wanting. On the one hand—and it speaks in their favor—new mechanists want to refer to both substance and process ontologies without favoring one or the other (they call themselves "ontological dualists"). They strive to depict and explain mechanisms in terms of both concrete entities with capacities or dispositions to act and activities in which these entities are engaged. On the other hand, however, the proponents of the new mechanical philosophy seem to struggle as they try to find a proper theory of parthood and wholeness, which is crucial for their project.[97]

Rejecting formal mereologies[98] (applicable in cases of simple aggregation), new mechanists are torn between the two horns of an ontological dilemma. On the one hand, they want to acknowledge the decomposability of a biological mechanism, whose constituents must have a certain level of robustness, characteristic properties, and reality—when taken apart from their places in a given mechanism. They also speak about "bottoming out" of nested hierarchical descriptions of mechanisms in lowest-level mechanisms.[99] On the other hand, they want to emphasize an irreducible character of each biological mechanism, requiring multilevel and upward-looking explanations. Arguing that components of biological mechanisms can be widely distributed and escape precise localization, and that their operations cannot be limited to the Cartesian machine-like push-pull dynamics, new mechanists seem to line up with the proponents of nonreductionism in scientific explanation. At the same time, however, they do not want to abandon a physicalist worldview, which seems to be a dogma for all natural sciences.[100]

6.3. Causal Mechanisms

Moreover, relating biological mechanisms to phenomena (the components in a mechanism are components in virtue of being relevant to the concrete phenomenon), new mechanists use the language of production, underlying, and maintaining.[101] They talk about mechanisms producing phenomena, defined in terms of (1) objects (e.g., production of an enzyme), (2) states of affairs (e.g., being oxidized), or (3) activities or events (e.g., digestion). Describing physiological phenomena (e.g., the Krebs cycle), they find it more appropriate to say that mechanisms underlie them. Finally, in some specific cases, such as homeostatic regulation of body temperature, they speak about mechanisms maintaining these phenomena. All three terms listed here belong to the category of causal claims, the metaphysics of which becomes a serious challenge for new mechanists. On the one hand, they acknowledge the shortcomings of the strategy favored by logical empiricists, who strove to explain phenomena (*explananda*) on the basis of laws of nature and the antecedent and boundary conditions (*explanantia*). Rather than looking for covering-law explanations, we see new mechanists arguing in favor of causal explanations in reference to structures that produce, underlie, or maintain phenomena. But there is no unity among them when it comes to the theory of causation they want to follow. Craver and Tabery[102] list four possible strategies discussed among new mechanists:

1. *Causation defined as transmission and propagation of marks or conserved quantities, such as mass-energy, linear momentum, or charge (i.e., Salmon's process view of causation).*[103] This view has been criticized because of the difficulty of its application in nonfundamental sciences such as biology, which do not make explicit references to conserved quantities (even if they assume their reality fundamentally). Moreover, biological mechanisms involve causation by omission, prevention, and double prevention (when a mechanism removes a cause, prevents a cause, or inhibits an inhibitor). These situations seem to escape description in terms of the process view of causation. Finally, the process view of causation implicitly assumes reduction of causation to physical interactions, which is an object of controversy among philosophers.
2. *Causation derived from the concept of mechanism (a view supported by Glennan).* It assumes that causal claims are simply claims about the existence of a mechanism. Hence—according to this theory—mechanisms become the hidden connection Hume sought between cause and effect. Naturally, this view of causation

can be charged with circularity, as the concept of mechanism contains necessarily a causal element.
3. *Causation understood in terms of productive activities, e.g., magnetic attraction and repulsion, or hydrogen bonding (an Anscombian theory followed by Machamer).* Those who follow this view claim that relying on science describing causal activities enables them to avoid the need to define more generally the concepts of cause and effect. This position has been criticized because of its failure to say what activities are, its inadequacy in accounting for the relationship between causal and explanatory relevance, and its inability to mark an adequate distinction between activities and correlations.
4. *Causation defined in terms of Humean counterfactual dependencies between causes and effects, with an inclination toward the manipulationist definition of causal relata.* According to this approach, making a difference to the value of one variable brings a change in other variables of a mechanism, influencing the phenomenon it produces. Even if this approach readily accommodates omissions, preventions, and double preventions, it is vulnerable to other challenges faced by the counterfactual view of causation, such as overdetermination and early and late preemption.[104]

None of these four definitions of causation discussed by new mechanists is satisfactory. This situation not only complicates their ability to delineate the exact character of causal work exercised by biological mechanisms. It also leads to a lack of clarity and precision in their attempt at specifying the ultimate ontological character of their philosophical position. Supporting the idea that biological systems are hierarchically organized into near-decomposable structures of mechanisms—within mechanisms—within mechanisms, new mechanists have difficulty identifying the ontological status of the particular stages of this hierarchy. Is the relation of the levels of complexity of biological mechanisms—producing phenomena—a relation of supervenience, which simply states that the higher-level properties of a system occur only if appropriate conditions are realized on the lower level? Or maybe what we are dealing here with is an example of EM?

6.4. Mechanisms and Downward Causation

In answer to this question, Craver and Bechtel argue that the analysis of mechanistically mediated effects (phenomena) shows that interlevel (in-between-levels) relations cannot be causal, but only constitutive, while intralevel (inside-of-level) dependencies can be and are causal. They claim that "the notion of 'level' involved in considering cases of emergence is not the same as the notion of level that is so ubiquitous in biology. Levels of mechanisms are constitutive levels; levels of strong emergence are not."[105]

This position has two serious problems. First, it does not give any positive explanation of what the actual character of levels in strong EM is and how they differ from constitutive levels of biological mechanisms. Second, it does not specify the metaphysical nature of the constitution of levels in these mechanisms. We have already learned that such constitution is not a simple aggregation—that is, a quantitative sum of the physical components of a mechanism. We are told it can be characterized as a unique spatial and temporal organization. But what is the source and nature of this organization? Does it bring a qualitative change in the components of a mechanism, or establish some new and unique qualitative features of the mechanism itself as a whole? If the levels are constitutive for a given mechanism, what is exerting the action of constituting them in a very specific way, and why is this process directed toward a particular and complex biological mechanism? Can we explain the phenomenon of the prolonged homeostasis of a simple living cell by the constitution of its causally active parts (levels of complexity)? How does this constitution make the cell as a whole change the biochemical activity of its parts in response to the external conditions? How do the bottom-up activities at basic levels of the constitution of the cell mechanism—classified in molecular biology as geometrico-mechanical, electro-chemical, energetic, and electromagnetic interactions engaging macromolecules, smaller molecules, and ions—explain causally changes in the cell as a whole?

It seems to me that the new philosophical mechanists emphasize too much the explanatory power of the constitution of parts into wholes, without providing a sufficient metaphysical clarification of the meaning of the very term "constitution." To say that it is not causal but organizational is definitely not enough. Their position lacks a serious ontological elucidation.

Moreover, going back to the problem of causation in complex living systems, we find Craver and Bechtel offering two major arguments why interlevel causation is impossible—both of which are rather flawed. First—following Salmon's process view of causation (see number 1 above)—they define it as some sort of a "physical connection"—that is, "a kind of cement, glue, spring, string, or some other physical transmission or exchange from one object, process or event to another through contact action or through a propagated signal."[106] They claim that this view does not accommodate interlevel relations between mechanisms and their parts because they are not distinct events, objects, or processes. Again, the argument weighs upon the concept of constitution, which is noncausal and supposedly explains the relation between mechanisms and their components—the concept which we have already classified as rather vague and metaphysically unclear. What is more, the argument implicitly assumes the reduced notion of causation in terms of physical interactions to be the only true and acceptable theory, which does not find a common approval among philosophers of today.

Second—following post-Humean reflection on causation—Craver and Bechtel claim that because mechanisms and their parts coexist in space and time, are not wholly distinct, and cannot intersect with one another or precede one another (i.e., they are symmetrical), they cannot be causally related. This argument is easy to refute in the context of anti-Humean arguments and examples of causal relations that are simultaneous both spatially and temporally (e.g., attraction and repulsion of two magnets joined together). I will say more on this topic in chapter 4.[107]

Nevertheless, building on their arguments against the possibility of an interlevel causation, Craver and Bechtel conclude that

1. all causation in biological mechanisms is intralevel;
2. it is constitution (not causal relations) that explains the relation between biological mechanisms (wholes) and their components (parts); and thus,
3. the idea of top-down causation (DC) is rather spooky and incoherent; although it can be used informally and as a shorthand, we need to understand that "the change at the higher level is

mediated by, and explicable in terms of, a mechanism. Nothing mysterious is lurking here. . . . Where there are mechanistically mediated effects, there is no need for the mysterious metaphysics of interlevel causation at all."[108]

This argumentation supports the view of Wimsatt, who—contrasting levels of biological mechanisms with levels of mere aggregation—describes the constitution of the former as a kind of EM, which can be classified as mechanistic or organizational EM.[109] Craver and Tabery relate it to the idea of weak or epistemic EM, which is simply a result of our "inability to predict the properties or behaviors of wholes from properties and behaviors of the parts." Moreover, they contrast mechanistic EM with what they call a spooky kind of EM, which involves "the appearance of new properties with no sufficient basis in mechanisms."[110] Their conclusion seems to suggest that there is a tendency among new mechanists to reject the possibility of the ontological (strong) kind of EM in biological systems. It is true that Craver tries to be more careful and nuanced, recommending at one point—together with Bechtel—a separation of "the question of whether strongly emergent properties are possible (in some sense of the word possible) from the question of whether top-down causation is possible."[111] But, as we will see, it is not entirely clear what strong EM would mean, when devoid of the reference to DC, which is usually regarded as its sine-qua-non aspect/condition. Again, Craver and Tabery do not sort this issue out for their readers.

6.5. Summary

My analysis of the new mechanical philosophy shows that the theory of ontological (strong) EM and DC challenges this important, if not predominant, methodological analysis of the way systems biology gathers its data.[112] We saw that the new mechanical philosophy cannot avoid metaphysical commitments, and that in this particular aspect of its agenda it needs further development and clarification. Otherwise, it will remain in the gray area, undecided between ontological reductionism and nonreductionism. But does the analysis of systems biology arguing in favor of the reality of ontological EM and DC avoid similar problems?

7. METAPHYSICAL ASPECTS OF EMERGENTISM

I said at the beginning of the previous section that the theory of EM in its contemporary version—and the concept of DC in particular—despite its novelty, attractiveness, and ontological robustness, raises some serious metaphysical doubts and questions. I still hold to my position. For what is the exact character of DC? How should we characterize it in the context of the reductionist understanding of causality in terms of physical (efficient) types of interactions, which seems to still predominate among scientists? Is DC reconcilable with Hume's causal skepticism and his regulatory and counterfactual views of causation, which define causal relata in terms of events rather than concrete entities and their natures? It appears to me that the proponents of strong EM and DC face a similar ontological dilemma that challenges followers of the new mechanicism in philosophy of biological sciences. (1) Those among them who want to remain faithful to the methodology of natural science tend to define DC in terms of physical interactions, thereby calling into question its irreducibility. (2) On the other hand, those who emphasize its novelty and distinctiveness from physical causes find it difficult to identify its ultimate character. They also risk violating the rules of physicalism and causal closure, which they want to follow as scientists and/or philosophers of science.

In view of the manifest inadequacy of both approaches, it has therefore been suggested that EM and DC can be saved only in the context of systemic causation which goes beyond efficient causation, reaching toward Aristotle's ideas of formal and final causes. This advice was followed by Terrence Deacon, who—in a number of publications, among which *Incomplete Nature* remains the most influential—offers a new version of EM theory, arguing in favor of a broader understanding of causation, which reaches back to Aristotelian concepts of form and teleology. Interestingly, in developing his model of emergentism Deacon rejects top-down mereological (whole-part) reasoning and suggests rethinking EM itself in dynamical terms. He introduces an intriguing notion of "constitutive absences" (constraints), understood as "possible features being excluded," as the core of his process view of EM, in which "what is absent is responsible for the causal power of organization and the asymmetric dynamics of a physical or living process."[113] In other words, the reduction of possibilities and options brings an increase in complexity and specialization, leading to the EM of the new features of inanimate and animate entities.

While fascinating and promising, Deacon's project—like those mentioned above—raises important metaphysical questions. Although he explicitly rejects eliminative reductionism, which assumes everything reduces to physical particles, Deacon does not side with the classical antireductionist positions of Aristotelian origin. Nor does he follow contemporary proponents of top-down causation. Rather, he suggests reinterpreting formal cause as a function, and final cause as an emergent outcome of basic mechanical physico-dynamic processes—a position which is still compatible with some form of limited reductionism and departs from the Aristotelian understanding of these types of causation. Moreover, his idea of the causality of absences seems to be philosophically counterintuitive, as it assumes that "what is not" can act on "what is." Finally, following many proponents of the scientific notion of emergentism, Deacon seems to reject the concept of hylomorphism (the view that things are composed of prime matter and substantial form), treating the explanation in terms of Aristotle's formal causes as homuncular.[114] He is also critical about the process metaphysic of Whitehead, which he rejects as another version of panpsychism. But at the same time, he does not seem to offer a fully developed alternative ontology for biological emergentism.

8. EMERGENTISM AND DISPOSITIONALISM

In the context of difficulties challenging both the top-down mereological and the dynamical versions of EM, I will propose in this book dispositional metaphysics and the corresponding view of causation as a possible solution to the ontological problems of emergentism. Formed within the analytic philosophical tradition, dispositionalism defines dispositions (powers) in things and organisms as intrinsic properties characteristic for natural kinds and explains causation as a manifestation of these dispositions. I think that this metaphysics is highly relevant and has significant explanatory power in the context of the debate on the reality and the character of EM, emergent properties, and DC. I find it helpful in solving many ontological puzzles challenging emergentism and the irreducible approach to systems biology. The goal of this volume is to prove it.

My option for dispositionalism is not accidental. I want to situate the whole debate on emergentism within the context of the most recent metaphysics developed in the analytic tradition. Among many theories of causation offered by contemporary metaphysicians, I choose the dispositional view of causation, which I find to be the closest to the robust Aristotelian notion of causal dependencies in nature. I believe that dispositionalism builds a bridge between ancient and contemporary metaphysics and ontology. Because it seems to support essentialism (the view that entities have essences decisive for their nature) and involves a possible retrieval of hylomorphism and final causation, dispositional metaphysics is rightly regarded by some philosophers as a neo-Aristotelian position. It shows that Aristotle's explanation—employing scientific principles as well as the notion of causes that admittedly lie beyond the bounds of science (but are not facile explanations [homunculi], but legitimate [natural] principles of a philosophy of nature)—is still valid and applicable in the context of contemporary science. I want to argue that it provides some crucial explanatory categories, fundamental for understanding and describing the nature of emergent phenomena. It also answers, in my opinion, the main difficulties challenging other accounts of EM that have been developed so far.

At the same time, however, I am not uncritically optimistic about dispositionalism. Acknowledging its explanatory power and the unique character of the view of causation it offers, I am aware that dispositionalism is not entirely identical with the ancient philosophy of nature and the metaphysics of Aristotle. Trying to avoid too hasty and unjustified connections between these two systems of thought, in the second part of this book I will carefully examine the Aristotelian legacy of dispositionalism. I will also address and answer some other problems raised by this metaphysical position.

Nonetheless, despite its difficulties, I believe that dispositional metaphysics opens the way to a more robust view of causation, which both avoids pure causal necessity and contingency and enables us to give a proper account of singular and plural cases of causal dependencies. Moreover, a dispositional view of causation proves to be attentive to cases of polygenic causation—where more than one cause is responsible for an effect—and to causal pleiotropy (i.e., causes contributing to various causal occurrences). It also helps us to deal with probability aspects of many causal situations, as well as the problem of the counterintuitive cases of causation of absences (my not doing A causes B to occur). Finally, dispositional theory of causation pays more attention to context-sensitivity and complexity of single causal events than other views and enables philosophers of science to ground ontologically the laws of nature. These advantages of the dispositional view of causation over other contemporary theories—especially those of neo-Humean descent—lead me to contend that it can serve as a suitable metaphysical foundation for emergentism. I will apply it both to the version of EM that emphasizes the importance of the top-down causation (DC-based EM) and to Deacon's dynamical-depth model of ontological EM, which emphasizes the role of "constitutive absences."

9. PLAN OF THE PROJECT

The research presented in this book is divided into two parts, each preceded by a short introduction. The first part critically analyzes the metaphysical aspects of the theory of EM. Beginning by describing the basic postulates of EM—in the historical context of their classical formulation (chapter 1)—our inquiry will give an account of the metaphysical challenges and weaknesses of the mereological top-down version of EM (chapter 2), to culminate by critically investigating Deacon's new version of emergentism (chapter 3). My examination of his dynamical model of EM will concentrate on the question of its causal nonreductionism and its references to Aristotelian four causes.

As a remedy for the metaphysical shortcomings of both the classical and Deacon's concepts of EM, the second component of the book will present dispositional metaphysics as a possible new ontological foundation of emergentism. The analysis of the main objectives of this metaphysics and its related view of causation (chapter 5) will be preceded by an overview of the six main theories of causation in analytic philosophy (chapter 4). Next, an investigation of the Aristotelian legacy of dispositional metaphysics and theory of causation will follow (chapter 6). The whole argument will conclude with my constructive proposal of an application of this new ontology—with its references to the classical Aristotelianism—in both the top-down and Deacon's dynamical depth models of EM, which will help to ground them in a more thorough metaphysical foundation (chapter 7).

PART 1

Metaphysical Aspects of Emergence and Downward Causation

When spontaneous processes like the complex adaptive functions of living bodies tend to produce increasing orderliness, complex interdependencies, and designs that are precisely correlated and matched to one another and the world, we can be excused for being just a little mystified.
—Terrence Deacon, *Emergence: The Hole at the Wheel's Hub*

To the extent that emergence denies to the whole any unity beyond the incidental arrangement of its parts, it fails to explain the being of the whole. And without some grounding in the being of the whole, emergence can only affirm but not explain the activity that proceeds from the whole. So long as emergence remains unable to account for the being of the whole, its tendency will be to slip back into a kind of reductionism, attributing to the part a more fundamental reality than to the whole.
—Michael Dodds, *Top Down, Bottom Up or Inside Out? Retrieving Aristotelian Causality in Contemporary Science*

The more advanced our scientific knowledge becomes, the more we realize the world is a messy place, resisting any naïve drive toward simplification and unification. The actual complexity of nature puts in question the reductionist agenda developed and pursued by many scientists and thinkers over the last three centuries of academic research and study. Today it becomes obvious that reductionism's ideal of clarity, precision, orderliness, and discipline is simply a myth, unable to account for the actual complexity of nature. But the reductionist paradigm has driven both scientific research and the philosophical reflection based on it long enough to make it difficult for many to acknowledge its shortcomings, while causing an enormous struggle for others who try to provide a suitable replacement for various types of reductionist declarations and dogmas.[1]

One of the main battlegrounds in this warfare is the field of biology, as it is concerned, by definition, with complex structures of living organisms. The rapid development of biochemistry and molecular biology in the past century had led many scientists to believe that their reductionist approach would prove to be the only valid method of biological research. Today, a growing interest in the systems approach in biology becomes a sign of a paradigm shift in the methodology of the life sciences. The widespread availability of high computational power, developments in mathematical and algorithmic techniques, and the development of mass data-production technologies (e.g., high-throughput data collecting) allow for collecting dense dynamical information from complex biological systems. The interest in systems becomes a viable alternative to traditional biological fields like molecular biology, which use experimental techniques to research and measure molecular properties, discover interactions, and build causal pathways. Rather than study biomolecules in vitro, systems biology studies in vivo the dynamic behavior and properties of entire biological systems that are formed by these biomolecules.

This new approach in biology shows the growing awareness among scientists that biological functions of particular components depend on their participation in complex systems, and it is only at this scale that we can construe predictively accurate models for theoretical and practical purposes. Moreover, it also helps them understand that qualitative properties of complex biological systems are not simply functions of quantitative aggregation of their physically simpler constituents.[2]

The holistic attitude of systems biology becomes one of the main inspirations of the recent revival of emergentism in both science and the philosophy of science. In its contemporary ontological (strong) version, the theory of EM acknowledges the reality of layered strata or levels of systems, which are consequences of the appearance of an interacting range of novel qualities. "[These qualities'] novelty is not merely temporal (such as the first instance of a particular geometric configuration), nor the first instance of a particular determinate of a familiar determinable (such as the first instance of mass 157.6819 kg in a contiguous hunk of matter)."[3] The qualities in question are novel, nonstructural, and fundamental. Moreover, in its most recent version, ontological EM is closely related to the theory of DC—that is, new, primitive, and top-down-oriented causal power, which is regarded as a decisive and most characteristic trait of emergent systems.

But what for many is the essence of EM and the foundation of the new ontological position called nonreductionist physicalism turns out to be a stumbling block and an obstacle for others, who acknowledge the metaphysical and logical inconsistencies of the EM theory based on the idea of DC. Their criticism inspires, in turn, a current development of a processual and dynamical version of EM. Its protagonists accuse the followers of the classical account of EM of being stuck in, and limited by, the mereological (part-whole) way of thinking. They see their own proposition as more consequent in applying a systems approach in natural science and philosophy of science. They also think that the whole endeavor shows the need for a new metaphysics that will give a better account of dynamical and processual changes in nature. Whether their project is capable of solving all philosophical puzzles and questions concerning EM, DC, and nonreductionist physicalism remains an open question, which is one of the main motives for the inquiry pursued in this part of my project.

I will begin by analyzing the most important metaphysical postulates of EM, including some necessary preliminary references to the historical context of their formulation (chapter 1). My investigation of the central dogma of EM will be then followed by an inquiry concerning its main metaphysical challenges, weaknesses, and flaws, showing the need and opening a way to its redefinition in terms of a more robust theory of causation (chapter 2). Finally, I will provide a thorough metaphysical analysis and a critical evaluation of the project developed by Terrence Deacon, in which he applies categories of causation related to those of Aristotle in his original dynamical depth model of EM (chapter 3).

CHAPTER 1

The Central Dogma of Emergentism

Even if some theorists of EM find it reasonable to look for its origins as early as in Aristotle, Plotinus, and Hegel,[1] I believe the concept has its roots primarily in the nineteenth-century analysis of the so-called composition of causes, developed by a group of philosophers seen as the protagonists of British emergentism. A thorough historical analysis of EM is not needed here. Since my interest is in its central metaphysical features and commitments, I will limit the historical inquiry to a short summary of the five main phases of the development of EM.[2]

1. HISTORICAL FACETS

The first phase of the historical development of emergentism was inspired by the dramatic advances in chemistry and biology in the nineteenth century that challenged conceptual bridges constructed earlier between those disciplines and physics. It is usually associated with the philosophy of John Stuart Mill (1806–73) and George Henry Lewes (1817–78), who analyzed the so-called compositions of causes and introduced the concept of emergent effects (Lewes). Mill's thought was picked up by the Scottish philosopher Alexander Bain (1818–1903), who spoke about new forces of nature caused by certain collocations of agents. After several decades, at the beginning of the twentieth century, emergentism reappeared in the philosophy of biology, in opposition to vitalism and mechanistic reductionism. Its major proponents were Samuel Alexander (1859–1938), Conwy Lloyd Morgan (1852–1936), and Charlie Dunbar Broad (1887–1971). The third phase of the debate on the concept of EM brought much criticism and skepticism about its relevance, due to the antimetaphysical agenda of logical positivism and analytical philosophy, in the middle of the twentieth century, in both continental and American contexts. The fourth phase, and the revival of emergentism, coincides with the debate on the mind-brain problem and with contributions to this debate, especially by Mario Bunge (b. 1919), Karl Raimund Popper (1902–94), Roger Wolcott Sperry (1904–94), and John Jamieson Carswell Smart (1920–2012). The fifth, current phase, besides the continuing research in the field of brain studies and philosophy of mind, includes the discovery and description of emergent properties in other branches of molecular and systemic biology. Moreover, research in emergent studies has been recently enriched by a broader analysis of the scientific, metaphysical, and theological aspects and implications of EM, which becomes the catalyst for our present discussion.[3]

2. CHARACTERISTICS OF EMERGENCE

The plurality of different realms of natural and human sciences referring to EM generates a multiplicity of meanings and classifications of different types of emergent properties, making it difficult to provide a universal definition of the term. Moreover, philosophical analyses and explanations of EM differ remarkably from the scientific ones. The former examine more speculative, ontological, and causal dimensions of the concept of EM, whereas the latter search more for its practical aspects and examples, limiting theoretical discussion to a minimum.[4] Following the first, more theoretical and speculative path, I will now try to list the most important philosophical characteristics of EM.[5]

2.1. Nonadditivity of Causes

The father of British emergentism, John Stuart Mill, distinguished (a) physical composition of causes and transition laws based on the vector or algebraic addition from (b) the chemical mode of combined action of causes and transition laws, where the product is not an algebraic sum of the effects of each reactant: "[The] difference between the case in which the joint effect of causes is the sum of their separate effects, and the case in which it is heterogeneous to them; between laws which work together without alteration, and laws which, when called upon to work together, cease and give place to others; is one of the fundamental distinctions in nature."[6] Mill called the mechanical (physical) type of effect a "homopathic effect" (governed by "homopathic laws"), as opposed to the chemical type of effect, which he named a "heteropathic effect" (governed by "heteropathic laws"). He claimed that heteropathic laws supersede the homopathic laws, providing an explanation for the need of special sciences, as it is impossible to deduce all chemical and physiological truths from the laws or properties of simple substances or elementary agents. Moreover, even if heteropathic causes can combine in accordance with the composition of causes to produce new homopathic effects, chemistry, which describes their instantiations, is still far from being a deductive science. It cannot be reduced to a small group of systematically well-integrated laws from which all other laws proper to it can be derived. But if such is the status and nature of chemistry, Mill concludes, then the laws of life will be all the more nondeducible from the laws of its chemical ingredients.

Lewes follows Mill's idea concerning nonadditivity of causes, while introducing new terminology distinguishing between "resultant" and "emergent" effects, the latter being incommensurable and irreducible to the sum of their components. He coins the very term "emergence" in saying: "There are two classes of effects markedly distinguishable as resultants and emergents. Thus, although each effect is the resultant of its components, the product of its factors, we cannot always trace the steps of the process, so as to see in the product the mode of operation of each factor. In this latter case, I propose to call the effect an emergent."[7] In more recent versions of emergentism we find those who relate nonadditivity of causes to the nonlinear features of complex systems— for example, being in the basin of a strange attractor—and treat it as motivating a nonreductionist version of physicalism.[8] Moreover, Jessica Wilson suggests that the proposition of Shoemaker, who distinguishes between "micro-manifest" and "micro-latent" powers of lower-level entities, can also be better understood in the context of the same idea of nonadditivity of causes.[9]

Nonetheless, going back to Mill and Lewes and their original formulation of the rule of nonadditivity of causes, we realize that the first definition of EM was given in reference to cause-effect relationships—that is, in the language characteristic of philosophy of causation. Indeed, as we will see, the causal aspect of EM theory proves to be crucial and indispensable for its relevance and plausibility. But before we explore it further, we should consider first some other philosophical features of EM.

2.2. Novelty of Complex Processes, Entities, and Properties

That the first definition of EM is given in causal terms requires from us a further reflection on the nature of reality grounding causal dependencies. Here we encounter a plurality of propositions concerning the metaphysical nature of emergents. Some of them are used interchangeably. Among the first emergentists, Samuel Alexander uses processual language when he talks about the "collocation of motions" possessing "a new quality distinctive of the higher complex" and "expressible without residue in terms of the processes proper to the level from which they emerge."[10]

Those who refer to chemical examples of EM (Mill, Lewes, Morgan), even if they acknowledge the processual character of chemical reactions, describe the emergent character of their outcomes in terms of entities and substances. Lewes, for instance, classifies liquid water or hydrogen as emergent. Broad contrasts pure mechanism, defined as the composition of all matter of the same stuff, with the emergent character of wholes. Both examples refer indirectly to the language of substance.

Probably the most popular among emergentists, however, is the language of emergent qualities and properties. Alexander describes them in terms of powers, dispositions, or capacities. Broad speaks of properties specific for emergent order and calls them "ultimate characteristics," in contrast with "ordinally neutral characteristics" and "reducible characteristics." He sees them as dependencies pertaining simultaneously among structures at different scales and distinguishes such synchronic aspects of EM from its diachronic features, which deal with the ways in which phenomena and properties develop over time:

Put in abstract terms the emergent theory asserts that there are certain wholes, composed (say) of constituents A, B, and C in relation R to each other; that all wholes composed of constituents of the same kind as A, B, and C in relations of the same kind as R have certain characteristic properties; that A, B, and C are capable of occurring in other kinds of complex where the relation is not the same kind as R; and that the characteristic properties of the whole R(A,B,C) cannot, even in theory, be deduced from the most complete knowledge of the properties of A, B, and C in isolation or in other wholes which are not of the form R(A,B,C). The mechanistic theory rejects the last clause of this assertion.[11]

2.3. Emergence and Supervenience

Jaegwon Kim notices that the property account of EM is related to the concept of supervenience (SUP). When describing it, Kim lists three putative components of supervenient properties: covariance, dependency, and nonreducibility. He claims that covariance (indiscernibility in respect to the base properties entailing indiscernibility in respect to supervenient properties) is metaphysically neutral, whereas dependency suggests an ontological and explanatory directionality. Although both are needed to describe supervenient relations, Kim thinks one should separate the covariation element from the dependency in SUP, because it helps to understand that covariation alone does not entail dependency. For him the question of what must be added to covariation to yield dependence remains an interesting and deep metaphysical query.[12] Paul Humphreys asks similar questions about the strength of necessitation involved in SUP and says that it can be interpreted as (a) logical necessity, (b) metaphysical necessity, (c) nomological necessity, or (d) conceptual necessity (we will say more about [c] and [d], respectively, in sections 2.5 and 2.8). He contends that they are subjects of considerable controversy.[13]

The nonreducibility component of SUP states that supervenient properties cannot be simply educed from their base properties. We will see that this is common to emergent properties as well. And yet we must not forget that EM is not the same as SUP. Humphreys claims that SUP is acceptable merely as a consistency condition, enabling attribution of concepts concerning properties characteristic for different levels of complexity, while EM provides an explanation of ontological relationships between them.[14] Hong Yu Wong, differentiating emergent (nonstructural) from resultant (structural) properties, claims that in the case of the former kind of properties, SUP is sui generis and does not suffice to establish their ontological status: "Emergent properties are *nonstructural* properties, in contrast to resultant complex properties which are *structural* properties; but both supervene on basal properties. Since emergent properties are nonstructural, supervenience in the case of emergent properties must be sui generis; it is not a matter of constitution, identity, realization, causation, or any of the usual relations that ground supervenience; it is a matter of fundamental, non-derivative emergent laws."[15]

2.4. Ontology of Levels

Humphreys's allusion to different levels of complexity turns our attention toward another characteristic feature of EM. The language concerning various stages of organization became typical for all proponents of EM throughout the history of the concept.[16] Samuel Alexander was the first to introduce it in the following passage from his *Space, Time and Deity*: "The higher quality emerges from the lower level of existence and has its roots therein, but it emerges therefrom, and it does not belong to that lower level, but constitutes its possessor a new order of existent with its special laws of behavior."[17] But even if the majority of emergentists followed Alexander's idea, it was only recently that a more detailed reflection concerning the metaphysics of levels of complexity was offered by Claus Emmeche, Simo Køppe, and Frederic Stjernfelt. They develop an ontological specification and classification of levels, emphasizing their physical interdependence and metaphysical "inclusivity," which goes beyond simple mereological composition. According to them, the higher level, being materially related to the lower one, does not violate its laws, while at the same time it cannot be simply deduced from it, which protects EM from falling into the pitfalls of dualism and eliminativism. In terms of ontological priority and posteriority, Emmeche et al. describe the status of higher and lower levels in this way: "A rational idea of levels must entail that the more basic levels are basic in the sense of the word that they are presupposed by the higher levels—but the word 'basic' does not entail any ontological priority. The higher levels are as ontologically pre-eminent as the lower ones, even if being presupposed by them, that is, they are defined by properties [and] by special cases of the lower levels. In this respect, levels are ontologically parallel, but non-parallel in so far as they coexist."[18]

After defining their ontological status, Emmeche et al. propose a basic classification of levels. They distinguish four "primary levels": the physical, the biological, the psychological, and the sociological; and they suggest that each can serve as a base for various sublevels. The biological level, for instance, can contain the cell, the organism, the population, the species, and the community sublevels. Emmeche et al. also clarify that the interlevel relations are "nonhomomorphic" in the sense that the EM of the biological from the physical level does not have the same complex of interlevel relations of dependence as the EM of the social and psychic levels from the biological, due to the continuous mutual conditioning and interdependence between emergent psyche and sociality.[19]

Although the ontology of levels proposed by Emmeche et al. brings some important clarifications, Jaegwon Kim asks further questions that still need to be answered: "How are these levels to be defined and individuated? Is there really a single unique hierarchy of levels that encompasses all of reality or does this need to be contextualized or relativized in certain ways? Does a single ladder-like structure suffice, or is a branching tree-like structure more appropriate? Exactly what ordering relations generate the hierarchical structures?"[20]

Humphreys asks similar questions and states that the criterion of special mereological inclusion of entities of a type A within entities of a type B is insufficient for establishing a hierarchy of levels. He suggests that a more promising account of levels adds the requirement that higher-level entities obey different laws than the lower-level entities. Nonetheless, he states that this criterion still depends on compositionality. For if we do not take compositionality into account, two types of simple fundamental entities, such as fermions and bosons, can be governed by different laws, which might erroneously suggest that they occupy different levels of complexity, while they belong to one and the same level of complexity of matter. Consequently, Humphreys suggests avoiding the language of levels and replacing it with the category of domains, which he finds neutral in terms of specifying the metaphysical relation among entities that belong to them and allowing for diachronic accounts of EM (unlike the levels approach, which is entirely synchronic).[21]

2.5. Emergent Laws

Humphreys's reference to the concept of emergent laws in order to salvage the talk of emergent levels of complexity/organization is not unique in the literature devoted to emergentism. As we have seen, when defining EM in terms of higher and lower levels of organization and complexity, Samuel Alexander takes the same path and mentions new laws ruling the emergents. Mill's idea is similar, when, contrasting homopathic and heteropathic effects, he concludes that the latter are governed by new, heteropathic laws. We can, therefore, classify new laws typical of higher levels of complexity as another characteristic feature of EM. Obviously, the status of the laws of nature is widely debated in the philosophy of science. We do not assign to them any kind of ontological character, for we see them as descriptive rather than prescriptive.[22] But even if they simply describe the results and effects of the activity of emergent beings and dynamical systems, they are real features of the world and thus might be considered as important indicators of EM. Those who define EM in reference to the role of novel scientific laws in emergent phenomena often speak of nomological EM.[23] Broad gives an interesting analysis of emergent laws, which he calls "trans-ordinal laws." His reflection is worth quoting at length, as it guides us toward three other characteristics of EM:

> We should have to recognise aggregates of various orders. And there would be two fundamentally different types of law, which might be called "intra-ordinal" and "trans-ordinal" respectively. A trans-ordinal law would be one which connects the properties of aggregates of adjacent orders. A and B would be adjacent, and in ascending order, if every aggregate of order B is composed of aggregates of order A, and if it has certain properties which no aggregate of order A possess and which cannot be deduced from the A-properties and the structure of the B-complex by any law of composition which has manifested itself at lower-levels. An intra-ordinal law would be

one which connects the properties of aggregates of the same order. A trans-ordinal law would be a statement of the irreducible fact that an aggregate composed of aggregates of the next lower order in such and such proportions and arrangements has such and such characteristic and non-deducible properties.[24]

2.6. Nondeducibility, Nonpredictability, and Irreducibility of Emergents

In the above passage from Broad and in his account of the synchronic aspect of EM quoted in section 2.2 we find him speaking about emergent properties and laws being nondeducible from, and irreducible to, the properties and laws characteristic of the lower orders of complexity. With regard to their nondeducibility he holds that the emergent, trans-ordinal laws are brute nomological and nonderivative facts that cannot be explained, as they do not reflect any radical "disunity in the external world." They must, he says, "simply be swallowed whole with that philosophical jam which Professor Alexander calls 'natural piety.'"[25]

While deducibility entails deriving a conclusion from something already known, predictability is based on envisaging future states of affairs. Karl Popper, who became one of the main proponents of defining EM in terms of nonpredictability, distinguishes between an absolute nonpredictability of events and nondeducibility of properties. He defines the former in terms of a thoroughgoing event-centered indeterminism: "[The] emergence of hierarchical levels or layers, and of an interaction between them, depends upon a fundamental indeterminism of the physical universe. Each level is open to causal influences coming from lower and from higher levels."[26] His idea of EM as nondeducibility of properties, on the other hand, is based on a property-centered indeterminism, expressed in his famous dictum: "[There] is the fact that in a universe in which there once existed . . . no elements other than, say, hydrogen and helium, no theorist who knew the physical laws then operative . . . could have predicted all the properties of the heavier elements not yet emerged, or that they would emerge."[27]

Kim softens Popper's radical view on the indeterminism and nonpredictability of emergents when he distinguishes between "inductive" and "theoretical" predictability. Knowing from experience (empirically) that the property E emerged from a certain lower-level property M of a system S at the time t, we are able to predict and formulate a general emergent law which says that whenever the system S instantiates the base condition M, the emergent property E will appear as well. This "inductive" predictability differs from a "theoretical" one. No matter how accurate and detailed our knowledge of S and M is, we cannot theoretically predict an EM of a new property E. This unpredictability, says Kim, is due either to our not having the concept of E before it actually occurs or to some possible changes in the microstructure of M that, transforming it into M^*, will cause the EM of E^*, instead of E.[28]

While nondeducibility and nonpredictability of emergents seem to support weak (inferential and/or conceptual) EM, the irreducibility thesis points toward strong (ontological) EM. "Irreducibility" refers, primarily, to both the simple intertheoretical (homogenous) and the bridge-laws-based (nonhomogenous) Nagelian types of ontological and causal reductionism. Both pathways of reduction mentioned here find their point of departure in Nagel's preliminary assertion that "a reduction is effected when the experimental laws of the secondary science (and if it has an adequate theory, its theory as well) are shown to be the logical consequences of the theoretical assumptions (inclusive of the coordinating definitions) of the primary science."[29] What differentiates them is that the former (simple intertheoretical reduction) covers homogenous cases, which do not require any reference to bridge laws, while the latter (bridge-laws-based reduction) covers nonhomogenous cases, which do need to implement the middle step engaging bridge laws.

Nagel offers these strategies as a necessary theoretical base for the project of the ultimate "identity" or "conservative" (but not eliminative) reduction of sciences dealing with complex phenomena to physics. A central difficulty of his project, however, is the question whether the bridge laws he argues for are true definitions rather than contingent identity claims or regulations. Kim thinks that they simply extend the linguistic or conceptual resources of the reduction base and that, therefore, they themselves require an explanation.[30]

Having said this, Kim proposes his own, functional model of reduction, which consists of three steps: (1) Emergent property E must be "functionalized"—that is, it has to be construed or reconstrued as a property defined by its causal/nomic relations to other properties of the system S. (2) E must have its realizers among the properties of S; in other words, a property M needs to be found that instantiates E. (3) A theory needs to be offered which explains how M-level realizers of E perform their causal task (e.g., a gene can be functionally characterized as "the mechanism in a biological organism causally responsible for the transmission of heritable characteristics from parents to offspring").[31]

Kim claims that functionalization of a property is both necessary and sufficient for reduction. However, Humphreys contends that Kim's model may fail in a most direct way in cases where a given higher-level property cannot be functionalized. He adds that this does not mean that the higher-level property in question would be causally isolated, as it could still enter into causal relations as an emergent property having its own, distinct causal powers. He concludes that the Kimian reduction does not rule out EM altogether, since "it is a contingent fact about the world how many properties exist that are resistant to the functionalization procedure."[32]

2.7. Emergence and Downward Causation

The reason Kim's functional model of reduction seems attractive to EM theorists is its concentration on the causal/nomic definition of properties. It reminds us that any description of a property at any level of organization of matter cannot ignore the question of its causal contribution to ongoing processes of the world. That is why, after describing several basic characteristics of emergent properties, we ask together with Kim one more question: What can emergents actually "do" after having emerged? For without having causal powers, emergent properties would simply turn out to be epiphenomenal and, as Samuel Alexander says, "undoubtedly would in time be abolished."[33] Addressing this question, Kim defines the last important feature of emergentism: "Emergent properties have causal powers of their own—novel causal powers irreducible to the causal powers of their basal constituents."[34]

These novel causal powers are usually given the name "downward causation" (some philosophers refer to downward- or macrodetermination). Unlike upward causation (an instantiation of a higher-level property by a lower-level property) and same-level causation, DC occurs when an emergent higher-level property causes (has an influence on) the instantiation of a lower-level property. An example of DC that is often brought and hotly debated in most recent philosophy of mind is the causal influence of mind on the brain—out of which it emerges—and on the activities of the entire human organism. Some thinkers give as an example of DC the phenomenon of symbiosis in biological systems. They emphasize that as an emergent phenomenon, it is not merely based on the simple addition of the contributions brought by the two different organisms. Its nature changes the microstructure and physiology of each one of them, often making their existence possible in an environment which otherwise would be unfavorable or even lethal for one or both of them.[35] Others apply the language of DC to the larger-scale description of nature. They speak about social influences acting down on individual human brain structures (social neuroscience) or about environment causally shaping biological development of individual organisms.

The term DC was introduced by an American social scientist, Donald Thomas Campbell (1916–96), in his paper delivered at the conference on the problems of reduction in biology in 1974. Acknowledging the above-mentioned theory of the world containing faculty-based hierarchy of levels (e.g., the stratification of molecules, cells, tissues, organs, organisms, populations, and species), Campbell formulates the emergentist principle in reference to biological evolution. He declares that in its exploration of segments of the universe, evolution encounters laws "which are not described by the laws of physics and inorganic chemistry, and which will not be described by the future substitutes for the present approximations of physics and inorganic chemistry."[36] Following this quite radical thesis is Campbell's definition of DC, in which he states that "all processes at the lower levels of a hierarchy are restrained by and act in conformity to the laws of the higher levels."[37] Again, in case of evolutionary processes, says Campbell, "where there is a node of selection at a higher level, the higher level laws are necessary for a complete specification of phenomena at both the higher level and also for lower levels."[38]

Campbell's ideas found support in Popper's and Roger Sperry's versions of EM. The latter, discussing mental phenomena, contrasts the reality of micro- and macro-properties, saying that "macro-determinism thus begins to be superimposed upon micro-determinism from the earliest stages onward and grows by a compounding process into increasing prominence as evolution progresses.... Microdeterminism is retained but is held to be incomplete, insufficient. The properties, forces and laws of micro-events are shown to be encompassed and superseded, not disrupted, by the properties, forces, and laws at macro-levels."[39]

An interesting—more recent and more general—account of DC can be found in the article on the rise and dissipation of molecular biology, coauthored by Powell and Dupré:

> When we attempt to generalise our thoughts about downward causation one idea that results is this: the influences, or causal potentials, that act at a particular location or region can come from multiple directions and distances, and the way the influences sum and act, together with the nature of whatever exists at that location, is determinative of what happens there. (By influences we typically mean forces arising from fields, whether acting over long or short ranges. At different scales, different fields predominate.) ... Conversely, parts are compatible—in the sense that they can retain their integrity or identity—only with particular contexts, within bounds beyond which they are changed. The conditions under which, and the manner in which, they are changed depend in complex ways on energetic and other factors.[40]

Moving beyond biological and neurobiological examples mentioned here, Michele Paolini Paoletti and Francesco Orilia suggest a list of phenomena across major contemporary scientific disciplines that could leave room for irreducible DC:

1. *physics*: quantum entanglement; high-temperature superconductivity; the arrow of time; the relationship between pressure, temperature and density in gas laws and molecular behaviour; Rayleigh-Bénard convection;
2. *chemistry*: symmetry breaking in virtue of configurational Hamiltonians; the chirality of complex macromolecules; more generally, the structure of macromolecules; the structures of atoms and their micro-physical effects in establishing chemical bonds;

3. *biology*: feedback control systems; natural selection; homeostatic processes; eukaryotic cells; the DNA code and the transcription of proteins; epigenome and the activation of genome; the behaviour of ants in colonies;
4. *neurosciences*: neural populations; neural mechanisms; neuroplasticity in the development of synapses (through learning and other activities); intentions and brain activities;
5. *psychology*: perception of wholes (according to Gestalt theories); beliefs; consciousness and mental states; qualitative feels (emerging from interconnected representations of certain objects and affecting those same representations); conceptual representations (emerging from memories and affecting them); language and individual speakers;
6. *sociology*: social institutions; values; norms; education; the value of money.[41]

At the same time, even though the concept of DC is usually defined in the context of the more contemporary philosophical reflection on the character of certain natural phenomena, we should not neglect some early contributions to its development brought by British emergentists. Arguing against epiphenomenal explanation of consciousness, Alexander declares: "The mental state is the epiphenomenon of the neural process. But of what neural process? Of its own neural process. But that process possesses the mental character, and there is no evidence to show that it would possess its specific neural character if it were not also mental. . . . A neural process does not cease to be mental and remain in all respects the same mental process as before."[42] Similar is Morgan's idea of the "involution" between different levels of properties complexity. In his *Emergent Evolution* he claims: "When some new kind of relatedness is supervenient (say at the level of life), the way in which the physical events which are involved run their course is different in virtue of its presence—different from what it would have been if life had been absent."[43] Both quotations clearly show an attempt of early emergentists at articulating an account of a new type of causality going from the top to the bottom level of complexity.

The power of the concept of DC lies in its providing an answer to the apparent insufficiency of the compositional explanation of complex higher-level entities encountered in nature and investigated by natural sciences.[44] At the same time, however, DC triggers a detailed and nuanced philosophical analysis and debate. We have already discussed in the introduction (section 6.4) the skepticism of the new philosophical mechanists about its reality. We will say more about this and other questions related to the notion of top-down causality in the next chapter. Nonetheless—regardless of the challenges it faces—DC remains for many proponents of EM the primary criterion of the distinction between its weak and strong versions, and a general sine-qua-non condition of the ontological reality of emergents.

2.8. Weak and Strong Emergence

As I mentioned in the introduction, weak EM is usually understood as a result of limits in our human analytic possibilities, and thus free from any ontological commitments. Consequently, we should expect that with the development of human knowledge and cognitive skills these epistemic (and often subjective) emergent properties will simply dissolve and disappear. Strong EM, on the other hand, is objective, as it assumes an ontological change, resulting in the occurrence of DC of emergent properties.[45] The change in question is usually described as transformation or fusion. The former refers to an individual entity or property, while the latter characterizes a change affecting a group of basal objects or properties.[46]

However, it is important to note that weak EM does not have to be radically separated from strong EM. This is the argument raised by Humphreys, who distinguishes within the domain of weak EM inferential and conceptual types of EM. He defines the former in reference to the situation in which "the process leading to the end-state of the system is computationally, and hence predictively, incomprehensible. That is, the only way to predict the behavior of such a system is to stimulate it using a step-by-step iteration of the dynamical process that leads to the emergent state."[47] The latter, conceptual EM maintains that "an entity, such as a state or a property, is conceptually emergent with respect to a theoretical framework F if and only if a conceptual or descriptive apparatus that is not a part of F must be developed in order to effectively represent that entity."[48] Most importantly, although the nature of inferential and conceptual types of EM is epistemological, "a tighter connection with ontological emergence can be gained in cases where we have good reason to believe that T [theory] is a correct representation of S [system]." Consequently, "if new, fundamental ontological features such as laws emerge, the predictive inadequacies would reflect the appearance of ontological diachronically emergent features."[49] In other words, a further development of scientific and theoretical explanation may lead not to a dismissal of weak EM but to its pointing toward/turning into a case of ontological EM.[50]

Clayton claims that "weak emergence is the starting position for most natural scientists."[51] He thinks it should not be a problem for contemporary science and philosophy of science to accept characteristics of relationships between lower and higher levels of organization of matter and the rule of supervenience, after they have rejected materialistic and mechanistic reductionism. We have seen that it is true in the case of the new philosophical mechanists. However—adds Clayton—the concept of strong EM and emergent properties exerting DC on the lower levels, from which they themselves have emerged, can be a challenge, if not a threat, to contemporary science, especially to those who want to avoid reductionism but keep a physicalist position (nonreductive physicalists). An awareness of this fact brings the matter of the present chapter to conclusion, opening the space for the discussion which will be the main concern of the next one.

3. SUMMARY

The challenges posed by DC to the consistency of the physicalist position in science that I have only alluded to here and will discuss in more detail below, as well as the questions DC raises in terms of the philosophical reflection on the nature of complexity and causation, become, in my opinion, a sign of some deeper metaphysical problems confronting the whole theory of EM. Looking at all the characteristics of EM, as they were described above, one may strive to see them as building up into a well-thought-out, coherent, and plausible theory, relevant to both the scientific picture of nature and a philosophical reflection based upon it. But the truth is that almost every one of these features of EM carries a bevy of metaphysical difficulties, many of them focused on the problem of irreducibility. The issue at stake is the question whether the ground for the irreducibility of emergents has a scientific, philosophical, or a mixed character. I will address this and other matters concerning the metaphysics of EM and DC in the following chapter, to which we now turn.

CHAPTER 2

Metaphysical Challenges for Emergence and Downward Causation

The theory of EM and DC pursued an astonishing career and has become—over the last few decades—probably the most popular conceptual tool used in the study of complex phenomena by both philosophically oriented scientists and scientifically informed philosophers. It is commonly used in our attempts at providing a viable explanation of the nature and character of global organization observed in various phenomena such as magnetization, laser light, crystallization, neuronal networks, living organisms, collective behavior, or ecological systems.[1] The central dogma of emergentism described in the previous chapter in terms of nonadditivity of causes, novelty, nondeducibility, nonpredictability; irreducibility of complex dynamical systems, entities, and properties; ontology of emergent levels of complexity and laws governing them; and DC is commonly accepted by scientists and philosophers of science. But what appears to them as a consistent, well-thought-out, and coherent theory, may seem in the eyes of an Aristotelian metaphysician to be nothing but a confusing welter of ideas, characterized by a maddening ambiguity of terms. An attempt to assess the ontological meaning and value of the expressions used to specify and define EM and DC reveals their ambiguity and imprecision. Both concepts raise some serious metaphysical concerns which cannot be ignored, and I will analyze them in the present chapter.

1. INSUFFICIENCY OF THE NEGATIVE DEFINITION OF EMERGENCE

Not without significance is the fact that the first reflection on EM and its first definition given by Mill and Lewes were formulated in mostly negative terms. This already shows the ambiguity which will be inextricably linked to the emergent theory ever after, making it difficult to specify its character and nature. Indeed, it seems relatively easy to say what EM is not—that is, which tools of mathematical description commonly used in scientific research are inadequate for describing the cases of global organization of entities and dynamical systems. At the same time, attempts at a positive description of EM are usually methodologically troublesome and confusing.

We saw Mill distinguishing and describing EM in terms of the nonadditivity of causes. His definition has a negative character. All that he notes is that in chemical reactions vector and algebraic addition of the effects of all reactants does not apply. The only positive aspect of his theory is an introduction of the concept of heteropathic laws, which govern the processes in question and are differentiated from homopathic laws, which are typical of mechanical processes and amenable to algebraic and vector addition of their effects. The reflection presented by Mill is highly insufficient. Why are the effects of chemical reactants not additive? Because they have features which are not of a physical nature? Or perhaps they are governed by a new type of causation, different from efficient causation studied in classical mechanics? If so, what is its nature? What is the character of heteropathic laws, and how do they differ from homopathic laws? In other words, does Mill's idea of nonadditivity of causes and heteropathic laws have an ontological aspect to it? And if it does, what is the subject of the ontological change?

In his own version of nonadditivity of causes, Lewes claims that an emergent feature is not simply a resultant effect of its components, as one cannot see in it the mode of operation of each factor that contributed to its instantiation. Again, this assertion has a purely negative character and is not sufficient to define the nature of EM. Nor does it tell us much about the ontological status of emergents. In fact, this insufficiency of the philosophical *via negativa* in emergent studies inspired a more advanced reflection on the metaphysics of emergents. Its closer analysis, however, opens a way to further doubts and questions, which I will address in the following sections.

2. ONTOLOGICAL STATUS OF EMERGENTS

As we have seen, with respect to the metaphysical nature of emergents, the most widespread tendency among emergentists is to define them in terms of qualities and properties. Stephan shares this general attitude and expresses it in his program of reduction of all possible candidates for "emergents" to properties:

> In the literature on emergence various types of things have been characterized as emergent: laws, effects, events, entities, and properties. However, we should easily agree to explicate the so-called emergent entities and events in terms of emergent properties: An entity is said to be emergent iff it has emergent properties. An event is said to be emergent iff it is an instantiation of an emergent property. Emergent effects, as mentioned by Lewes, have to be interpreted as emergent entities (substances) or properties. Thus, what remains are laws and properties. A law may be called emergent iff it contains emergent properties. Laws may do their work either in the context of transition theories explaining changes or in the context of property theories explaining the instantiation of properties by complex systems.[2]

Stephan's opinion raises some serious ontological questions, as the very term "properties" is multiply ambiguous: (1) some properties seem to have semantic work to do, while others have a causal role to play, (2) some seem to be physical, while others are abstract, (3) some theorists find them playing a metaphysical role as universals pointing toward natural and unnatural kinds, while others see them as tropes (particular instances) building up entities. In classical Aristotelian metaphysics, properties are always "attached to" or can be regarded as "parts of" something—that is, they are properties of concrete entities. No properties float freely in nature. They fall into the category of accidents, which do not have existence by themselves (*per se*) but are always mediated in substance. Hence, it is the substantial form of the entity that decides about the ontological status of its properties. In other words, a new property is always a sign and a function of a new substantial or accidental form of the entity that mediates it.

Naturally, one may question the attempt to explain the nature of properties in terms of classical metaphysics and prefer to analyze their status in the context of the contemporary ontological debate. Here, we need to realize that analytic metaphysics distinguishes between (1) relational and (2) constituent ontologies. The former (1) insists that properties are not metaphysical parts of substances, which are best described as particulars standing in a fundamental relation of exemplification to the universals that ground their character. The latter (2) sees substances exemplifying properties by having them as parts and comes in two versions: (2a) bundle and (2b) substratum theories. The problem of the relational ontology (1) is that it seems to suggest that substances have no intrinsic properties, which makes the attempt to define emergents in terms of emergent properties implausible. Those who follow the bundle theory (2a) tend to see substances as either bundles of tropes (individualized or particular properties) or bundles of universals, grounding their character. The problems of specifying the principle of unification of tropes (so-called metaphysical glue), the possibility of accidental changes (because all tropes seem to be equally important for the character of substance, it cannot change in any way), the identity of indiscernibles, and the principle of individuation seem to provide an argument in favor of the substratum theory (2b). In its most popular version, which assumes the reality of Locke's bare substrata, the theory faces the problem of assuming some similarity and character of supposedly propertyless substrata, with regard to their common ability to exemplify different universals. Its alternative version—that is, the modular substance theory—assumes that substrata do have one dimension of character determining the nature of the substrate's substance, which leads Koons and Pickavance to suggest that substrata are "meant to be thinly characterized with substance kinds."[3] This argument brings the substratum theory closer to the robust Aristotelian view of substantial form and hylomorphism, which I find still preferable based on its clarity and explanatory power.[4]

In this context, I find Stephan's suggestion, that the nature of emergents should be defined in terms of properties, troublesome and questionable. When we hear him saying that "an entity is said to be emergent iff it has emergent properties," we need to ask about the ontological priority and the metaphysical character of the entity or a group of entities in question. To give an example, is water emergent because it has the properties of liquidity, wetness, viscosity, and so on? Or rather, does it have these properties because it is emergent? This would mean that its very nature is ontologically different because of the new substantial or accidental form of the conglomeration of H_2O molecules—the phenomenon that makes them exhibit new characteristics which are not present in H_2O molecules taken separately. If the latter is true, then the emergent character of entities is not reducible to emergent properties characteristic of them.[5]

Similar are the challenges concerning the idea of the reduction of the emergent character of events. In a purely Humean spirit, events are often regarded as causal in their nature. But assigning to events the quality of causation does not give them an independent ontological status. Their occurrence needs to engage concrete entities and becomes a function of cause-effect relations between them. Consequently, Stephan's claim that "an event is said to be emergent iff it is an instantiation of an emergent property" raises again a metaphysical query concerning the ontological ground of the property in question. It must be a property of something, and its emergent character requires an ontological base. Therefore, strictly speaking, it is not an event that instantiates an emergent property, but rather concrete entities that are constitutive for this event, providing an ontological ground for emergents.

Stephan's attempt to describe new emergent laws in terms of emergent properties seems to be even more confusing. In his claim that "a law may be called emergent iff it contains emergent properties," he seems to argue in favor of attributing to the laws of nature an ontological character. Such is at least the flavor of his idea of the containment of emergent properties in a law. As I have said, in Aristotelian philosophy properties must be rooted in substances (concrete beings). They do not have an ontological status on their own. To describe the relation between properties and substances one can use the category of containment, but to suggest a containment of emergent properties in laws of nature is to implicitly assign an ontological status to the latter which is philosophically dubious. However, this might not necessarily be the position embraced by Stephan. Continuing his reflection, he states: "Laws may do their work either in the context of transition theories explaining changes or in the context of property theories explaining the instantiation of properties by complex systems." On the basis of this assertion we may assume Stephan assigns to laws of nature a descriptive rather than a prescriptive character and locates the ontological basis of emergent properties in complex systems. If so, however, his own rule of reducing all candidates for emergents to properties is endangered, as complex systems seem to have emergent qualities prior to the properties that originate from them.

The same question concerning the ontology of emergents applies to the theory of different levels of complexity and various stages of organization, which I have listed among the main characteristic features of EM. When introducing the ontology of levels, Alexander speaks about "order[s] of existence" characterized by "special laws of behavior."[6] But again, the orders of existence in question must be orders of existence of concrete beings, to which the laws of behavior mentioned by Alexander might apply. As we have seen, in their proposition of the basic classification of levels Emmeche et al. specify in more detail the character of the biological level. They list its various sublevels, such as the cell, the organism, the population, the species, and the community. Their analysis shows all the more that emergent properties have to be grounded in concrete beings (substances), which should be given the status of emergents prior to qualities that are mediated in them. Even if we define EM in terms of global organization as described in the phenomena of laser light, crystallization, embryo development, or the immune system, their emergent character must be mediated in concrete entities governed by a principle of unity and codependence. But if this is the case, then the theory of EM faces a great ontological difficulty, which seriously questions its plausibility.[7]

3. PROBLEMS OF NONREDUCTIONIST PHYSICALISM

Once we acknowledge that the description of EM in terms of emergent properties is insufficient and needs to be ontologically grounded in concrete entities, we enter a metaphysical struggle between emergentism and physicalism. The former emphasizes the nondeducible, nonpredictable, and, above all, irreducible character of emergents. The latter sees the nature of the entire universe as physical and explainable in reference to some fundamental physical laws, defined in mathematical terms. The problem of the majority of emergentists is their desire to stay in both camps. They claim to follow the rules of physical monism, which assumes that, at the end of the day, all chemical, biological, psychological, moral, or social aspects of reality are either physical or supervene on the physical. At the same time, they want to acknowledge the irreducibility of emergents. Therefore, they classify their ontological position as "nonreductionist physicalism" (NP).[8]

But in my opinion, the very concept of NP—metaphysically speaking—is self-contradictory. Why? Simply because physicalism, by definition, assumes either reductionism or eliminativism, or both. Thus Kim, when commenting on how the term gained quick currency in the late 1960s and early '70s, does not mince his words and says its ontology is physicalistic, while its "ideology" is dualistic.[9] As long as we define the first component of NP in negative terms, arguing for irreducibility and nonidentity of emergents to physical properties and relations, the whole term seems plausible. Once we try to give a positive account of how it is related to physicalism, we encounter insurmountable problems. Extending Kim's analysis of the mental to the general account of EM, we can list at least five main attempts at the formulation of the positive part of the definition of NP:[10]

1. Davidson's negative definition of the "anomalism of the mental" (the doctrine that there are no precise or strict laws relating mental events to other mental events and to physical events) can be extended to all emergents and translated into a positive rule of "anomalous monism." The rule calls for an acceptance of the crude fact that although all events in the universe have physical properties, some of them have also emergent qualities. However, any further questions are pointless, as we cannot indicate laws relating the physical to the emergent. Besides the fact that such a claim does not seem to contribute much to the positive account of NP, through its denial of laws linking physical and emergent, anomalous monism leads to a denial of any causal role of emergents, and consequently becomes another form of eliminativism.[11]

2. Putnam's "multiple realizability" (or "compositional plasticity") of psychological phenomena can be extended to the realm of all emergents and defined as a positive account of NP. Although he supports physical monism, Putnam defends a nonreductionist character of psychological event-types. He argues they can be physically realized, instantiated, or implemented in endlessly diverse ways, and thus cannot be identified with any "single" kind of physical state. But even if his theory defends emergents against a uniform or global reduction to the physical base, says Kim, it does not protect them from specific or local reductions. In other words, multiple realizability is not nonreductionist, as it allows for multiple local reductions. Moreover, it implicitly assumes that emergents are realized through certain configurations of events at the physical level without introducing any qualitative change, which makes them even more vulnerable to reductionism.[12]
3. Another argument in defense of NP comes from the theory of SUP. It defends physicalism in that it does not introduce any qualitative change in emergents which merely supervene on the physical. The same relation of SUP, however, entails not only metaphysically neutral covariance between physical and emergent, but also their nonidentity, dependence, and directionality, all three having an ontological dimension. But some have objected that SUP simply adds mystery to mystery, as it does not specify the nature of dependence between supervenient and subvenient. Nor does it define the ontological status of the former. Moreover, Kim notes the difficulty in meeting simultaneously the two main requirements of SUP: nonreducibility and dependence. Knowing that the relation between subvenient and supervenient properties needs to be weak enough to avoid their identity, which would entail reducibility, and strong enough to provide for their dependence, he states: "The main difficulty has been this: if a relation is weak enough to be nonreductive, it tends to be too weak to serve as a dependence relation; conversely, when a relation is strong enough to give us dependence, it tends to be too strong—strong enough to imply reducibility."[13]
4. Shoemaker distinguishes between manifest and latent properties. The former are present and detectable at the basic level, while

the latter are present and grounded but not detectible at the same basic level. They become realized when the system in which they are embedded reaches a required complexity.[14] Naturally, what becomes problematic in Shoemaker's theory is the ontological status of the latent properties. If they are merely physical, they have a status of hidden variables and are ultimately reducible to basic manifest properties. If they share a new ontological character, they violate the rule of physicalism.

5. The last and the most popular argument supporting NP is formulated in strictly causal terms. This is the argument from DC, which became, as we noted in the last chapter, a sine-qua-non condition of strong (ontological) EM. It finds support among many thinkers, whose thought we will analyze in the following section.

4. METAPHYSICS OF DOWNWARD CAUSATION

Hulswit's seemingly obvious and intuitive definition of DC as "a situation in which certain aspects of the active elements of an active physical whole can be adequately described only by referring to some causal influence of certain aspects of the active whole"[15] is by no means free from metaphysical problems. Closer analysis reveals a great difficulty and ambiguity among emergentists in their attempt at specifying the following: (1) What is the cause in DC?, (2) What is being acted upon?, and (3) What is the nature of DC? These questions bring us back to the problem of the ontological status of emergents, opening a way to a similar plurality of answers, which we need to investigate carefully.[16]

4.1. What Is the Cause in Downward Causation?

Many answers have been proposed to the question of what is a cause in DC. Strangely enough, a number among them concentrate on a rather vague idea of "general principles" as having causal powers. When we return to Donald Campbell, whom I referred to in the previous chapter as the first to employ the concept of DC in the 1970s, in the context of evolutionary biology, we will see that for him, "a cause" in DC means a kind of law, regularity, or principle, which can be understood as a general disposition of a "whole," such as a *"molecule, cell, tissue, organ, organism, breeding population, species,* in some instances *social system,* and perhaps even *ecosystem"*: "Where natural selection operates through life and death at a higher level of organization, the laws of the higher-level selective system determine in part the distribution of lower-level events and substances."[17]

The recurring difficulty of assigning to the laws of nature an ontological character can be mitigated by the proposition of those who emphasize dynamical aspects of EM and DC and define the causal aspect of the latter in terms of "boundary conditions," "constraints," or "patterns of organization." Michael Polanyi, who introduced the term "boundary conditions" in life sciences, understands them as higher-level principles controlling lower-level processes.[18] Arthur Peacocke uses "boundary conditions" language as well when defining whole-part constraint (his name for DC).[19] Nancey Murphy, George Ellis, and Paul Davies hold similar positions.[20] Davies and Ellis widen the terminological base, speaking of "complexity thresholds" or "higher-level structural relations." Alicia Juarrero introduces the concept of "context-sensitive constraints" in complex processes, which connect objects with their environments (systems), strengthening their embeddedness. She redefines the concept of constraints, used formally in Newtonian mechanics, and presents them not only as reducing alternatives but also as producing new possibilities.[21] The idea of patterns of organization appears in Deacon's definition of the second-order EM, when he writes, "The interaction dynamics at lower levels becomes strongly affected by regularities emerging at higher levels of organization."[22] Probably the most detailed description of causal potency of "patterns of organization" is offered by Robert van Gulick, who defines their properties as follows:

1. Such patterns are recurrent and stable features of the world.
2. Many such patterns are stable despite variations or exchanges in the underlying physical constituents; the pattern is conserved even though its constituents are not (e.g. in a hurricane or a blade of grass).
3. Many such patterns are self-sustaining or self-reproductive in the face of perturbing physical forces that might degrade or destroy them (e.g. DNA patterns).

4. Such patterns can affect which causal powers of their constituents are activated or likely to be activated. A given physical constituent may have many causal powers, but only some subset of them will be active in a given situation. The larger (i.e. the pattern) of which it is a part may affect which of its causal powers get activated. . . . Thus the whole is not any simple function of its parts, since the whole at least partially determines what contributions are made by its parts.
5. The selective activation of the causal powers of its parts (4) may in many cases contribute to the maintenance and preservation of the pattern itself (2, 3).[23]

Another set of answers given to the question of the causal factor in DC concentrates on concrete events and entities, rather than boundary conditions or patterns of organization. Nancey Murphy returns to the thought of a philosophical theologian, Austin Farrer. Although he does not use the term DC, Farrer talks in his 1957 Gifford Lectures about systems in which "the constituents are caught, and as it were bewitched, by large patterns of action."[24] Hence, concrete actions (events) have causal power in his description. As we will see in the next chapter, Terrence Deacon, when forming a definition of dynamic third-order EM and its causal factors, seems to develop a similar way of argumentation. Using event-descriptive terminology, he speaks about "a higher-order stochastic process extending across time that—like the limited stochastic processes of thermodynamics and morphodynamics—is capable of both cancelling and amplifying biases."[25] Again, we have here events (stochastic processes) that can be regarded as causal factors.

When she comments on Farrer's work, Murphy goes further and discusses "entities that exhibit new causal powers (or, perhaps better, participate in new causal processes or fulfill new causal roles) that cannot be reduced to the combined effects of lower-level causal processes."[26] Emmeche et al. seem to follow her ideas. Describing a strong and medium DC, they employ a concept of an "entity or process," ascribing to it causal properties.[27] Similar is the position of Sperry and Kim, who define certain properties of matter on higher levels of its organization as being causal with respect to properties of entities on lower levels. Thus, they side indirectly with others who perceive concrete emergent entities

Table 2.1. Various interpretations of the causal factor in DC grouped in four categories

Causal Factor in DC	Supporting Thinkers	Metaphysical Status
general principles, laws, regularities, wholes	Donald Campbell	unspecified
boundary conditions	Michael Polanyi, Arthur Peacocke, Nancey Murphy, George Ellis, Paul Davies	unspecified
context sensitive constraints	Alicia Juarrero, Terrence Deacon	unspecified
patterns of organization	Robert van Gulick, Terrence Deacon	unspecified
processes	Austin Farrer, Nancey Murphy, Terrence Deacon, Emmeche et al.	events
entities	Nancey Murphy, Emmeche et al., Jaegwon Kim, Roger Sperry	entities

as causes in DC, acknowledging thus implicitly that emergent properties have to be embedded in concrete entities.

As we can see, we are faced with many different ways of understanding the nature of a causal agent in DC. Either they concentrate on the concept of laws, general principles, and patterns of organization, or they refer to events and entities with their properties. I have summarized all the responses to the question of the causal factor in DC in table 2.1. This plurality of definitions of the active causal factor in DC has deep metaphysical consequences. But before we turn toward them, we must address the problem of the other end of causal relation and ask what is being acted upon in DC.

4.2. What Is Being Acted upon in Downward Causation?

An attempt at defining what is being acted upon, or is the object of causation in DC, is no less troublesome. One group of thinkers talks about entities—that is, parts of a whole—and constituents or units of a system as being acted upon in DC. Among them we find Paul Davies, Austin

Farrer, and Arthur Peacocke. Others classify objects subject to DC as lower-level properties of basal constituents (Roger Sperry and Jaegwon Kim). Still others see them as events—namely, component constituent dynamics, lower-order interactions, actions on the lower levels, or microevents (Terrence Deacon, George Ellis, and Robert van Gulick). There are also those who present a mixed opinion. They see the objects of downward causal influence as lower-level events and substances, entities or processes on a lower level, molecules in self-organizing processes, or lower-level conditions, structures, or causal processes (Donald Campbell, Emmeche et al., Alicia Juarrero, and Nancey Murphy).

Once again, we can see the difficulty among proponents of DC in specifying unambiguously the metaphysical nature of the things being acted upon in DC. See the summary of their views in table 2.2.

As we can see, our careful analysis of various descriptions of DC in terms of both causal factors and things being acted upon does not provide any clear and precise explanation or definition of this type of causality. Moreover, the plurality of answers concerning the metaphysical

Table 2.2. Various interpretations of the object of DC grouped in four categories

Object of Causal Influence in DC	Supporting Thinkers	Metaphysical Status
parts of a whole constituents of a system units of a system	Paul Davies Austin Farrer Arthur Peacocke	entities
lower-level properties properties of the lower-level basal constituents	Roger Sperry Jaegwon Kim	properties
component constituent dynamics, lower-order interactions action on the lower levels microevents	Terrence Deacon George Ellis Robert van Gulick	events
lower-level events and substances entities or processes on a lower level molecules in self-organizing processes lower-level conditions, structures, or causal processes	Donald Campbell Emmeche et al. Alicia Juarrero Nancey Murphy	mixed

status of agent and patient in DC indicates a deeper problem. It shows a tremendous difficulty in understanding and defining the very nature of DC. Standing at the intersection of the interdisciplinary endeavor engaging science and philosophy, DC faces both the challenge of the causal closure of the physical and the requirement of nonreductionism that is an intrinsic feature of EM. We need to look more closely at this problem.

4.3. What Is the Nature of Downward Causation?

The concept of DC as defined by Campbell and Sperry is intrinsically connected with natural science and its methodology. After all, it has been defined in the context of recent discoveries in physics, chemistry, and systems biology in particular that have led us to discover complexities and properties that escape the description of basic-level constituents of any given natural system. On the other hand, however, we cannot forget that the notion of DC is also, strictly speaking, philosophical in its character, as it is specified in reference to the language of causation, entities, and their ontological status. Here we encounter the same hindrance and difficulty that has already been mentioned with the analysis of emergents. The same obstacles that hamper an attempt to reconcile physical monism with the irreducible status of emergents impede the alleged harmony between DC and the causal closure of the physical, which is sometimes defined as the causal inherence principle, stating that causal powers of emerging properties are the product of the causal powers of the "basal" properties.

From a scientific point of view, any philosophical reflection on physical causation, if permitted, restricts and constrains causality to efficient cause, based on the physical principle of action and reaction and amenable to a mathematical description. This position has been predominant in natural sciences since modernity. Those who want to follow this line of argumentation and to think about DC in terms of the causal closure of physics face the difficulty of defending its irreducibility. Those willing to accept a unique ontological character of DC, on the other hand, run the risk of violating the basic rule of the causal closure of physics.

My previous considerations demonstrate, in fact, a strong inclination among many proponents of EM to understand DC in terms of efficient causality. This is evident in the case of those who define both sides participating in DC as concrete events or entities with their properties. Those who believe laws, general principles, and patterns of organization to be causal seem to be alluding to efficient causality as well. For in what way can a certain law, a general principle, or a pattern of organization of matter have a causal influence on the lower-level constituents? What kind of causation are we dealing with here? Many of the emergentists listed above leave this question unanswered.

The general problem of EM and DC understood in terms of efficient causality and in accord with the principle of the causal closure of physics was described by van Gulick as follows:

> The challenge of those who wish to combine physicalism with a robustly causal version of emergence is to find a way in which higher-order properties can be causally significant without violating the causal laws that operate at lower physical levels. On one hand, if they override the micro-physical laws, they threaten physicalism. On the other hand, if the higher-level laws are merely convenient ways of summarizing complex micro-patterns that arise in special contexts, then whatever practical cognitive value such laws may have, they seem to leave the higher-order properties without any real causal work to do.[28]

To avoid this problem, van Gulick proposes a new model in which higher-order patterns involve the selective activation of lower-order causal powers. But the question of the nature of DC remains still unanswered. If selective activation is a physical efficient cause, the whole argument falls into reductionism. If it is not a physical cause, the principle of physical causal closure is violated.

Nancey Murphy struggles with the same problem. Thinking about ontological aspects of EM and trying to define causal factors of emergent properties, she avoids the concept of new causal forces over and above those known to physics. Postulating them would again violate the causal closure of physics. She suggests instead approving the idea of new causal powers that cannot be reduced to the summary of lower-level processes.[29] But what is the nature of these causal powers? If they are not physical, how can they act upon physical constituents of lower-levels? What is the "causal joint" between high and low levels of organization of matter? The nonphysical aspect of the causal powers in question seems to contradict the principle of physical causal closure, emphasized in nonreductive physicalism. Thus, the question remains unanswered.

Is the ontological status of DC, therefore, impossible to determine, and is the very concept of ontological EM that weighs heavily on it consequently incoherent and self-contradictory? Should we agree with the sharp criticism offered by Menno Hulswit, who states, "The concept of 'downward causation' is muddled with regard to the meaning of causation and fuzzy with regard to what it is that respectively causes and is caused in downward causation"?[30] Even if the status of EM and DC is problematic and uncertain, I want to argue, following the suggestion of some philosophers of science (including Emmeche et al. and others), that the difficulty of specifying their nature is not inextricable. Quite the contrary, it can be solved if we expand our notion of causality, reinterpreting DC in terms of the Aristotelian notion of formal and final causes. But before we discuss this strategy in more detail, we must address one of the most important arguments against DC, which was brought by Kim and is related to the major difficulties of the theory we have discussed so far. Kim's argument gained much attention among emergentists and philosophers of science over the last few years.

5. KIM'S ARGUMENT FROM CIRCULARITY AND CAUSAL EXCLUSION

In analyzing EM and DC, Jaegwon Kim emphasizes the fact that in DC higher-level properties must have a real causal influence on their microconstituents. Thus understood, "reflexive downward causation" (a term coined by Kim) is combined with upward causation. Following Sperry, he gives an example of an eddy, which comes into being if, and only if, each and every molecule constituent to the pool of water begins to move in an appropriate way. At the same time, the eddy is moving all molecules around "whether they like it or not." But here comes the question: "How is it possible for the whole to causally affect its constituent parts on which its very existence and nature depend? If causation or determination is transitive, doesn't this ultimately imply a kind of self-causation, or self-determination—an apparent absurdity?"[31] Kim adds that his reasoning implies a tacit acceptance of a metaphysical principle, which he calls "the causal-power actuality principle": "For an object, x, to exercise, at time t, the causal/determinative powers it has in virtue of having property P, x must already possess P at t. When x is caused to acquire P at t, it does not already possess P at t and is not capable of exercising the causal/determinative powers inherent in P."[32] Kim embraces this principle, which is an expression of the transitive character of causation, and argues that the circularity and incoherence threatening the plausibility of the whole concept of DC can be avoided only if we understand and define reflexive DC as diachronic.[33] An introduction of a time delay between the cause and effect in DC is indispensable for him if we want to make it metaphysically plausible. He rejects synchronic DC, saying its coherence is doubtful. He finds support, at this point, in Humphreys and Richard Campbell.[34]

However, diachronic self-reflective DC does not solve all the problems of EM. Applying it to mind-body causation, Kim still finds it questionable, and claims it eventually collapses into physicalism. He formulates his famous causal exclusion argument (or an argument from systematic overdetermination) and analyzes it by means of a model of a hypothetical emergent mental property. We can illustrate his argument using a concrete example of a mental state of pain and escape reaction to it. See figure 2.1.[35]

Kim tries to salvage DC and suggests giving it a "conceptual" interpretation. We can think about higher levels as levels of concepts and descriptions, or levels characteristic of our representative apparatus, but not as levels of real phenomena and properties in the world. DC would then mean a cause described in terms of higher-level concepts or higher-level language, although it could be representable in lower-level concepts and language as well. This approach will not save real DC—Kim admits—but it will save downward causal explanation, and maybe we should not expect anything more.[36]

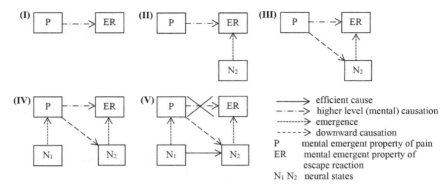

Figure 2.1. An example of the collapse of DC into physical causation, according to Jaegwon Kim

(I) An instance of an emergent property of pain, P, causes another emergent property of escape reaction ER to instantiate, as an effect of same-level causation. **(II)** ER, being an emergent property, must have a basal (physical) property, a neural state N (N_2 in particular) from which it emerges. Moreover, as long as N_2 occurs, ER will be instantiated, whether or not ER's purported cause, P, occurs at all. **(III)** The only way to save the claim that P caused ER appears to be to say that P caused ER by causing N_2. In this case, the same-level causation from P to ER entails DC from P to N_2. **(IV)** Now, P, as an emergent, must itself have an EM base property—that is, a neural state N_1. **(V)** Here Kim asks a critical question: If an emergent property of pain, P, emerges from basal neural state N_1, why cannot N_1 displace P as a cause of neural state N_2? If causation is understood as nomological (law-based) sufficiency, N_1 as P's emergent base is nomologically sufficient for P, and because P is nomologically sufficient for N_2, it follows that N_1 is nomologically sufficient for N_2, and can be regarded its cause. Moreover—adds Kim—a causal chain from N_1 to N_2 with P as an intermediate causal link is not possible because the EM relation from N_1 to P is not properly causal. If P is to be retained as a cause, we are faced with the highly implausible consequence that the case of DC (from P to N_2) involves causal overdetermination (since N_1 remains a cause of N_2 as well). But if DC goes—Kim concludes—emergentism goes with it.

6. A RESPONSE TO KIM

The argument developed by Kim seems to be devastating for all enthusiasts of DC in philosophy of mind. Moreover, Erasmus Mayr contends that "as several philosophers have noted, Kim's argument, if successful, not only applies to the relation between mental and underlying physical events. It is a completely general argument-scheme, which can be applied to the relation between any two layers of reality, whenever one is held to be more fundamental or basic than the other, and the latter supervenes on the former."[37] Hence, it does not surprise that Kim's view on DC has become a subject of a wide conversation and critique offered by scholars working across several academic disciplines, ranging from philosophy of mind, to metaphysics, to philosophy of science, to philosophy of particular sciences. I will now mention and discuss some of the most important positions in this debate.

6.1. Alwyn Scott

Alwyn Scott rejects Kim's concept of diachronic DC, showing that from a nonlinear point of view the problem of circular causal loops and time in DC does not exist at all. According to Scott, an emergent structure does not pop into existence at time t. It "begins from an infinitesimal seed (noise) that appears at a lower level of description and develops through a process of exponential growth (instability). Eventually, this growth is limited by nonlinear effects, and a stable entity is established." Using Kim's notation applied in his "causal-power actuality principle," Scott depicts DC using nonlinear differential equations:

$$\frac{dx}{dt} = F(x, P), \qquad \frac{dP}{dt} = G(x, P),$$

where x is an object, P is an emergent property, and F and G are nonlinear functions of both x and P. The emergent structure is not represented by $x(t)$ and $P(t)$, but by x_0 and P_0, according to the following equations:

$$0 = F(x_0, P_0), \qquad 0 = G(x_0, P_0).$$

If x_0 and P_0 are an asymptotically stable solution of the given system, we may assume that

$$x(t) \to x_0, \qquad P(t) \to P_0.$$

Because $t \to \infty$, what we are dealing with here is a dynamic balance between upward and downward causation. "Thus, Kim's causal-power actuality principle"—says Scott—"is a theoretical artifact stemming from his static analysis of a dynamic situation."[38] He then gives two examples of simple and complex causal loops in the feedback mechanisms to explain the relation between upward and downward causation. See figure 2.2.

Scott's critique of Kim seems reasonable. It offers an answer to the problem of circular causal loops in DC, as well as an argument against Kim's idea of diachronic reflexive DC. In terms of nonlinear causality, the problem of time simply disappears, and the whole concept of diachronic DC is unnecessary.

On the other hand, however, Scott's criticism does not address the key problem, which is the nature of DC. In his explanation of the causal loops in figure 2.2 Scott discusses entities which belong to different levels of both the biological and cognitive hierarchies. They are entangled in a complex causal network. But what is the nature of causation

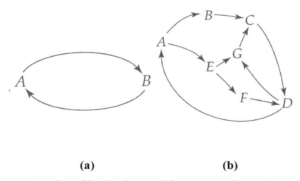

(a)   (b)

Figure 2.2. Examples of feedback causal loops according to Scott

(a) A simple loop. (b) A graphic representation of the organization of a complex network. "In this diagram, the node **A** might represent the production of energy within an organism, which induces a muscular contraction **B**, leading the organism to a source of food **C**, which is ingested **D**, helping to restore the original energy expended by **A**. Additionally, **A** might energize a thought process **E**, which recalls a positive memory of taste **F**, further encouraging ingestion **D**. The thought **E** might also induce the generation of a digestive enzyme **G** which also makes the source of food seem more attractive. Finally, ingestion **D** might further induce generation of the enzyme **G**. In this simple example—which is intended only as a cartoon—the network comprises the following closed loops of causation: **ABCD**, **CDG**, **AEFD**, and **AEGCD**, where the letters correspond to entities at various levels of both the biological and the cognitive hierarchies."

Source: Scott, *Nonlinear Universe*, 288–89, by permission of Springer Nature.

between mental and biological entities? What is the causal joint between cognitive and physical structures? It is not clear whether Scott is thinking exclusively in terms of efficient causality or suggests a different mode of causation when presenting an example of a complex causal loop, in which various biological and cognitive properties are entangled. This fact makes his explanation insufficient.

6.2. Timothy O'Connor and Hong Yu Wong

Another dynamical model of EM undermining Kim's charges of both circularity and systemic overdetermination of DC comes from Timothy O'Connor and Hong Yu Wong and was delineated in their paper "The Metaphysics of Emergence."[39] Their main argument—holding that emergent properties are nonstructural properties of composite individuals—is depicted and described in figure 2.3. It presents the original diagram developed by O'Connor and Wong and explains it in terms of my example of pain and escape reaction.

O'Connor and Wong argue that if we resolve each physical state into two substates—a state generating the emergent property and a different state that is affected by it—we can avoid Kim's charge of causal circularity. Moreover, they claim that their dynamical model evades his causal exclusion argument, as it shows that "the distinctive potentialities of emergent properties do stem indirectly from the total potentialities of the basic physical properties. But they do not determine the emergent effects (or fix the emergent probabilities) independently of the causal activity of those emergents."[40] Thus, the relation between emergents and their physical "base" conditions is "diachronic and causal, rather than as a sui generis variety of synchronic supervenience," which does not allow Kim's argument to "get off the ground against the dynamical model of emergence."[41]

More recently, Wong offered another argument against Kim. He notes that Kim, in his line of reasoning, does not invoke the completeness of physics. Wong argues this is permitted because emergentists (who do not endorse NP) need not be committed to it. In Kim's own words, "Most emergentists will have no problem with the failure of the physical causal closure; although they may have to tinker with their doctrines somewhere to ensure the overall consistency of their position, they are not

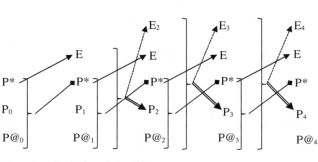

P*	complex physical configuration
P_n	remaining aspect of S's intrinsic physical state at t_n
$P@_n$	summation of physical factors in S's immediate environment at t_n
E	emergent state E
⟶	upward causation of baseline emergent state E
---▶	upward causation of superemergent states E_n
⟶◆	maintenance causation of emergence-sustaining configuration P*
⟹	wide horizontal causation (including DC)

Figure 2.3. O'Connor and Wong's dynamical EM model of the evolution of system S over time

A neurophysiological system S comes to have a certain complex physical configuration P* at t_o, giving rise to the disposition to E, which is pain awareness at t_1. This dispositional state, in conjunction with the total physical state of S at t_1 ($P_1 + P@_1$), and in the presence of the physical pain stimulus, causes the awareness of pain at t_1, an escape reaction (E_2), and an aspect P_2 of the subsequent physical state at t_2. As the physical pain stimulus is removed and this information is encoded in S's physical state, E_2 gives way to E_3 which inhibits the escape reaction, as well as influences the perceptual information encoded in P_3. The whole process progresses to t_4 and so forth. Notice that the physical state of S and its environment ($P@_n$) at t_n suffice, apart from E, to maintain the critical structural feature P*. At the same time P* is of a sufficiently general type as to persist through constant changes over time. It therefore cannot be equated with the total physical state of the system at any given time. In other words, P* only partly determines the underlying physical state of S at t_n. Consequently, E at given t_n—e.g., t_1—will help to determine the physical state of S and the occurrence of another emergent state, E_2, at the subsequent moment, t_2, but it does not determine S's continuing to exhibit P* (an argument against circularity of DC).

Source: This is a replica of the original image that can be found in O'Connor and Wong, "Metaphysics of Emergence," 666.

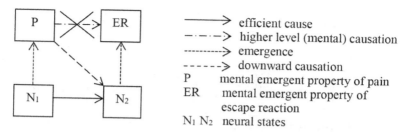

Figure 2.4. The final step of Kim's causal exclusion argument

likely to shed any tears over the fate of the closure principle. For many emergentists that precisely was the intended consequence of their position."[42] Moreover, careful observation of his argument shows that Kim believes the relation between basal physical properties and the emergent properties related to them is noncausal (the relation between N_1 and P and between N_2 and ER in figure 2.4). But if this is the case then there is no causal chain going from N_1 through P to N_2. Because the completeness of physics is not in play, and the downward causal powers of P are emergent ones, even if N_1 has powers that are always manifest and ready to bring N_2, it is still plausible for the emergentist to describe the scenario as one in which P emerges from N_1 (synchronically) and causes N_2. Wong finds this plausible even if P and N_1 together overdetermine N_2. He concludes that the whole argument presented by Kim does not stand and gives us no reason to think that emergent properties are epiphenomenal.[43]

6.3. Cynthia Macdonald and Graham Macdonald

Another argument against Kim's causal exclusion, developed by Cynthia Macdonald and Graham Macdonald, is simply founded on their claim that because the supervenient and base properties are instantiated in a single instance, "there is *no distinction of levels of instances*, only levels of properties; at the level of instances, the world is flat."[44] Therefore, to return to my example of pain and escape reaction, even if on the level of instances the instantiations of N_{1i} and N_{2i} are identical with the instantiations of P_i and ER_i (see figure 2.5 A), on the level of properties we can still argue for the distinct emergent character of properties P and ER (see figure 2.5 B). Macdonald and Macdonald claim that Kim construes his

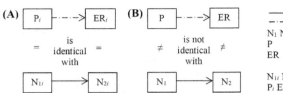
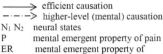

Figure 2.5. Macdonald and Macdonald's refutation of Kim's causal exclusion argument

(A) Identity of instances on physical and emergent levels. **(B)** Nonidentity of properties on physical and emergent levels.

argument along the causal inheritance principle, which states that the instance of the higher-order property inherits all its causal power from the instance of the lower-level property, a principle that "is not obviously derivable from the less controversial claim *that identical instances have identical causal powers*."[45]

6.4. Carl Gillett

Trying to face the same challenge of Kim's argument from causal exclusion, Carl Gillett suggests redefining DC in reference to metaphysical terms that are noncausal. He contends that in cases described as examples of DC, "we have non-causal and non-compositional determination of components by composed entities through something like role-*shaping* or role-*constraining*, rather than role-*filling*, in the composed entity determining that the components have certain differential powers." He continues: "We therefore also need to coin a new term to mark this kind of determination relation, and I have consequently made the following suggestion. Combining the Greek words 'macro' and 'chresis,' where the latter is roughly the Greek for 'use,' we get the terms '*machresis*,' and '*machretic determination*,' for the general phenomenon of composed, or 'macro,' entities that non-causally, and non-compositionally, determine the nature of their components through causal role-shaping."[46] At the same time, Gillett strives to save the reality of DC. He does that by distinguishing between what he calls "direct" DC—referring to "a composed entity that directly causally acts upon its own components"—and "mediated" DC—referring to "a composed entity that bears a causal relation to lower-level entities that are *not* its components and where these lower-level entities are effects of components of [another higher-level] composed entity."[47]

I think that what Gillett describes as a noncausal and noncompositional determination that can exist "in a downward direction alongside a compositional relation of role-filling upwards"[48] may come close, in some respect, to what I will describe as an Aristotelian formal cause understood in the context of contemporary science. Unlike Gillett, however, I will not hesitate to classify it as being a cause, based on a retrieval of Aristotle's plural notion of causation offered in the following parts of this book. At the same time, I find Gillett's idea of "mediated" DC somewhat trivial and escaping the problem of the causal notion of DC, rather than solving it. For what is at stake in DC is not its causal action on basal constituents of some other entity(-ies) but its own constitutional base.

6.5. Francesco Orilia and Michele Paolini Paoletti

Editors of the recently published collection of papers on DC, Francesco Orilia and Michele Paolini Paoletti, argue that the problem of Kim's position is his treating causal relata as events, which can be defined as follows:

> Events (i) are structured entities involving as constituents a property or relation understood as a universal, one or more objects (concrete particulars, at least in typical cases), and a time . . . ; (ii) have an existence condition according to which they exist inasmuch as the object exemplifies the property, or the objects exemplify the relation, at the time in question; (iii) have this identity condition: $[x_1, P_1, t_1] = [x_2, P_2, t_2]$ if, and only if, $x_1 = x_2$, $P_1 = P_2$, and $t_1 = t_2$, where in general a structured symbol such as "$[x, P, t]$" stands for the event which results from the object x's exemplifying the property P, called the *constitutive property* of the event . . . , at time t.[49]

Orilia and Paoletti claim that Kim's argument holds, and DC is untenable and reducible to causal action of basal constituents of complex entities as long as causal relata are treated as events. They offer three solutions to this problem, of which I will mention the first and the last.[50] The first solution refers to those who conceive causal relata not as events but as tropes.[51] As I mentioned in section 2 of this chapter, according to a bundle theory of substance, each entity consists of tropes (individualized or particular properties). Some of the tropist ontologists distinguish between properties qua characterizers of objects (simple tropes), and properties qua unifiers of objects (classes of resembling tropes—i.e., entities that allow us to group together various objects under general terms). Orilia and Paoletti suggest that in this approach, the distinction between mental (emergent) and physical properties—required by the layered picture of reality—is best seen as regarding properties qua unifiers. This way, they claim, multiple realizability (which they find important for there being EM and DC), as well as distinctness of levels and supervenience, can be saved, and the possibility of DC granted. However, this solution encounters at least some of the major problems faced by its foundational (tropist) metaphysics (once again, see section 2 of this chapter), including the possibility of accidental changes, identity of indiscernibles, and the principle of individuation. Moreover, mere defense of the reality of DC in terms of properties qua unifiers does not explain its character and nature.

The third solution offered by Orilia and Paoletti anticipates, to some extent, my own position, developed in part 2. They emphasize that the Kimian picture "neglects the possibility that what is upwardly determined are properties understood as causal powers and events consisting in the exemplifications of such powers, which we may call *power events*."[52] They suggest distinguishing powers—which have corresponding supervenience bases but are not themselves causes—from "*exercise events*," which are causal, yet devoid of subvenience bases that would compete with them as causes. They conclude that according to this solution, "free acts of will are not . . . supervenient on corresponding physical bases. On the contrary, we can perhaps take them as causally producing the neural events constituting a causal precondition of the ensuing action, insofar as such neural events are not causally determined by preceding physical events—something which can be granted on the assumption, seemingly licensed by current physics, that there is room for indeterminacy in the physical world."[53]

Most importantly, Orilia and Paoletti associate this picture with a special form of causation—namely, agent causation, which they are willing to extend to all entities, speaking thus about "*substance* causation." This return to substantial language is intriguing. However, it requires a careful examination and contextualization with reference to the complex contemporary debate on causation in analytic philosophy, as well as the classical Aristotelian thought. I offer such analysis in part 2. Commenting on Orilia and Paoletti's third solution, I need to add that I find their reference to the indeterminacy of quantum mechanics rather puzzling. It does not seem necessary to refer to its premises in order to defend the reality of DC.

6.6. Erasmus Mayr, Anna Marmodoro, Rani Anjum, and Stephen Mumford

Apart from Orilia and Paoletti, the reference to powers in defense of the reality of DC inspires a whole group of analytic philosophers, whose thought further develops the background for my argument. Erasmus Mayr—distinguishing (after Orilia and Paoletti) emergent powers from "*exercise* events" (which he describes as "characteristic change processes," CCPs)—pays attention to a key flaw in Kim's argument, which he thinks consists in

> conceiving of the power and the sequence of events involved in the CCP as factors which compete with one another for causal influence, such that the causal influence of the one leaves less room for the causal influence of the other. For the realist about powers, this conception must be wrong. The events involved in the CCP are themselves necessary for the power's full manifestation (if they, or similar events, did not occur, the power wouldn't be fully manifested, either), without, however, fully constituting this manifestation on their own, since they must also be due to an exercise of the underlying power.[54]

Anna Marmodoro, Rani Anjum, and Stephen Mumford strive to offer an account of strong EM and DC that is stated in positive terms and not merely in negative ones (as we have seen it above while discussing the classical definitions of both terms). Marmodoro defends the reality of strong EM and DC by arguing in favor of the "complexity in the nature of powers," i.e., the reality of "powers whose natures are mereologically complex by composition of powers (and not reducible to relations to other powers)." She also suggests distinguishing physical unification ("physical structure *unites*") from metaphysical unification ("metaphysical structure *unifies*") and describes the latter as re-individuation in terms of the whole, which "results in a change in the [given] power's ontological status." Consequently, she differentiates a "structural power" (i.e., the structure of powers constituting an emergent entity—for example, mass, charge, spin, and space-time constituting an electron) from a "*substantial power*" (i.e., a metaphysical unification of the physical structure of powers into a single individual).[55] Following a similar way of reasoning and speaking of emergent causal powers, Anjum and Mumford introduce a new term to specify their nature: "Emergent powers can then act on their parts, and this is what we mean by downward causal influence. It might be useful to think of this as, to coin a phrase, demergence. Emergence is where there are new powers of wholes in virtue of causal interactions among their parts; demergence is where there are subsequent new powers of the parts in virtue of the causal action of the whole upon them."[56]

All of the proponents of powers ontology mentioned here claim that their position enables them to save the reality of EM and DC. I think that this might be true. Nevertheless, to defend an argument of this kind, one needs to clarify much more about the fundamental base of its background metaphysics. Since it is being classified as a version of the new Aristotelianism, it also requires a careful examination of its premises in reference to the ancient thought of Aristotle. Without such analysis—similar to the positions of Scott, O'Connor and Wong, Macdonald and Macdonald, and the first solution offered by Orilia and Paoletti—we might end up with an appealing attempt at defending the reality of EM and DC, devoid of a clear-cut specification of the latter's character and nature.

However, before entering the conversation on contemporary analytic views of causation—including the powers (dispositional) view of causation and its foundational metaphysics—I should go back to my former suggestion that most likely the best way to overcome the difficulty in defining DC requires our rethinking the very notion of causality, extending it in reference to Aristotle's classical typology of causes. We will see how this contention, which preceded the reference to powers ontology in defense of EM and DC, has led to an interesting new account of EM offered by Terrence Deacon.[57]

7. EMERGENCE AND DOWNWARD CAUSATION REDEFINED

In addition to arguments defending DC discussed in the previous section, I would now like to propose one more line of reasoning. This will not only help to avoid Kim's criticism, but also suggest a response to the question of the nature of DC, which poses a challenge for all emergentists and, in a way, remains unanswered in their explanations analyzed so far.

Looking again at Kim's argument from systematic overdetermination, depicted in figure 2.1, I want to argue that even if we acknowledge that the downward-causative feature of a mental property of pain (P), "standing" between two neural states (N_1 and N_2) and embedded in a regular efficient causal chain, is truly superfluous, leading to a problem of overdetermination, the whole argument collapses once we assert that DC is not an efficient cause and cannot be understood in terms of this type of causality. Such a position, however, becomes truly revolutionary, as it leads to a breakdown of causal monism and requires a departure from the commonly accepted account of causation that both science and the philosophy of science gradually embraced and have defended since the eighteenth century.

I want to argue in favor of this revolution in the philosophy of causation (rather than trying to redefine DC in noncausal terms). Even if the analysis of causality in terms of efficient cause is still predominant in analytic metaphysics (as I will show in chapter 4), the anti-Humean turn in philosophy of causation and the revival of teleology in the philosophy of evolutionary biology—which is a part of a more general revival of classical philosophy of nature—shows that the time is ripe to go beyond efficient causation in defining EM as well.[58]

7.1. Beyond Efficient Cause

One of the first suggestions to broaden the notion of causation in emergent studies came with the trifold division of DC formulated by Emmeche et al. In their article published in 2000, Emmeche, Køppe, and

Stjernfelt present an original classification of different types of DC, based on a distinction between efficient and formal causality. The first type of DC in their classification is a strong downward causation (SDC). Its definition assumes an efficient causal influence of higher-level entities on lower-level entities. Moreover, it understands the former as substantially different from the latter, and thus entails a substance dualism allowing for a violation of the assumption of the inclusivity of levels (see chapter 1, section 2.4). According to Emmeche et al., SDC is erroneous not only because it defines DC in terms of efficient causation, located in space and time, but also due to its accepting the idea of exchange of energy between higher and lower levels. Their argument follows my criticism of DC understood as efficient cause presented above. Pointing out the problem of the causal joint between higher and lower levels, Emmeche et al. show that not only SDC but also strong upward causation raises similar questions. For just as it is difficult to think of the biological level inflicting purely physical effects, the idea of physical cause having nonphysical efficient influence on higher emergent levels is no less problematic. Thus, Emmeche et al. reject SDC and suggest we should look for an understanding of DC beyond the idea of efficient causality.[59]

At this point in their analysis, they introduce the concept of medium downward causation (MDC). To describe it they refer to the mathematical and physical language of differential equations describing the dynamics of systems. Using terms such as "boundary conditions" or "constraining conditions," they define MDC in terms of higher-level entities as constraining conditions for the emergent activity of lower levels. In other words, "the higher level is characterized by *organizational principles*—lawlike regularities—that have an effect ('downward,' as it were) on the distribution of lower level events and substances."[60]

This definition reminds one of the whole group of authors considered in the fourth section of this chapter, who use the same language of boundary conditions and lawlike regularities in their description of the causal factor in DC (Juarrero, Peacocke, Murphy, Ellis, Davies, and others). But a fundamental difference separates them from Emmeche et al., who emphasize that "medium DC does not involve the idea of a strict 'efficient' temporal causality from an independent higher level to a lower one, rather, the entities at various levels may enter part-whole relations (e.g., mental phenomena control their component neural and biophysical subelements), in which the control of the part by the whole can be seen as a kind of functional (teleological) causation, which is based on efficient, material as well as formal causation in a multinested system of constraints."[61] What we find here is an attempt to retrieve the Aristotelian fourfold division into material, formal, efficient, and final causality. Certainly, the position presented by Emmeche et al. needs further clarification and poses many questions (e.g., the nature of the teleological and formal causation and the metaphysical status of a substance as a whole— in reference to the status of its components). Nevertheless, the very fact of breaking the causal monism is significant, and that is why I strongly support Emmeche et al. when they argue: "There is a place for a rational concept of downward causation (in some version) in science and philosophy, but only with a broader framework of causal explanation. Very often 'causality' is implicitly equated with the usual notion of efficient causality, but if downward causation is regarded as an instance of efficient causality it will form a 'strong version' of the concept, which . . . is not a plausible one. The notion of causality should therefore be enlarged to make sense of downward causation."[62]

Finally, Emmeche et al., using phase-space terminology, define weak downward causation (WDC), in which the higher level is the form into which the constituents of a lower level are arranged. In other words, the higher level is not a concrete substance, but an organizational phenomenon. The formal cause in WDC can be understood by analogy with the concept of an "attractor" in a dynamical system—that is, a steady state toward which the system may evolve. The application of formal causation in WDC is another example of a wider understanding of causality by Emmeche et al.[63] Their position is recalled by C. N. El-Hani and A. M. Pereira, who, in search of a middle road between reductionism and radical dualism, emphasize the importance of the entanglement of matter and form, the role of higher-order structures constraining lower ones, and the role of functional causation.[64]

7.2. Aristotelianism and Emergentism

The example of Emmeche et al. suggesting a return to the Aristotelian notion of causality in emergent studies is not an isolated one. Several authors among emergentists have taken a similar position, although the majority of them do not develop this idea more broadly. Michael Silberstein, for instance, mentions at one point (with no further explanation) that "systemic causation means admitting types of causation that goes beyond efficient causation to include causation as global constraints, teleological causation akin to Aristotle's final and formal causes, and the like."[65]

A. Moreno and J. Umerez also go back to Aristotle, arguing for the acceptance of a new type of causation in biological systems, which "is 'formal' in a sense that it infuses forms, i.e., it *materially restructures matter according to a form*."[66] But at the same time they claim their position differs from the classical Aristotelian understanding of material and formal cause: "In Aristotle, both formal and material causes are intrinsic, whereas efficient and final ones are extrinsic. In our view of formal cause being efficient and being intrinsic do not exclude each other. In a sense, formal cause is intrinsic inasmuch as it is inherently generated in the very system which becomes an autonomous complex system. Anyway, formal cause can also be extrinsic with respect to some levels or subsystems (or even systems) which allow for relatively autonomous kinds of description."[67]

Alwyn Scott, in his introduction to *The Nonlinear Universe*, presents a description of all four of Aristotle's causes. He also adds that, in modern terms, Aristotle's material and formal causes are put together and classified as "distal causes," his efficient cause is called a "proximal cause," while the final cause is given the name "teleological cause." Although he lists the same four causes, we can see that his position differs from the one offered by Moreno and Umerez.[68]

In his analysis of emergents, George Ellis refers to the idea of causally effective goals. He sees them as central factors in feedback control systems and ascribes to them causal properties which are the result of the information about the system's desired behavior or responses they embody. For Ellis, living systems are "teleonomic"—that is, goal-seeking. Moreover, these goals are not the same as material states, although they are expressed and become effective in a material way. Ellis believes goals can be either inbuilt (as homeostasis), learned, or consciously chosen. Although he does not refer to Aristotle, his notion of nonmaterial causally effective goals resembles the Aristotelian account of final causation.[69]

We could probably find more examples of authors writing on EM who allude, more or less directly and openly, to the classic Aristotelian understanding of causation, and to formal and final causes in particular.[70] Although this tendency seems to be very promising and inspiring, preliminary allusions of this kind are by no means sufficient or complete. Quite the contrary, they require further development that would provide a more robust analysis of the specific mechanisms of causation in emergentism and ground the whole theory in a clearly defined metaphysics. And here we encounter a difficulty. For Aristotle's philosophy of causation is founded on his hylomorphism and substance metaphysics, the use of which in the contemporary discussion on EM and DC has already been criticized and questioned.

7.3. In Search of Proper Metaphysics

The attempt of Emmeche et al. to bring Aristotle's notion of causality and natural philosophy back into conversation has met with a rather skeptical reception and evaluation by Menno Hulswit. His comments seem to apply to other authors who favor explaining EM and DC in terms of formal and final causality as well. Hulswit begins his criticism by saying that Emmeche et al. show a strong bias toward thinking in terms of substances and substantial forms without presenting any arguments to explain and support their position. He emphasizes that "the mere claim that certain phenomena ought to be explained in terms of formal (and/or final) causation, without providing a clue on how that should be done, amounts to putting the cart before the horse."[71] He becomes even more radical when he talks about "Western 'substance addiction,'" which needs to be overcome by an analysis of DC in terms of interactions which "should do full justice to the primacy of processes and events, along the lines of suggestions made by C. S. Peirce and A. N. Whitehead."[72] To support his thesis, Hulswit refers to Bickhard and Campbell, who argue that the Aristotelian substance ontology is "an inadequate metaphysics for relationships and process, most especially open process," and claim that we should substitute it with a process metaphysics.[73]

The problem of Hulswit's own position, however, is that his suggestion to rethink DC in processual terms is not actually followed by an attempt at such an analysis. Thus, we may use his own words of criticism, saying that now he himself is putting the cart before the horse in his own project.

Nonetheless, Hulswit continues his critical evaluation of Emmeche et al., pointing toward the inconsistency in their description of MDC. He notes that, when formulating its definition, they identify higher- and lower-level entities as substances or substantial forms, but later, when giving concrete examples of MDC, they refer to thoughts constraining neurological states, and mental phenomena controlling their component neural and biophysical subelements. According to Hulswit, neither of these can be treated as substances. What is more, the same criticism refers to processes and interactions.[74]

Hulswit is also skeptical about the summary of their exposition of MDC and WDC, in which they use the language of supervenience: "The point of departure in both cases is the assumption of formal causality. As higher-level entities (e.g., a cell) supervene on lower order entities (molecules), formal causality on the higher level supervenes on the efficient causality of the lower level. This can be interpreted as the selection—from a very large set of possible (efficient) interactions—of a small set of realizable (efficient) interactions on the lower level, on which the higher level then (formally) supervenes. In any case, in our view this is the only non-contradictory version of downward causation possible."[75] The explanation of the nature of DC based on the theory of supervenience is somewhat unfortunate. As I mentioned earlier (see chapter 1, section 2.2), DC is more than simply a relation of dependency between supervenient and subvenient properties, a relation that entails that there cannot be any change on the higher level without a change on the lower level and vice versa. It assumes an ontological relationship between these levels. The whole discussion shows once again that it is not only causal relata but also the nature of DC understood as formal and final causation that must be explained in more detail.

In his article, Hulswit presents one more critical argument, which is of great importance for the whole debate. He contrasts the classical Aristotelian concept of formal cause with the modern concept of natural laws. He claims that, although they have the same function—that is, to explain the apparent stability of the world—they differ significantly, for "the formal cause was meant to explain the stability of the world in terms of the structure of things, whereas natural laws explain the stability of the world in terms of the dynamic relations between events."[76] At this point I strongly disagree with Hulswit. His differentiation between the formal cause and natural laws presupposes a static interpretation of the Aristotelian metaphysic of substance. Defining Aristotle's substantial form as a structuralizing principle, rather than a basis and source of properties and causal powers of entities, is highly irrelevant. As I showed in the introduction, Aristotle's *Physics* and *Metaphysics* have their origin in his reflection on both stability and the dynamic change of entities (substances) observed in the natural world. His idea of formal causation is thus by no means static. I believe this fact responds to the criticism of Hulswit, showing that an attempt to formulate a plausible account of EM and DC explained in terms of formal and final causation is still possible, provided Aristotle's principal metaphysical assumptions and the dynamic aspect of his thought are rediscovered in the context of contemporary science.

The problem is that none of the authors mentioned so far in this section has attempted such an endeavor. Even if they alluded to Aristotle's four causes in explaining EM and DC, they did not present a thorough analysis of the concepts in question in terms of Aristotle's metaphysics and philosophy of nature. Interestingly, an attempt of this kind has been recently pursued by biologist and EM theorist Terrence Deacon. Although he approaches the problem of EM from the scientific point of view, his analysis leads him to rediscover Aristotelian categories of causation, which he applies to a concrete and thorough model of the growing complexity of nonliving systems, the origin of life, and consciousness. In the following chapter, I will present a detailed philosophical analysis and evaluation of his project, trying to answer the question to what extent it is truly Aristotelian.

CHAPTER 3

Dynamical Depth and Causal Nonreductionism

The scope and breadth of Deacon's project are remarkable. The primary objective of explaining the difference between nonliving and living structures and processes becomes for him a point of departure for analyzing the origin of consciousness, meaning, purpose, significance, value, and other related phenomena, which he classifies under a general category of "ententional" (the term he coins).[1] The same effort of specifying the borderline between living and nonliving dynamics leads Deacon to develop his original and unique theoretical model of EM. He defines it in terms of the three dynamic modes or levels of what he calls, together with Spyridon Koutroufinis, "dynamical depth"[2] and describes in reference to an idea of the causal influence of "constraints"—that is, "specifically absent features" and "unrealized potentials." He claims this new approach brings about a paradigm shift in EM studies.

The interdisciplinary character of Deacon's project, embedded in natural science, philosophy of science, philosophy of nature, and metaphysics, determines its originality and attractiveness, while making it vulnerable to critical questions and evaluations. Though he is praised for offering "the conceptual tools necessary to finally move emergence theory out of philosophy and into sciences,"[3] Deacon's reintroduction of teleology, as a consequence of his dissatisfaction with the reductionist theories explaining the origin of life, can hardly be regarded as a suggestion purely scientific in its character and meaning. He openly returns to Aristotle's idea of final causality, which is, strictly speaking, a metaphysical principle. On the other hand, though his version of emergentism is applauded as a promising proposition of philosophy of nature "both neo-Aristotelian and well integrated with contemporary science,"[4] Deacon's rejection of substance metaphysics raises serious questions about the Aristotelian character of his project. Though he openly challenges the causal reductionism of modern science and goes back to the classical fourfold notion of causality, Deacon's reinterpretation of all four Aristotelian causes departs considerably from the original and may be at risk of falling into the same pitfall of scientific eliminative reduction. Though he is praised for his "rejection of a substance metaphysics in favor of a process metaphysics,"[5] Deacon seems to avoid Whitehead, the founder of the philosophy of process, without proposing any thorough metaphysics to replace it. Finally, although he makes the causality of absences a cornerstone of his project, the very idea of the causal character of constraints raises some serious metaphysical doubts and questions.

These difficulties, however, by no means undermine the importance, plausibility, and promise of Deacon's project. Quite the contrary, as is typical of interdisciplinary endeavors, they do not prevent it from being a fascinating proposition, bearing a significant potential of bringing natural science and philosophy into a fruitful conversation. In what follows I will first analyze the originality of Deacon's ideas in the context of the current scientific theories of information, dynamical systems, dissipative structures, and self-organizing processes. Next, I will summarize the three levels of dynamical depth. Then I will carefully investigate this new model of EM in terms of its reference to the plural Aristotelian notion of causality. Finally, a general philosophical evaluation of Deacon's project will follow, showing how it needs a more specific metaphysical base, grounding all three levels of dynamical depth.

1. TELEOLOGY MATTERS IN LIFE THEORIES

"The recovery of a non-eliminative concept of information and a non-trivial concept of 'selfhood' are both critical to a complete theory of the physics of biological organism," says Deacon in his article written with Tyrone Cashman, criticizing the common tendency to reduce biological problems to engineering problems.⁶ He thinks the classical Shannonian notion of information, reducing it to copying or conveying physical patterns and identifying it with entropy (a measure of disorder and unpredictability), ignores entirely issues of reference, function, and significance and does not explain the causality of processes involved in its creation, transmission, and reception. While it proves very useful in the area of computational and communicational technology, Shannon's theory of information cannot explain the functional and adaptive significance of the various genetic substrates which form the molecular basis for biological inheritance. Yet the inheritance, preservation, replication, and transmission of information are precisely what define goal-directedness, which distinguishes—together with persistent self-maintenance of a system in a far-from-equilibrium state—living from nonliving dynamics. Acknowledging this, Deacon finds the teleological factor crucial for explaining the property of life, which he is not afraid to define in terms of a primitive biological "self" and "self-directedness." He thus sees the query concerning the origin of life boiling down to "an explanation of how a biological self can emerge from non-self components."⁷

The same teleological question makes Deacon dissatisfied with the current concepts of life's origin developed in reference to dynamical systems, dissipative structures, and self-organization theories. He does not question the important observation that persistently far-from-equilibrium systems "can spontaneously develop toward more orderly global organization as they are forced to more efficiently dissipate an incessantly imposed disturbance."⁸ Nor does he ignore the role of intrinsic dynamical constraints characteristic of these systems and compensating for the external imposition of constraints, in accordance with the "maximum entropy production" principle, which states that "a dynamical system will tend to organize itself in such a way that it maximizes the generation of entropy, given the available pathways for dissipation (i.e., within given constraints), even if it is persistently perturbed far from equilibrium."⁹ But at the same time, Deacon states that because it develops in opposition to the persistent displacement from equilibrium (by decreasing its local entropy in a way that increases global entropy), a self-organizing process has "no capacity for persistence beyond the extrinsically imposed gradients that it develops in response to. When such external perturbation ceases, the acquired regularity of the local system dissipates and the system re-approaches equilibrium following the Second Law of thermodynamics."[10] What is needed for the phenomenon of life to emerge is, therefore, a self-propagating inherent formative power which is not susceptible to mechanistic reduction.

For this reason, Deacon emphasizes that autocatalysis, as a special case of self-organization in which two or more molecular types reciprocally catalyze the synthesis of one another, thus increasing constraints locally and maximizing entropy globally, and which has motivated the coining of the term *autopoiesis* ("self-forming" or "self-producing"), should not be too easily regarded as a fundamental property of life. He claims that "ascribing 'means' and 'ends' properties to these reciprocal catalytic processes is an externally imposed teleological interpretation. A purpose or end is not merely a consequent state of things."[11] Since autocatalysis does not fundamentally differ from typical chemical reactions, an interaction of arbitrarily distributed molecules in solution does not constitute a higher-order entity except as abstractly conceived. Introducing the requirement of containment does not solve the problem of teleology being rather a regulative assessment than a constitutive property of autopoietic systems. Therefore, Deacon and Cashman boldly conclude that "with no particular unitary beneficiary—no self—for the sake of which these processes take place and no localizable source for the generation of these processes there is no actual (as opposed to merely descriptive) teleology."[12] This strong teleological statement brings us to the presentation of Deacon's model of EM.

2. DYNAMICAL DEPTH—A TRUE FACE OF EMERGENCE?

Emergent phenomena grow out of an amplification dynamic that can spontaneously develop in very large ensembles of interacting elements by virtue of the continuing circulation of interaction constraints and biases, which become expressed as system-wide characteristics.[13]

This hypothesis, formulated by Deacon in 2006, in his article "Emergence: The Hole at the Wheel's Hub," opens a way to his new understanding of the mechanism of EM, which he defines as "the formation of novel, higher-order, composite phenomena with coherence and autonomy at [a] larger scale."[14] Referring to the complex debate concerning the notion of downward or top-down causation, Deacon claims his version of EM can do well without any reference to this concept, reframing EM in terms of topological influences and constraints.[15] Moreover, he states that the process conception of EM he proposes can solve all the conceptual problems raised by former attempts at specifying the mechanism of EM, based on the description of whole-part organization, supervenience, and causal inflection.[16] Departure from these ways of defining EM, says Deacon, enables us to avoid implementing ideas such as "abstract formal properties" or "abstract ideal forms" to categorize higher levels of organization.[17] Instead of referring to these concepts, he proposes a new approach to systems dynamics in which "it's not so much what *was* determined to happen that is most relevant for future states of the system, but rather what *was not* cancelled or eliminated. It is the negative aspect that becomes most prominent. This is the most general sense of *constitutive absence*: something that is produced by virtue of determinate processes that eliminate most or all of the alternative forms. It is this, more than anything else, which accounts for the curious 'time-reversed' appearance of such phenomena."[18]

The notion of "constitutive absences"—that is, "possible features being excluded"—becomes the core of Deacon's process concept of EM, in which he departs from a descriptive notion of form to show "how what is absent is responsible for the causal power of organization and the asymmetric dynamics of a physical or living process."[19] In other words, "emergent properties are not something added, but rather a reflection of something restricted and hidden via ascent in scale due to constraints propagated from lower-level dynamical processes."[20] In his article written with Koutroufinis, Deacon thus describes the nature of constraints characteristic of particular levels of complexity scale:

> The term "constraints" refers to all factors that reduce the number of the possible states of a system, so that its behavior resides only in a limited part of its state space. Since entropy increases with the number of a system's possible states, it follows that constrained systems are not in the state of their maximum possible entropy and therefore are able to perform work. Constraints can be imposed from outside

on a system or can be generated internally, *i.e.*, by a system's own dynamics. We describe the former as *extrinsic* and the latter as *intrinsic* constraints.[21]

2.1. Homeodynamics

Deacon identifies three levels of dynamical depth, which involve "the nesting of stochastic dynamical processes within one another." The processes in question are distinguished by "differences in the ways they eliminate, introduce, or preserve constraints."[22] He begins with the most basic, first-order EM, which includes higher-order linear thermodynamic phenomena, characterized by their spontaneous tendency to eliminate constraints and increase global entropy, thus reaching thermodynamic equilibrium. Deacon calls this level of EM homeodynamics and notes that the term can be applied to higher-order properties of stochastic systems such as liquid properties (e.g., laminar flow, surface tension, viscosity). Although statistical thermodynamics and quantum mechanics offer a remarkably precise theory of how the properties of separate water molecules can provide for the liquid properties in an aggregate, the latter are nonetheless not applicable to the description of water molecules in isolation. It turns out that liquid properties are relational (as opposed to intrinsic molecular properties of mass, charge, configuration of electron shells, etc.) and are not symmetric across levels of description. Consequently, "it is only when certain of the regularities of molecular interaction relationships add up rather than cancel one another that certain *between*-molecule relationships can produce aggregate behaviors with ascent in scale."[23]

To better explain the phenomenon of homeodynamics Deacon distinguishes between "orthograde" changes, which he defines as natural and spontaneous changes in the state of a system (regardless of external interference), and "contragrade" changes, which are extrinsically forced and run counter to orthograde tendencies. He claims the fundamental reversal of orthograde processes in contragrade changes is a "defining attribute of an emergent transition."[24] In the given example of water molecules, their repeated microscopic interactions lead to a specific distribution of their unique features, such as charge, geometry, orientation, momentum, internal vibration, and so on. These features (responsible for orthograde changes) cancel one another in an aggregate (contragrade changes), bringing about the higher-order state with higher-level properties.[25]

2.2. Morphodynamics

Second-order EM (morphodynamics) is characteristic of a dynamical organization of diverse classes of phenomena which have a spontaneous regularizing tendency to become more organized and globally constrained over time. The amplification and propagation of specific constraints happen due to constant perturbation, which reverses the typical thermodynamic orthograde tendency. The critical feature distinguishing morphodynamic processes from homeodynamic processes is the ability of the former to generate locally new constraints (spatiotemporal patterns). These intrinsic constraints are produced to dissipate the imposed extrinsic constraints. Although such systems are usually called "self-organizing," Deacon suggests they should be described as "self-simplifying," since the internal dynamics of their constituents diminishes in comparison to "being a relatively isolated system at or near thermodynamic equilibrium."[26]

As mentioned above, dynamical systems of this kind remain stable so long as the countervailing processes of extrinsic constraint introduction and intrinsic constraint elimination tend to do work that organizes system dynamics in a way that matches the rate of constraint dissipation to the rate of constraint imposition. Once the supportive externally imposed gradient ceases, balancing of input and output stops with it, which shows that the internally generated constraints of these systems are in no way self-preserving. Because they depend on the persistent presence of extrinsic constraints, they cannot contribute to their own persistence. All they can do is degrade these extrinsic constraints. That is why Koutroufinis describes them as cases of "self-organization without self."[27]

Each morphodynamic process is an orthograde process, because "it is an asymmetric orientation of change that will spontaneously re-form if its asymmetry is somehow disturbed and so long as lower-order dynamics remain constant." It arises from and supervenes on (in a non-mereological sense) the lower-level orthograde tendencies, due to a "higher-order orthograde attractor," which characterizes an emergent dynamical transition—a "form-begetting-form."[28]

Deacon lists several examples of morphodynamic systems: the formation of regular spiral whorls of plant structures (called spiral phylotaxis), the formation of geological polygons, the formation of an eddy in a stream, the formation of Bénard cells in a heated liquid, the growth of snow crystals, the generation of laser light, and autocatalytic reactions.[29] In each of these examples "we find a tangled hierarchy of causality, where micro-configurational particularities can be amplified to determine macro-configurational regularities and where these in turn further constrain and/or amplify subsequent micro-configurational regularities."[30]

2.3. Teleodynamics

Although morphodynamic processes generate order, they lack representation and functional organization, which is characteristic of the third-order EM. It is called teleodynamics because of its end-directedness and consequence-organized features, which allow dynamic systems not only to resist entropy increase but also to persistently decrease it, within themselves and their progeny over the course of evolution. Deacon and Koutroufinis list five reasons to ascribe the category of "selfhood" to teleodynamic systems emerging from precisely complementary and interdependent morphodynamic systems:

1. The end-directedness of a teleodynamic system differs from the activity of an "attractor" characteristic of morphodynamic systems. While the latter reflects a constraint on dynamic possibilities in reaction to external environmental influences, the end-directedness of a teleodynamically organized system is generated internally as its higher order intrinsic constraint, preventing the disruption of the synergy between the component morphodynamic processes that provide for its unity. This teleodynamic target state is, thus, a "higher order attractor."
2. The organization of teleodynamic systems enables them to preserve and regenerate their own essential constraints and to interact with their environment in a way that sustains supportive external influences and compensates for unsupportive or destructive ones (by self-reparation or reproduction). It provides for their

inner coherence, which can be defined as self-representation and is lacking in morphodynamic processes, which are intrinsically self-undermining.
3. In contrast to morphodynamic attractors, the higher-order attractors play a critical role in their own generation. Because their production is a form of dynamical compression, teleodynamic systems can be considered self-compressing systems. We can describe this second-order compression (compression of the relation between compression processes) as irreducible self-reference, which is not amenable to description by formalisms adequate for nonliving dynamical systems.
4. In reference to their intrinsic ends, self-coherent teleodynamic systems represent a distinct self-relevant environment, regulating their internal dynamics in relation to the external environment. In other words, teleodynamic systems have an *Umwelt* ("self-centered world") and not just surroundings (as merely reactive morphodynamic systems do).
5. Self-compression, self-representation, and representation of the environment are three aspects of the system's integrated material-energetic dynamics, but not three different activities. The unity of the system in question is a function of a higher-order intrinsic constraint.[31]

"Selfhood" in this sense is understood as a dynamical system minimizing the probability of its own disintegration through maintaining supportive conditions, especially those less likely to occur spontaneously. It is *"a dynamical form of organization that promotes its own persistence and maintenance by modifying this dynamics to more effectively utilize supportive extrinsic conditions."*[32] Deacon and Koutroufinis discuss a "curious causal-circularity linking living processes to the production of their own preconditions," which makes living complexity escape "nondynamical conventional measures of complexity." Summarizing their list of arguments in favor of ascribing "selfhood" to teleodynamic systems, they declare:

> What we term *dynamical depth* then, is this hierarchic complexity and irreducibility of constraint-generating dynamics, such as distinguishes teleodynamics from morphodynamics and morphodynamics

from homeodynamics. Each of these transitions is characterized by the *generation of intrinsic constraints* on the relationships between processes at lower levels and as a result with increasing autonomy from extrinsically imposed constraints. Since constraints are a prerequisite for producing physical work, the increasing autonomy of constraint generation with dynamical depth also corresponds to an increasing diversity of the capacity to do work. Thus the flexibility with which a dynamical system can interact with its environment also increases with dynamical depth.[33]

Deacon sees teleodynamics as a dynamical realization of final causality with respect to the specific self-generating and end-directed attractor dynamics that processes develop toward—"*a consequence-organized dynamic that is its own consequence.*"[34] Because it is typical of systems exhibiting developmental and/or evolutionary features, this kind of EM is most appropriate in the description of the living dynamical systems, in which "the orthograde increase in entropy is effectively reversed by virtue of contragrade processes that generate order and new structural components at the expense of a net entropy increase in their surroundings."[35]

Deacon emphasizes the dependence of teleodynamics on morphodynamics and morphodynamics on homeodynamics, which constitutes "a three-stage nested hierarchy of modes of dynamics, which ultimately links the most basic orthograde process—the second law of thermodynamics—with the teleodynamic logic of living and mental processes."[36] According to Deacon, to describe the forms of representation implicit in teleodynamics, we must introduce a combination of multiscale, historical, and semiotic analyses (analyses based on the relations between signs), which entails reference to concepts such as representation, adaptation, information, and function.[37]

2.4. Autogenesis

Deacon offers a theoretical model of teleodynamics, which he calls "autogenesis"[38] and distinguishes from simple *autopoiesis* (self-production). Together with Cashman, he claims that the model they offer meets Stuart Kauffman's requirements for "autonomous agency" and "acting on its own behalf": (1) self-reparation (self-reconstruction) and self-production (self-replication), and (2) carrying out at least one thermodynamic work cycle in the process.[39]

Deacon exemplifies the distinctive features of an "autogen" (or "autocell") by ascribing to it a specific form of reciprocal coupling between morphodynamic processes of autocatalysis and an enclosure-generating process called "self-assembly." Taken separately, these two processes are inevitably self-limiting and self-undermining. They continue as long as substrate molecules and free energy are available in immediate proximity, depleting these resources exponentially. Moreover, since autocatalysis is a transient non-equilibrium process, whereas self-assembly is an entropy-increasing equilibrium process, autocatalysis and self-assembly are to some extent mutually exclusive. At the same time, however, autocatalysis can replenish the local concentration of self-assembling molecules, while self-assembly is able to provide conditions conducive to autocatalysis.[40]

Consequently, it may occur that one or more of the side products of catalysis tend to self-assemble, forming an enclosure for the autocatalytic activity, similar in its shape to a polyhedron (e.g., a virus shell) or a hollow tube (e.g., a microtubule formation). Such a process gives an origin to an autogen—a self-generating system. While normally the rate of autocatalysis must be equal to or exceed the diffusion rate (otherwise the process will cease), in autogenesis diffusion is impeded by physical barriers to molecular movement generated in the process of self-assembly. This produces what Deacon calls a "negentropy ratchet effect," in which full dissipation of constraints is never completed.[41] Although such molecular containment and enclosure can end autocatalysis, it gives an origin to *"a unit structure that contains within itself the very set of constraints that are necessary and sufficient to re-create these same constraints in a new system in a supportive environment."*[42] See figure 3.1.

Deacon claims this self-limiting higher-order constraint on maximal entropy production characteristic of autogenesis demarcates the boundary and an emergent transition between teleodynamic and morphodynamic processes. It also becomes the threshold zone between living and nonliving systems, as well as the ultimate locus of selfhood.[43] He emphasizes three crucial characteristics of autocells: nonlipid and nonpermeable containment shell, lack of self-replicating template components, and reciprocal cosynthesis of all components. He then contrasts

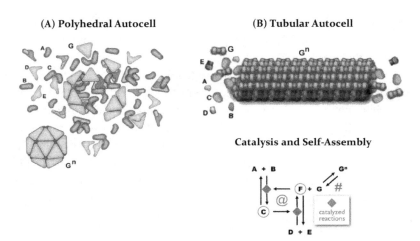

Figure 3.1. Two general classes of autocells depicted as geometric constructions

An autocell produced by polyhedral containment is depicted in (A), and an autocell produced by spirally elongated tubular containment is depicted in (B). Both are minimal autocells, products of a reciprocal catalytic cycle @, depicted with an arrow diagram. The cycle engages two catalysts C and F. One of the molecular by-products (G) tends to self-assemble (#) into a closed structure G^n, encapsulating the ensemble of reciprocal catalysts. Both catalysts are depicted as synthesized from two substrate molecules in each case (A and B, D and E). A polyhedral autocell completely encloses the complementary catalyst, achieving structural closure and allowing no further growth. A tubular autocell does not completely enclose its interior; instead, contained molecules are retained by viscosity of van der Waals interactions with the inner walls. The tubular autocell also retains the ability to continually elongate. Reproduction in both cases depends on extrinsic forces to break containment.

Source: courtesy of Terrence Deacon. See Deacon, "Reciprocal Linkage," 141.

his theoretical model of the minimal components necessary for bacterial cell life with other theories, such as Koonin's "minimal cells," Gánti's "chemoton" and "self-reproducing spherule," and various versions of "protocell." Deacon finds problematic an implicit assumption of the great majority of these approaches, which claim that replication and transcription of molecular information, as embodied in nucleic acids, is a necessary and fundamental characteristic for any system capable of autonomous reproduction and evolution. The primitive cell turns out to be extremely demanding in terms of the critical support mechanism for processes involving such molecular-based information. In contrast to these systems, which he calls *semeota* (from *seme* = sign), as referring to semiotic-based organism replication and identity, Deacon classifies autocells as *morphota* (from *morphe* = form), that is, morphology-based systems. He emphasizes that their maintenance, regulation, self-integrity, and reproduction are embodied holistically (not in specialized molecules such as DNA or RNA): "Autocell dynamics demonstrate that these fundamental attributes of life can emerge without the presumptive critical and ubiquitous role of separate information-bearing molecules; for example, nucleic acid polymers functioning as templates."[44]

Moreover, while an autopoietic process need not have a tendency to reproduce and evolve, the self-reparation and self-replication tendency of an autogen in response to physical damage of its shell makes it a potential candidate for the operation of natural selection. Deacon notices that autocell systems fulfill three abstract criteria for evolvability: (1) self-reproduction with fidelity, guaranteed by necessary synergies between their components (without template-based replication); (2) competition for resources between autocell lineages; and (3) heritable variation occurring via partial breakup and reenclosure of autocells, augmenting reproduction of autogens with changes favorable in the surrounding molecular and energetic environment.[45] Thus, according to Deacon, an evolving autogen becomes not only a "negentropy ratchet" but also a "ratchet of life."[46] In the summary of their description of autogenesis Deacon and Cashman say: "It is not merely a self-propagating formative process, but a reciprocally reinforcing system of self-propagating formative processes, which together form a cohesive integral unit with the potential to actively maintain this unity of processes against dissociation. Only this is sufficient to generate autonomous agency—a physical self that is the locus of an autonomous intrinsic end-directed dynamic, pointing to the meaning of the term teleodynamics."[47]

What is crucial from my point of view is Deacon's recognition of and emphasis on the teleological character of autogenesis. He highlights the fact that individuation (the origin of biological selfhood) is not simply a function of growth in size of an autogen. For him it is a "disposition to produce a specific unrealized potential—an end, which is a type of condition rather than any specific bounded structure."[48]

Finally, what is emphasized in Deacon's classification of three levels of dynamical depth is that they are irreducible. Teleodynamic processes require self-amplifying morphodynamic processes to create their necessary conditions, which in turn require self-amplifying thermodynamic processes to create their necessary conditions. In other words, although third-order emergent processes have their origin in the most basic first-order emergent interactions, one cannot reduce teleodynamics to thermodynamics with no reference to morphodynamics.[49] This notion of irreducibility in Deacon's model of EM brings us to the first point of my evaluation of his project.[50]

3. CAUSAL NONREDUCTIONISM OF DEACON'S PROJECT

An attempt to specify the theoretical and philosophical background of Deacon's model of dynamical depth turns us toward the fundamental assertions of causal nonreductionism. In his works Deacon recognizes, and briefly describes, the change which occurred at the beginning of the Renaissance, when Bacon, Descartes, and Spinoza questioned and then rejected final causality. He shows how this process led to "restricting the conception of causal influence to the immediate pushes and pulls of physical interaction" and, in effect, caused scientists "to replace these black boxes and their end-directed explanations of function, design, or purposive action with mechanistic accounts."[51] Deacon is also critical of the mechanistic explanation of evolution and natural selection, in which end-directedness is replaced by accidentally formed mechanisms favored by environmental conditions. When taken to the extreme, this way of argumentation brings about Dawkins's concept of evolution as a Blind Watchmaker.[52] Deacon believes this metaphor to be deeply influenced by certain assumptions associated with the designing of human artifacts, which renders the machine metaphor a misleading oversimplification: "Organisms are not built or assembled; they grow by the multiplication of cells, a process of division and differentiation from prior, less differentiated precursors. . . . The machine metaphor of the world implicitly begs for a watchmaker, even as it denies his or her existence."[53]

Deacon criticizes eliminative materialism for assuming that concepts like information, representation, and function are ultimately reducible to mechanistic accounts. He disagrees with the claim that, even if an attempt to explain goal-directed concepts in the language of basic physical properties and interactions of things is clumsy and arduous, nonetheless, these concepts are nothing more than just a sum of these properties and interactions. As we have seen, he finds concepts such as information, function, purpose, meaning, intention, significance, consciousness, and value intrinsically irreducible.[54]

What is especially interesting from my perspective is that, when criticizing causal reductionism in science, Deacon presents a counterexample: the plural notion of causality in Aristotle. He lists Aristotle's four causes, finding in his philosophy "the most sophisticated early recognition of a distinction between ... different modes of causality."[55] However, we must ask and analyze carefully to what extent Deacon is really following Aristotle's philosophy of nature and where he departs from him, reinterpreting or simply rejecting his thought as inadequate in the context of contemporary science. I believe such analysis opens the way to a more general evaluation of the metaphysical aspects of Deacon's model of EM, which is strategic for the purpose of this book. I will now address these issues in more detail.

3.1. Material Cause

Deacon admits he ignores Aristotle's material causation consciously and intentionally in his project.[56] He directly alludes to material cause only occasionally and appears to understand matter simply as stuff out of which things are made. In Deacon's example of a carpenter's work, "material cause is what determines the structural stability of a house."[57] Deacon's reading of Aristotle at this point is similar to that offered by other scientists and philosophers of science who are more familiar with Aristotle's notion of causality. Alwyn Scott, for instance, when speaking of Aristotle's material causality, gives several examples: "Atoms of iron are necessary to produce hemoglobin, obesity in the United States is materially caused by overproduction of corn, water is essential for Life."[58] Among philosophers of science, Claus Emmeche et al., while explaining the material cause, use concepts such as "consisting of" and "made of." These concepts are usually applied to a description of relations between wholes and the parts that construct them.[59] At first sight, Deacon appears to limit his inquiry to this type of explanation as well.

My analysis of Aristotle's notion of material cause in the introduction clearly shows that such an interpretation is not attentive to his idea of matter as an underlying principle (πρώτη ὕλη), a principle of potentiality which persists through all the changes and processes a given substance is exposed to, a metaphysical category that goes beyond what is perceptible in sensual experience. An attempt to define matter simply as stuff, amenable to mathematical and physical description, reduces its philosophical meaning, which is at the heart of Aristotle's explanation.

The truth is, however, that Deacon is by no means so indifferent to the metaphysical aspects of material causality as he claims to be. Quite the contrary, a careful reading of *Incomplete Nature* shows he presents a very strong position in the debate on the nature of matter. He begins his criticism of substance metaphysics in the chapter on EM with a rather bold metaphysical statement, based on the Copenhagen interpretation of quantum mechanics. He claims that because science is not able to specify properties of elementary particles on the quantum level, except by describing them statistically, "at this presumed lowest level, discrete parts do not exist. The particulate features of matter are statistical regularities of this dynamical instability, due to the existence of quasi-stable, resonantlike properties of quantum field processes. This is why there can be such strange unparticlelike properties at the quantum level."[60]

And yet, after delineating this metaphysically flavored reflection on matter, Deacon immediately adds he does not strive to base his explanation of emergent properties in reference to "quantum strangeness," nor does he propose to "draw quantum implications for processes at human scales."[61] Indeed, his model of dynamical depth is based on the analysis of micro- and macroscopic rather than quantum-level phenomena. But Deacon's making this allusion to indeterminacy of quantum mechanics is crucial. It proves he intuitively admits his dynamical model of EM needs to be grounded in an account of the principle of potentiality in nature. He implicitly strives to locate this model in the description of reality in terms of quantum field processes, despite the model's difficulties in explaining how quantum events translate into macroscopic phenomena (a problem Deacon does not address in his project).

I want to emphasize that, whether or not he is aware of it, the metaphysical aspect of his explanation of matter and his search for the principle of potentiality in quantum mechanics bring Deacon close to the

Aristotelian notion of primary matter. One of the contemporary interpreters of Aristotle's philosophy of nature, William Wallace, notices that with the advance of the sciences of physics and chemistry scientists realize that the ultimate matter, the stuff of which things come into being and pass away, becomes—paradoxically— somewhat mysterious and elusive: "It has been said that matter has been 'dematerialized' in our generation. What this means is that it can no longer be thought of ultimately as little hard chunks of stuff. Better to think of it as the matrix, the underlying principle from which all natural forms emerge, more like an indeterminate or potential energy that grounds every change going on in the universe."[62] Obviously, we cannot think mass (a conserved quantity), energy, and quantum field processes—the contemporary equivalents of matter in natural sciences—identical with primary matter, which is *sensu stricto* a metaphysical principle.[63] But we cannot ignore an important correspondence between these two approaches at defining the principle of potentiality in nature. This is one reason I will argue, in section 3.5 of this chapter, why Deacon's rejection of substance metaphysics, with its notion of potentiality, is too hasty. In the meantime, I must analyze the retrieval of the other three causes in his dynamical model of EM.

3.2. Formal and Efficient Causality

Moving toward the formal cause, Deacon describes it together with efficient causality, claiming they are intrinsically interrelated. At one point, in the chapter on EM in *Incomplete Nature*, Deacon seems to follow Aristotle accurately in his understanding of substantial and accidental form. He contrasts Sperry's analogy of a thing being affected by the geometrical shape and configuration of the whole, which does not change the local properties of the parts (e.g., parts of a wheel), with Humphreys's idea of fusion, which affects intrinsic properties and brings changes in the very constitution of the parts of a whole.[64] This contrast is reminiscent of a genuine Aristotelian notion of substantial versus accidental form and of substantial versus accidental changes in nature. Moreover, in the chapter on homeodynamics, Deacon follows Aristotle in distinguishing between spontaneous and forced change. He acknowledges that spontaneous changes are intrinsic to each substance and are not an outcome of efficient causality, which is responsible for forced changes. According to Deacon, spontaneous changes occur in virtue of the formal cause.[65]

Nonetheless, approaching the core of his explanation of the very mechanism of EM understood in terms of constraints and orthograde and contragrade changes, we find Deacon at risk of reducing formal cause to merely geometric properties. He states that orthograde changes are functions of the geometric properties of a probability space and happen spontaneously, irrespective of anything else. Their occurrence is "an unperturbed reflection of the space of possible trajectories of change" for a given system. He then concludes, saying: "I take this to be a reasonable way to reinterpret Aristotle's notion of a formal cause in a modern scientific framework, because the source of the asymmetry is ultimately a formal or geometric principle."[66]

Formal cause redefined in such a way plays an even more significant role in one of Deacon's recent publications, coauthored with Srivastava and Bacigalupi, in which he introduces terms such as "*forms* of the constituent dynamical processes," "formal reciprocity constraint," "formal 'semiotic' constraints," "formal source of regulation," "'formal' disequilibrium," and "formal relationship" of morphodynamic processes. What is being emphasized in this terminology is the idea of the formative aspects of causality of absences (constraints). According to Deacon, it is not only what is present and analyzable in terms of parts of the system depicted in the probability space, but also what is absent, that provides for the formal identity of dynamic processes.[67]

Such an approach seems to reduce the positive aspect of formal causality to the accidental formal cause (that is, elements of a system depicted in the vector probability space), while ignoring the substantial formal cause (that which makes a thing to be what it is and so to have the properties it exhibits). The negative aspect of formal causation, expressed in Deacon's theory of the absential nature of constraints, is rather foreign to the Aristotelian definition of substantial form as well. Consequently, Aristotle's emphasis on the fundamental relation between substantial form and primary matter—the core characteristics of hylomorphism—is replaced by Deacon's idea of an inherent relation between efficient and formal causes, where the latter appears to be redefined in terms of the geometry of the vector probability space and seen as a function of efficient causal processes.[68]

Even within his own terminological framework, Deacon's theory leads to further questions. The idea of spontaneous orthograde changes begs the question of the source of this regularity of a given dynamical system. If form is just the behavior of a system, developed in response to the causal influence of constraints and described in terms of geometric properties of a probability space rather than in reference to an intrinsic principle of unity (characteristic of inanimate things, organisms, and processes), then the question of why they have the properties that they have and why they tend to do what they do when entering dynamic processes remains unanswered. Moreover, similar is the case of efficient causality, which brings contragrade changes in Deacon's system. "Forcing change away from what is stable and resistant to modification," efficient cause is defined as "the juxtaposition of different orthograde processes," which "can produce complex forms of constraint."[69] But again, what is the source or cause of this juxtaposition of orthograde processes? Does it happen by chance, or is it a function of the intrinsic properties of their constituents? If the latter is true, then form has to be something more than just a set of geometric properties of a phase space. Otherwise, everything is reducible to basic physical interactions between particles or basic dynamical systems, which brings Deacon's project of replacing philosophical terminology with scientific at risk of falling into eliminative reductionism.[70]

That Deacon regards the traditional philosophical concept of form as another homunculus becomes clear from his analysis of Aristotle's notion of entelechy. As we have seen in the introduction (section 2.1), Aristotle sees entelechy as form's state of completion and fulfillment. Deacon seems to misunderstand him when he says that entelechy is like a purpose or agency that is present only potentially in the earliest stages and fully realized at the end of development of an organism.[71] As we have seen in sections 2.1 and 2.2 of the introduction, purpose for Aristotle is the final cause, and agency is the efficient cause. Entelechy is a final realization of formal cause. Formal cause is thus present—fully, not only potentially—from the very beginning of an organism's existence. It causes an organism to be what it is at any stage of its development. Entelechy does not realize itself; it is the organism that reaches maturity (entelechy) in virtue of its formal cause.[72]

Deacon rightly notices that although his concept of entelechy (or better: form) was influenced by Plato's notion of ideal forms existing independently of any specific embodiment and expressed imperfectly in beings, Aristotle parted from his mentor, insisting that form needs to be embodied. He is also correct in his observation that the more Platonic—that is, disembodied—conception of form understood as a self-actualizing living potential gave rise to the twentieth-century version of biological vitalism supported by Hans Driesch. It is true that, similar to strict preformationism, strict vitalism is homuncular.[73] But this criticism does not apply to an Aristotelian understanding of formal causation. As an inherent feature in-forming primary matter, form does not come from the outside and is not disembodied. It cannot be treated as a general template for the final result of a development toward entelechy. Nor is it a mysterious formative potential or force. It is defined as a principle of actualization and unity, which is intrinsic to a nonliving or living thing, fully present and active at any stage of its development. Despite this fact, however, Deacon seems to view Aristotelian hylomorphism as homuncular. He reduces form to emergent constrained dynamics, depicted by the geometry of a phase space, and claims that as a "formal reciprocity constraint" in a reciprocal interaction of morphological processes, it opens the way to teleology—a new and decisive form of causation, introducing qualitative differences and central for the theory of the origin of life and all entential phenomena.[74]

3.3. Teleology

The question of natural teleology stands in the center of Deacon's argument in *Incomplete Nature*. This is probably the reason why John Farrell suggests the book might have been titled *Aristotle's Revenge*.[75] As mentioned above, Deacon emphasizes repeatedly that concepts such as information, function, purpose, meaning, intention, significance, consciousness, and value are real and irreducible. Following Aristotle, he defines a teleological phenomenon as that for the sake of which something happens, and he stresses that the concept of teleology does not assume future goals have a mysterious causal influence on the present.[76] He describes all phenomena that are intrinsically incomplete as "entential."

Nevertheless, although the point of departure of Deacon's account of teleology is consistent with the explanation found in Aristotle, his next steps bring about ideas that are rather peculiar from the point of view of Aristotelian metaphysics. Trying to describe teleology in scientific terms, Deacon finds it necessary to explain the way in which teleological properties emerge from nonteleological. He claims that teleodynamics—a dynamical realization of final causality—can be described in quasi-mechanistic terms.[77] While searching for an explanation of the ways in which the causal dynamics of teleological processes emerges from simpler, blind, mechanistic systems, Deacon notes that ententional phenomena are defined by their fundamental incompleteness; they are reaching out toward something they are not.[78] This makes him suggest that absences have causal efficacy and that the development (or EM) of teleology occurs due to the growing limitation of the degrees of freedom which is an outcome of the operation of absences in constraint-generating dynamical systems. He then concludes that the traditional motto of complex-systems biology that states that the whole is more than the sum of its parts must be replaced by an opposite assertion—namely, that the whole is less than the sum of its parts.[79]

These ideas are problematic from the point of view of Aristotle's philosophy of nature, in which final causality never arises from mechanistic or any other type of causation, as there is no transformation of causes in his philosophical view of nature. In opposition to this rule, Deacon states: "Until we explain the transformation by which this one mode of causality [physical] becomes the other, our sciences will remain dualistically divided, with natural science in one realm and the human sciences in the other."[80] On another occasion we find him saying: "Hierarchically organized constraint relationship provides a plausibility proof for the emergence of *telos* (a form of Aristotelian 'final causality') from mere physico-dynamic processes (Aristotelian 'efficient causality')."[81]

Again, this reasoning is radically foreign to Aristotle's understanding of final causation. Even if teleology is regarded in his philosophy, together with efficient causation, as an extrinsic principle of living and nonliving things, it nevertheless has an intrinsic aspect to it. Although the end or goal may be extrinsic to nonliving and living things, the directedness to it is intrinsic in each one of them, as it finds its source in their proper natures (defined in terms of formal causation). Consequently, even if we agree with Deacon that hierarchically organized complex dynamics proper to chemical and biological systems do exhibit novel and higher modes of teleology, we will find questionable his idea that they emerge from a nonteleological—that is, efficient—causation.

The difference between Aristotle and Deacon at this point becomes evident. Unlike Aristotle—for whom teleology is deeply related to formal and efficient causes of each natural entity and organism, but never emerges from them—Deacon strives to prove the reality and irreducible character of "ententional phenomena" without introducing the classical concept of substantial form. We have seen that he regards it as a homuncular category. What gives rise to an emergent whole with its characteristic teleological features, in his theory, is not a substantial form but specific absences, defined in purely negative terms. What is not present has a causal influence on what is. The source and nature of such causation, however, becomes another major problem of his model of EM.

3.4. Absences as Causes

It is true that, when explaining that one and the same thing may be the cause of contrary results, Aristotle brings an example of the captain (pilot) whose absence is the cause of the sinking of the ship.[82] What we are dealing with in Aristotle's example, however, is a complex situation in which the structure of the ship and its crew was set up so that in case of the absence of the captain he would be "missed," and so be a "missing" cause. He would then cause something by "not" being there. But the absence itself cannot be a cause for Aristotle, which might be summarized as "no causality without actuality."

Apart from the example of the ship and its captain, one may try to define incompleteness and absences as a privation of form. But although it is a principle of change for Aristotle—as the change proceeds from it—privation is never an actual cause, for nothing comes from nothing. "Whatever comes to be is always complex. There is, on the one hand, (a) something which comes into existence, and again (b) something which becomes that—the latter (b) in two senses, either the subject or the opposite."[83] By "the opposite" Aristotle means a mode of nonbeing. Hence, when clay is shaped into a ball, the round comes from the nonround or the privation of roundness. Still, nothing comes from nothing, since the round comes from the nonround not *per se* but only *per accidens*, insofar as the nonround is not (simply) nonround but is clay. Clay in this case is the principle of potency from which the round comes to be *per se* (insofar as it is potential) and not merely *per accidens*. In other words, we can predicate causation about privations only analogically (*per accidens*), bearing in mind that as forms of nonbeing, they are defined in terms of deficiencies, gaps, imperfections, and so on, in actualities, which are *per se* sources of causal interactions.

Unlike Aristotle, in his explanation of EM, Deacon ascribes causal efficacy to the sheer "missing-ness" (lack of presence) characterizing various degrees of freedom. He speaks about "absence-based causality," "absential influence," or "efficacy of absence."[84] But what kind of causation is this? What can an absence actually do? Does it play the role of an efficient cause, acting specifically at various levels of dynamical depth? Or maybe we should treat it more like a formal or final cause in the Aristotelian sense? Another answer would be to suggest that Deacon defines a new and unique type of causation, neither known nor studied in philosophy of causation hitherto. But would the definition of causal absences in purely negative terms suffice?

In his rhetorical device of the hole at the wheel's hub Deacon emphasizes the importance of the lack of physical components of the wheel at its center for its proper functioning. However, his reasoning begs a question. Why ascribe the critical importance for the proper operation of a wheel to the absence of physical components at its center, rather than to the presence of the rim, spokes, and the hub, which together form the structure with an empty space at its center, causing the wheel to be what it is (in Aristotelian categories, they provide for an accidental form of the wheel)? There would be no hole at the wheel's hub if not for the hub and all other parts of the wheel, in the first place. Moreover, how can "the hole's feature of being non-material" be "in relation to the material components of the wheel?" What kind of relation is it? How can "the lack of physical components of the wheel at that position" ("being non-material") be "physically extended ... in space and time?" In what way is it "an actuality" and not "a mere possibility"? All these questions refer to assertions made by Deacon in his response to Peter Bokulich.[85] I find them metaphysically wanting. The adjective "non-material" suggests that a constitutive absence is something, while it should be regarded by definition as nothing. Hence, instead of calling it "non-material," to be true to his intention, Deacon should have called it "nonbeing," to avoid confusion. But if a constitutive absence is "nonbeing," it cannot be "physically extended ... in space and time" or be described as "an actuality" and not "a mere possibility." Nor can it be assigned a causal power.[86]

It is important to note that in one of his more recent papers, coauthored with Cashman and presented at the conference organized by the Center for Theology and the Natural Sciences in Berkeley, California, Deacon seems to change his strategy radically. He says—in response to an early draft of this book—that his idea of the "efficacy of absence" is just a "rhetorical trope," which actually "undermines the point that it intends to emphasize," and that "absences themselves don't do work, nor do they resist work."[87] And yet we find him suggesting, in the same paragraph, that absences in fact do something—that is, "they enable and channel the outcome of energy release"; thus he still ascribes to them causal powers.[88] In response to my critical comment on this point, presented at the same conference, Deacon and Cashman explain that "absences are the result of physical constraints. Physical constraints are contragrade—that is, part of the efficient causality complex. However, the absences that result from them are formal. Since absences are required for the controlled results of a process, they might be called formal 'causes.' But these are nothing like efficient causes—and nothing like the formal causality of Aristotle's *eidos* or the *forma substantialis* of St. Thomas. They relate more to Aristotle's *morphe* than to his *eidos*."[89]

Technically speaking, Aristotle's *eidos* (εἶδος) is not identical to the *forma substantialis* of Aquinas. As could be seen in the introduction (section 2.1), *eidos* (εἶδος = "outward appearance") and *morphē* (μορφή = "shape") have similar meanings, which should be contrasted with Aristotle's primary term for form—namely, ὁ λόγος τοῦ τί ἦν εἶναι = "the statement of the essence." It is only the latter that can be translated as Aquinas's *forma substantialis*. Nevertheless, aside from this terminological imprecision, what we encounter in Deacon and Cashman's statement is their embracing of the reduced version of Aristotelian form, understood as the external physical shape (appearance), which can be imposed from the outside (e.g., the shape of a statue or a rock) or acquired by a living organism in the process of its development. They refer this notion of form to the nature of "absences" in their metaphysical model of autogenesis, which I find rather questionable. For the only thing that the classical philosophical notion of form in its reduced version tells us is how a given entity appears to us in its external physical qualities. It does not tell us why it is what it is in itself, and why it possesses or takes this particular shape. This kind of philosophical knowledge is acquired only in reference to the robust notion of form as "the statement of the essence," or *forma substantialis*. Hence, from the Aristotelian-Thomistic point of view, Deacon and Cashman take the reduced, geometrical version of Aristotle's form, trying to make it do the same "metaphysical work" that is usually assigned to the more robust version of form as the intrinsic ὁ λόγος τοῦ τί ἦν εἶναι—the meaning of which they find homuncular.

Moreover, thinking about the nature of formal causation, we find—in the same response to their interlocutors—a rather peculiar analysis in which Deacon and Cashman claim that "the causal function of Aristotle's form is 'to be desired.'" In opposition to this view they state, "In our understanding of absence as form, form is not the goal of the matter. There is no desire or reaching after form. Absence as form (*morphē*) is an allowing, an enabling, an opening up for."[90] This explanation is metaphysically dubious. First, if it is possible to say that primary matter "desires" to be in-formed, it does so not because form caused (made) it to do so. Rather, "desiring" form belongs to the character of primary matter as the principle of potentiality. Consequently, the causal function of Aristotle's form is to in-form primary matter, providing for the very nature of an entity that is "composed" of primary matter and substantial form, and not to cause primary matter to "desire" form.

Second, trying to find an answer to the question of the exact nature of absences, we see Deacon and Cashman once again using causal terms (such as "allowing," "enabling," and "opening up for") in their definition of absence as form. Hence, we are back to square one and the question about the precise nature of these terms. We might assume they want us to read them analogically, but that leaves us with a rather puzzling picture of nature. We have learned from Deacon and Cashman that "absences are the result of physical constraints, [which are] part of the efficient causality complex." What they seem to argue for is thus the reality of efficient (physical) causation of constraints, resulting in absences, which are neither physical nor causal in the efficient sense. I find this explanation rather peculiar.

In *Incomplete Nature* Deacon claims that "to argue that constraint is critical to causal explanation does not in any way advocate some mystical notion of causality."[91] He sees constraint simply as a reduction of options for change in one process that can account for a further reduction of options in another process which depends on the first. But this explanation inevitably raises questions of how something absent can affect what is present, how constraints actually produce the reduction of options for change, why the reduction of options occurs at all, and why it appears to be goal directed in the case of teleodynamics. More fundamentally, we should ask what provides for the options for change in the dynamic processes in the first place. It seems that possible degrees of freedom are functions of the very nature of the beings and processes in question. If it were not for their characteristic features, there would be no options for change, nor possibilities of constraints. Once again, my argument proves that there can be no constraints without a reference to what actually exists and determines possible degrees of freedom, as well as possible constraints.

In other words, according to Aristotelian metaphysics, it is because an entity A is A—that is, it is a substance in-formed by a particular substantial form—that it is not $B, C, D, \ldots, Z$. Deacon's metaphysics has it in reverse. It is because it is not $B, C, D, \ldots, Z$ that A is A. But the latter approach seems to require from us specifying all possible degrees of freedom, all possible forms of entities in the universe, and all degrees of freedom that were reduced in order for A to become A. Knowing that this is practically impossible, a follower of Deacon's "metaphysics of incompleteness"[92] may say that we have to limit our description to the most proximate and selected range of possibilities. Such a strategy proves very useful and effective in natural science, but it is highly unsatisfactory in philosophy, which searches for ultimate answers to the questions it asks.

In any case, the whole conversation on the causation of absences leaves—in my opinion—the question about their very nature unanswered. On the one hand, we find Deacon and Cashman stating that absences do not do any work. On the other hand, they find it difficult to avoid using causal terms such as "allowing," "enabling," "channeling," or "opening up for" in reference to absences. In their most recent position they do attribute causal efficacy to absences, but define it not in terms of efficient causation, or the primary notion of formal causation in Aristotle ("the statement of the essence" or *forma substantialis*), but in terms of the reduced notion of form characterized as shape (μορφή). However, we have seen that formal cause, understood in this way, is responsible merely for the way entities appear to us in their external, physical qualities. It does not explain why they are what they are in themselves.

What is, then, the conclusion of the whole debate at its present stage? From the context of my investigation, it becomes clear that—despite difficulties in providing an exact explanation—Deacon does want to assign causal importance to absences. The attempt to withdraw completely from his earlier position on this topic seems to undermine Deacon's own project, which defines emergent transition in terms of new levels of dynamical organization, originating due to the impact of absences resulting from constraints. Thus, in what follows I will remain faithful to the well-documented and widely discussed earlier version of Deacon's argument, in which he does assume that absences do causal work. At the same time, I will keep in mind all the reservations and qualifications concerning the nature of this causation he recently introduced, together with Cashman.

3.5. Substances versus Processes

Having analyzed Deacon's reinterpretation of all four types of Aristotle's causes and his concept of the causation of absences, I shall now address the problem of his rejection of substance metaphysics in favor of the dynamic (process) view of reality. First, Deacon's understanding of substance metaphysics does not appear to be entirely correct. Aristotle does not assume that "the properties of things inhere in their material construction."[93] He sees them rather as a function of a substantial form, which, as a source of unity, in-forms primary matter. He would also entirely disagree with the statement of Mark Bickhard (cited by Deacon), who claims that the substance metaphysics assumption requires that, at base, "particles participate in organization, but do not themselves have organization."[94] Bickhard contrasts this assertion with contemporary quantum physics's emphasis on the fact that particles are the somewhat indeterminate loci of inherently oscillatory quantum fields, which are irreducibly processual and thus intrinsically dynamic and organized. To that I answer that nothing in Aristotle's philosophy of nature would prevent fundamental particles from having organization of some kind.[95] However, when they are about to enter into a construction of a whole that will have a new substantial form, the primary matter that was in-formed by forms of those fundamental particles is now disposed to receive a new, higher form of a new thing or being. That is why we should emphasize that Aristotle's substance metaphysics is not based on the notion that the actuality of the very small accounts for the whole. It rather begins with the whole of any size (at either the micro or macro level of analysis), its substantial form, and natural finality. Comparing Aristotle's substance metaphysics with Deacon's notion of the process organization of matter, we can say that just as "processes don't have other processes as their parts,"[96] so the form of a substance does not have other forms as its parts. It is an irreducible principle, educed from the potentiality of primary matter. If this is the case, then what Deacon is arguing against in the *Incomplete Nature* is not classical substance metaphysics but the materialistic reductionism of modern science, which rejected the basic principles of the natural philosophy and the metaphysics of Aristotle.

The problem with Deacon's analysis is that he associates the predominant method in contemporary science—that is, the mechanical explanation of whole-parts relationships—with the notion of substance metaphysics, which he classifies as mereological. He does not realize that Aristotle's substance metaphysics is founded on hylomorphism rather than on mereology. Following his mereological assumption about substance metaphysics, Deacon claims it is the main target of the criticism of Kim. As we have seen, however, Kim's criticism is valid as long as the idea of causation is reduced to the efficient cause. Without doubt, the reductionist mereological explanation, as defined by Deacon, does not escape this criticism. But the case of substance metaphysics is different. Bringing into consideration a plural notion of causality, and beginning its argumentation from the description of a whole, being one in virtue of a substantial form, substance metaphysics remains free from the problems raised by Kim's analysis.

Nevertheless, even if we assume Deacon is right in his rejection of substance metaphysics, he does not seem to offer a solid metaphysical theory to replace it in his project. Although in his explanation of the material aspects of the universe Deacon refers to some philosophical aspects of the Copenhagen interpretation of quantum mechanics, he does not offer an explanation of how quantum field processes provide for macroscopic events. Moreover, it becomes evident from reading his

Incomplete Nature that Deacon does not side with the process metaphysics of Whitehead, which was inspired by the impact of the new science of quantum. He regards this as another version of panpsychism (pansubjectivism), in opposition to its proponents, who define it rather as panexperientialism. Deacon says the assumptions of process metaphysics do not explain "why the character of physical processes associated with life and mind differs so radically from those associated with the rest of physics and chemistry—even the weird physics of the quantum."[97]

This lack of a solid metaphysical background raises several important questions which I have already addressed and will further develop in my general philosophical evaluation of Deacon's project. Before I present it, however, I need to conclude my analysis of his contribution to the revival of Aristotle's theory of four causes.

3.6. How Aristotelian Is Deacon?

> In many ways, I see this analysis of causal topologies as a modern reaffirmation of the original Aristotelian insight about categories of causality. Whereas Aristotle simply treated his four modes of causality as categorically independent, however, I have tried to demonstrate how at least three of them—efficient (thermodynamic), formal (morphodynamic), and final (teleodynamic) causality—are hierarchically and internally related to one another by virtue of their nested topological forms. Of course there is so much else to distinguish this analysis from that of Aristotle (including ignoring his material causes) that the reader would be justified in seeing this as little more than a loose analogy. The similarities are nonetheless striking, especially considering that it was not the intention to revive Aristotelian physics.[98]

I believe this statement from the summary of Deacon's "Emergence: The Hole at the Wheel's Hub" applies to the argumentation in his *Incomplete Nature* and some more recent articles as well. Putting aside his assertion that Aristotle "treated his four modes of causality as categorically independent," which can be seen as inaccurate in the context of the analysis of Aristotle's theory of causation presented in the introduction (section 2.3), the question remains whether Deacon offers in his works a truly "modern reaffirmation of the original Aristotelian insight about categories of causality," or rather simply "a loose analogy" to it.

The answer to this question is complex. On the one hand, in some places in his works Deacon presents Aristotle's concepts accurately, extending and enriching the view of nature presented in contemporary natural science. On the other, however, a more detailed analysis shows Deacon's account of the Aristotelian philosophy of nature appears to be a reinterpretation of his thought rather than a revival of his original argumentation, a reinterpretation which, although inspiring and intriguing, may be in danger, at some points, of falling into a blind alley of eliminative reductionism. We have seen, for instance, that, although Deacon's description of orthograde and contragrade changes accounts for both efficient and formal causes, he tends to reduce the latter to a merely geometric principle describing behavior of dynamical systems. He sees form as "shape" or "appearance" (μορφή or εἶδος) rather than "the statement of the essence" (ὁ λόγος τοῦ τί ἦν εἶναι). Aristotle's final cause is redefined in Deacon's model of EM as an outcome of the causal efficacy of absences in constraint-generating dynamical systems—a theory which is rather obscure and alien to Aristotle's metaphysics. When speaking of material cause, Deacon does not take into account Aristotle's concept of primary matter, which is crucial for his hylomorphism. Finally, Deacon generally distances himself from substance metaphysics, without which it appears difficult, if not impossible, to argue for the revival of Aristotle's theory of causality. This seems to suggest that what we find in Deacon's *Incomplete Nature* is an interesting reinterpretation of the natural philosophy of Aristotle, in which he departs considerably from the Philosopher's original thought.

Having said this, we need to remember that in the quotation opening this section Deacon himself says openly that he did not intend to bring about a revival of Aristotle. What he wants to argue for is the antireductionist character of "ententional" phenomena. Fair enough. But such a position begs the question of the foundation of his antireductionism. Deacon claims, at one point, that he does not regard himself as an antireductionist in the traditional meaning of the term. Quite the contrary, he says, "there can be little doubt that reductionistic science is fundamentally sound," and "it would be pointless to even imagine that it is somehow misguided."[99] Leaving aside the question of a definition of the traditional antireductionism of which he speaks, I infer from these statements that Deacon is fully comfortable with the methodology of natural science, which is reductionist by definition. This insight leads me to the conclusion that the antireductionist character of his model of EM must be grounded at the level of a philosophical (metaphysical) interpretation of reality, with scientific data in the background. If this is the case, however, my analysis shows that Deacon's work needs a more precise definition of the metaphysical presuppositions of his theory of EM. This assertion becomes an introduction to my general philosophical evaluation of his project.

4. INCOMPLETENESS OF DEACON'S MODEL OF EMERGENCE

Deacon's model of dynamical depth becomes undoubtedly the most important alternative to the traditional understanding of EM in terms of SUP and DC. It is also the first systematic emergent theory arguing in favor of causal nonreductionism. Its character and scope are truly distinctive. At the same time, however, we have seen that Deacon does not simply argue for a retrieval of Aristotle's notion of causation. He suggests a thorough reinterpretation of all four causes and supplements his analysis with a new type of causation—namely, causation of absences. He incorporates Aristotle's causes into his theory of the "machinery" providing for the growing complexity of biological dynamical systems. Therefore, in order to fully appreciate Deacon's project, we need to acknowledge both the strengths and weaknesses of his endeavor. I believe my analysis offered in the previous section, and strengthened by the present evaluation, will enable me to offer a constructive proposal for Deacon's dynamical model of EM in the following part of this project.

What stands at the center of Deacon's theory is an attempt to explain the transition from nonliving to living systems. Here I want to emphasize once again the distinctiveness of his model of autogenesis among current theories of the origin of life. I strongly agree that the reference to dissipative structures and dynamical autopoietic systems, with no capacity for persistence beyond extrinsically imposed gradients of energy and constraints, does not answer the question of the phenomenon of life.

Deacon is right when he says that autocatalysis does not fundamentally differ from a typical chemical reaction and that the introduction of the requirement of containment does not provide by itself for the development of what he calls a "primitive biological self." I wholeheartedly follow his recognition that what is at stake in the process of the EM of life is the origin of specific intrinsic teleological features of a system, enabling it to develop the ability of self-reparation and self-replication and making it a candidate for the operation of natural selection. Finally, I acknowledge the originality of Deacon's characterization of an evolving autogen as a "negentropy ratchet" and a "ratchet of life."

These advantages notwithstanding, I have already shown that what becomes problematic in Deacon's theoretical model of EM, from an Aristotelian point of view, is his tendency to reduce formal causation to geometric properties of a probability space and to regard the classical interpretation of Aristotle's form as homuncular. I have also found questionable Deacon's suggestion concerning the EM of teleology from mere physico-dynamic processes classified as a contemporary notion of Aristotle's efficient causation. I find these suggestions reductionist not only scientifically but also metaphysically.[100] What matters in the end in Deacon's model of EM is (1) an interplay between regularity and spontaneity of orthograde (physical) changes and (2) their juxtaposition in contragrade changes. Their reversal brings new constraints, and thus becomes a decisive attribute of emergent transitions at various levels of EM. But the whole theory begs the question concerning the very source of both spontaneous and forced changes and concerning the behavior of entities and dynamical systems that are subject to these changes. This general question can be specified and related to each level of Deacon's model of dynamical depth.

To begin at the top of the ladder of ascent of emergent properties, why does the reciprocal coupling between morphodynamic processes of autocatalysis and processes of self-assembly, a coupling which produces higher-level constraints and gives an origin to an autogen, take place at all? Why do the by-products of autocatalysis show a spontaneous tendency to self-assemble and to form an enclosure? What makes the two processes complementary and interdependent? What makes them develop together spontaneously toward a "higher order attractor"—that is, a teleodynamic target state? Even if we assume that the behavior of these two processes is a reaction to thermodynamic gradients of entropy in the environment, we still need to ask what makes them show this kind of response to these external conditions. Moreover, the laws of thermodynamics, like any other natural laws, are descriptive rather than prescriptive. They do not have any causal power themselves but reflect and describe the nature and causality of all dynamical systems and beings building up into a universe we know. Sherman and Deacon are right when they speak about "the thermodynamic universe."[101] The laws of thermodynamics are not unrelated to the content of the universe. They both come from and apply to a concrete set of entities and dynamical systems forming it. Hence, the energy gradient responsible for the reciprocal coupling of morphodynamic processes of autocatalysis and processes of self-assembly needs to be explained in terms of the nature of entities and dynamical systems that form it. Similar is the case of the self-reparation and self-replication characteristics of an autogen. As a reaction to thermodynamic changes in the environment and physical damage, these processes cannot be explained exclusively in terms of physical conditions. A reference to intrinsic teleology characteristic of an autogen is needed.[102]

Descending to the level of morphodynamics, we find Deacon repeatedly emphasizing the spontaneous character of the regularizing tendency of various classes of phenomena to become more organized and globally constrained over time. This spontaneity becomes the main reason to classify such systems as "self-organizing." But the theory raises the same question concerning the source of the spontaneity of these processes. What is the source of externally imposed gradients of energy? What makes the molecules of water react in a specific way in characteristic thermodynamic conditions, forming an eddy in a stream? What makes the molecules of a heated liquid form Bénard cells in response to certain energy gradients? What is the source of the spontaneous tendency of these processes to regenerate their intrinsic constraints in case of random perturbations? Why do their constituents bother to do what they do in reaction to thermodynamic gradients of entropy? Again, I claim that an attempt at answering these questions requires a reference to the metaphysical nature of entities entering these morphodynamic processes.

Going further down to the lowest level of homeodynamics, we encounter a similar emphasis on the spontaneous tendency of higher-order linear thermodynamic phenomena to eliminate constraints, thus increasing global entropy and reaching the state of equilibrium. But what is the source of this spontaneous tendency? Why do relational properties of water molecules bring about the higher-level liquid properties such as laminar flow, surface tension, and viscosity? Again, a reference to thermodynamics is not enough. The intrinsic nature of water molecules needs to be taken into account as well.

In Deacon's narration about EM all higher-level descriptions are possible thanks to constraints that reduce the degrees of freedom of the underlying microphysics and that are eliminated, introduced, and preserved in dynamical systems. This makes him concentrate on constraints and ascribe to them causal powers. Aside from the difficulty in specifying the exact nature of the causation of absences, I find that this strategy does not pay enough attention to dynamic systems and their constituents, which are actively engaged in eliminating, introducing, and preserving constraints. I want to argue that Deacon needs a metaphysics that will give an account of, and explain the spontaneity of, the behavior of these systems, so often emphasized and thus crucial for his theory of the ascent of complexity of emergents. I want to highlight the fact that it is not only the absential nature of constraints that escapes the reductive analysis, but also the actual nature of dynamical systems and their components, which is responsible for eliminating, introducing, and preserving constraints. Deacon emphasizes repeatedly that the nature of these systems, characteristic of all three levels of dynamical depth, is active and dynamic rather than passive. His theory of EM, therefore, needs a metaphysics that can adequately account for this dynamism.

Another set of questions refers to the nature of what can be classified as emergent structures, dynamical systems, and properties of various levels of dynamical depth. Does Deacon's theory truly do well without any reference to DC? To begin with teleodynamics, how does the "higher order constraint" prevent the disruption of the synergy between the component morphodynamic processes? In what way does the organization of teleodynamic systems enable them to preserve and regenerate their own essential constraints and to interact with the environment in a way that sustains supportive, and compensates for unsupportive, influences? What is the nature of self-compression and self-reference of teleodynamic systems? All these properties of biological "selfhood" are delineated by Deacon and Koutroufinis in implicitly causal terms, which are reminiscent of the language of DC used by proponents of ontological EM in its classical version. The metaphysical aspects of both descriptions are equally unclear.

Similar questions apply to the lower levels of dynamical depth. What exactly is the nature of a spontaneous regularizing tendency to become more organized and globally constrained over time, which is exercised at the level of morphodynamic systems? Or the nature of a spontaneous tendency to eliminate constraints, increasing global entropy and reaching thermodynamic equilibrium, exercised by homeodynamic phenomena? What is the causal character of these tendencies, which seem to be exercised by complex dynamical systems or wholes and to affect their constituents? At this point, Deacon's attempt to avoid the language of DC is not entirely successful. Therefore, once again, I want to argue that his theory needs a suitable metaphysics that will enable him to answer these and other possible questions concerning causal aspects of his dynamical depth model of EM. I will offer a constructive proposal fulfilling these requirements in reference to both the old and the new Aristotelianism in the concluding chapter of the second part of this book. To that part we shall now turn.[103]

PART 2

Dispositions/Powers Metaphysics and Emergence

It is one of the aims of physical science to discover and to describe the inherent causal powers of things. For these powers are the truth-makers for those laws of nature that are generally known as causal laws. Causal powers thus have a very important role in the world, for they are the sources of the immanent causal laws of nature.

—Brian Ellis, *Scientific Essentialism*

The qualitative characteristics of things are held to be a real part of ontology, not mere epiphenomena of, or expressions of, or reducible to, the underlying quantitative characteristics of things given by a mathematical theory, no matter how predictively and explanatorily successful the mathematical theory may be.

—David S. Oderberg, *Real Essentialism*

Who would have thought that analytic philosophy, which was deeply influenced by the "ideal-language" or "formalism" postulate raised by Wittgenstein and influenced by the empirical methodology of logical positivism—both strongly antimetaphysical in their presuppositions—would rediscover, in the second half of the twentieth century, the need for ontological reflection in any comprehensive description of reality? The revival of metaphysics came with the works of David Lewis and David

Armstrong, theorizing on the topics of possibility and necessity, universals, and causation. Their ideas were followed by Willard Quine's attack on the popular distinction between analytic (deductive and linguistic) and synthetic (inductive and extralinguistic) statements about reality. He rejected this as "an unempirical dogma of empiricists, a metaphysical article of faith,"[1] claiming that no truths are grounded only in meanings, independent of facts, just as no truths are grounded only in facts, independent of deductive analysis. Metaphysical reflection inspired by Lewis, Armstrong, and Quine was further developed in modal logic and possible-worlds semantics introduced by Saul Kripke, who also speculates on the existence of essences.

The contemporary revival of metaphysics in the analytic tradition is all the more interesting as one of its main objectives is to provide an ontology that is sound and responsive to the current scientific picture of the universe. This relation, however, does not go in one direction only. It is not merely science that influences philosophy. Ontological questions reappeared in natural sciences along with the theories of special and general relativity, quantum mechanics, quantum field theory, neuroscience, and the neo-Darwinian theory of evolution. They concern the problems of the nature of time and space, the interpretation of the statistics of quantum events, the phenomenon of mind and its influence on the brain, the nature of natural selection, the reality of teleology, and many other issues debated in natural sciences.

One of the growing interests and major areas of research in current analytic metaphysics is the ontology of dispositions/powers and their manifestations. It finds its origin in recognition that within the domain of scientific evidence for causal terms it is not enough to provide "that-evidence" (Humean description of resultant regularities). Questions of the nature of things entering into causal relations ("what-evidence") and the origin and character of these processes ("how-evidence") move the reflection on causation from the analysis of causal events—to which Humean metaphysics attributes causal powers—back to the investigation of living and nonliving things as potential sources of causal activity.[2] That this attitude brings a recapitulation of Aristotelian ontological views becomes evident for many. The connection, however, is not obvious and leads to some challenges, which I will address in this part of my research.

Nevertheless, despite its difficulties, I find in the project of dispositional/powers metaphysics a fascinating and promising conceptual tool, capable of providing a plausible ontological foundation for contemporary scientific research, including the theory of EM and DC. In what follows I will first discuss the six major theories of causation in the analytic philosophy (chapter 4). This overview of contemporary causal theories will help us both contextualize and appreciate the original and unique character of the dispositional approach to the question of the nature of causal dependencies, which is the main subject of the following chapter 5. My analysis of the dispositional view of causation in this chapter includes an investigation of the main objectives, assumptions, and claims of dispositional/powers metaphysics. In chapter 6 I will look for the answer to the question concerning the Aristotelian legacy of the new metaphysics in question. Finally, I will offer a new constructive proposal for applying dispositional/powers metaphysics—with its references to classical Aristotelianism—in both DC-based and Deacon's dynamical depth models of ontological (strong) EM (chapter 7). I believe that my grounding of the powers ontology (and theories of EM and DC) in Aristotle's thought makes my project unique and that it clarifies and develops the defense of the reality of the irreducible complexity of natural entities and processes they enter.

CHAPTER 4

Theories of Causation in Analytic Metaphysics

As I have said, early developments of analytic metaphysics are closely related to the philosophical analysis of causation. The continual interest in the character of causal dependencies among philosophers coming from this tradition has led them to offer a number of causal theories. I have already referred in the introduction (section 6.3) to Hume's counterfactual theory of causation, and to the process and the manipulationist views of causal dependencies in nature. In order to fully appreciate the uniqueness of the dispositional view of causation—which was mentioned indirectly in chapter 2, on the occasion of my analysis of the responses to Kim's argument from causal exclusion—I need to revisit causal theories I have already mentioned, situating them in the context of a more systematic evaluation of all major notions of causation in analytic metaphysics. This is the object of my interest in what follows.[1]

1. THE REGULARITY VIEW OF CAUSATION (RVC)

One of the most popular lines of inquiry into the nature of causation goes back to Hume and his first definition of causality, which I mentioned in section 4.2 of the introduction. Formulated in analytic terminology, it states that an event C causes an event E iff C is spatiotemporally contiguous to E, E succeeds C in time, and all events of type C are regularly followed by (or constantly conjoined with) events of type E.[2] Causation is thus a matter of spatiotemporal contiguity, succession in time, and regularity of events. Unlike Hume, however, those who follow RVC take regularities as ontologically real and mind-independent. Nevertheless, they face several objections posing a real challenge to their theory. Since it is based on a constant conjunction of types (kinds) of events classified as causes or effects, it finds it difficult to specify what makes a particular C to cause a particular E. Moreover, RVC seems to need to identify the ground of regularity, which otherwise becomes a matter of chance. Some are willing to refer to powers, others to a force-based productive relation, still others to the concept of thick (Platonic) laws of nature stating that, ultimately, laws are not merely regularities but have an ontological status.[3] Followers of RVC, however, firmly deny any need to appeal to a different ontological category to explain the presence of regularities, which makes their theory incomplete, in my opinion.[4]

1.1. INUS-Conditions View of Causation (INUSConVC)

Since it does not acknowledge any need for a deeper metaphysical story about causation, RVC is agreeable to many empiricists. It sides with John Stuart Mill, in particular, when he states that there is no difference between a cause and an antecedent condition of an event and that the cause is, in fact, the sum total of all positive and negative conditions of an event taken together (he speaks of a "causal field"). However, most contemporary followers of RVC, though positive about Mill's theory, acknowledge the need to distinguish between conditions that are strictly causal and those that are only causally relevant (providing causal circumstances). Otherwise, we end up with counterintuitive examples of regarding as causally related any two temporally ordered and regularly following events, such as the sequence day-night or the growth of hair on babies and the growth of their teeth. Thus, it has been argued that causes can be defined under RVC as

1. *Necessary conditions*: an event C is the cause of an event E iff C and E have obtained and C was, under the circumstances (all other causally relevant conditions), necessary for E (e.g., striking a match is a necessary condition for its lighting under regular

circumstances, such as the presence of oxygen, the match being dry, the lack of other sources of fire, etc.).
2. *Sufficient conditions*: an event C is the cause of an event E iff C and E have obtained and C was, under the circumstances (all other causally relevant conditions), sufficient for E.
3. *Necessary and sufficient conditions*: an event C is the cause of an event E iff C and E have obtained and C was, under the circumstances (all other causally relevant conditions), both necessary and sufficient for E.
4. *INUS conditions (after the initial letters of the main terms used in definition)*: an event C is the cause of an event E iff C and E have obtained and C was an Insufficient (a wider set of conditions was required), Nonredundant (without it the complex condition would not have been able to cause E) part of a set of conditions that was itself Unnecessary (E could have come into being under other circumstances), but Sufficient for E.[5]

The INUS interpretation is quite useful, allowing us to describe causal relations in many scientific contexts. Sometimes finding conditions that are just necessary or sufficient (a claim weaker than INUS) is also satisfactory. On the other hand, however, all four approaches listed here raise a number of problems and questions. To begin with, it seems that the theory requires some general specification of what counts as a "condition." Losee notes that "a sufficient condition is some 'arrangement' such that the effect in question must occur in its presence. But what sorts of 'constituents' are relevant to the arrangement?"[6] He says that clearly, properties would qualify (e.g., dryness of the match and the roughness of the surface scratched in the example of lighting the match). But, as I have stated before, there are no free-floating properties in nature. Defining "conditions" in terms of properties requires a theory of properties and their relation to entities that bear them.

This observation leads to another question. The method of listing and characterizing INUS conditions for particular occurrences—which allows us to classify them as causal dependencies—seems to be fairly limited in its application to other similar and dissimilar contexts. In other words, as a version of RVC, INUSConVC faces the problem of generalization of its claims. How many events following one another regularly do we have to observe and investigate before we will be able to say that they are related causally? Does mere observation of regularities of events in nature suffice for attributing to them ontological reality and distinctiveness as causal relata?

Moreover, all four definitions of causes in INUSConVC listed here raise other queries. The first one (1) fails in cases of causal overdetermination, when many causes produce the same effect. Neither one of them is necessary because the effect would occur in its absence as well. The second (2) suffers from the undeterminative sufficiency problem—that is, situations in which the conditions of a sufficiency thesis are satisfied, and yet clearly, they are not causes (e.g., the fact that the table is square is sufficient to explain why it is not round, but is not a cause of it not being round). The sufficiency thesis is also problematic in cases of probabilistic causation (see below). In addition to problems related to (1) and (2), (3) suffers from the problem of directionality, making it difficult, if not impossible, to distinguish between causes and effects (if C is necessary and sufficient for E, then logically E is sufficient and necessary for C). Option number (4) is vulnerable to objections similar to those that threaten (2) and (3). Finally, all four propositions may be misguided in classifying as causal dependencies (C causes E) joint effects of common causes (C and E are both effects caused by C^*). This objection is classified as the problem of epiphenomena.

1.2. Inferability View of Causation (InfVC)

Philipp Frank, in his *Philosophy of Science*,[7] advocated a variant of the neo-Humean RVC, which can be classified as an inferability view of causation (InfVC). He interpreted causal relation as a functional relation which permits an inference about the (future) state of the physical system in a particular region of space-time, provided we know the state of that system in some other region of space-time. Frank suggested causal laws are differential equations of the form

$$\frac{d\xi_k}{dt} = F_k(\xi_1, \xi_2, \xi_3, \ldots, \xi_n) \ (k = 1, 2, 3, \ldots, n),$$

where $\xi_1, \xi_2, \xi_3, \ldots, \xi_n$ are variables determining the state of the system in question, and F_k is a mathematical function of these variables.

Provided we are able to determine experimentally the value of manageably small and specific variables describing a state of a given system, we should be able to propose a mathematically simple function F_k that enables us to calculate the value of the same variables describing another state of the same system.[8] In other words, as a variant of RVC, InfVC "connects causality and explanation by identifying causes and effects with states of physical systems, causal laws with functional relations of a certain form between these states, and causal explanations as deductive arguments from premises that include causal laws and initial state descriptions."[9]

InfVC proves practically useful in the domain of natural science, allowing its practitioners to classify important scientific laws, such as Newton's law of motion (crucial for deduction of the spatial location of planets, billiard balls, etc. as a function of time) or Mendel's laws of inheritance (important for deduction of features characteristic for succeeding generations from information about given populations), as causal relations. At the same time, however, InfVC is not applicable to all scientific laws. As an example, we may think of Boyle's law, stating that at constant temperature the volume of a given mass of gas is inversely proportional to its pressure ($PV = k$, where P is the pressure of the gas, V is the volume of the gas, and k is a constant representative of the pressure and volume of the system). Even if Boyle's law expresses a functional relation between the states of the system, it does not enable us to predict the variation of the state of the gas in terms of spatial location or time.[10]

Moreover, InfVC faces another limitation, which challenges RVC more generally. Similar to the proponents of INUSConVC—who emphasize the need to distinguish between conditions that are strictly causal and those that are only causally relevant—those who follow InfVC acknowledge the difficulty in specifying the number of state variables and forces that should be considered in formulating the algorithm of the mathematical function F_k. An absolute precision of our causal assessments would require from us a description of the entire state of the universe at times t and Δt. Apart from the practical impossibility of providing such a description, it would be rather useless. For it turns out that the more complete our specification of the state of the system at t, the less possible is a recurrence of a precisely similar state in the future, which makes the principle of causation tautological (it ceases to be a statement about physical reality). By contrast, if we decide to characterize a system by a small number of state variables—which enables us to identify the system's recurrent states—we pay the price of a certain imprecision and inadequacy between the mathematical and the actual empirical description of determined magnitudes. In other words, when we accept InfVC, "we have a choice between making the principle of causality precise and tautological or vague and factual."[11]

1.3. RVC, Laws of Nature, and Natural Kinds

Stathis Psillos points toward the struggle of the followers of RVC to define the character of the laws of nature. They are tempted to say these laws are merely regularities, but since they are hesitant to treat all regularities as being lawlike, they state that laws of nature are in fact regularities plus the property of lawlikeness, where the regularities that constitute the laws of nature are those that are expressed by the axioms and theorems of an ideal deductive system of our knowledge of the world. They claim that thus understood, the property of lawlikeness avoids introducing any metaphysically distinct and deeper kind of entities—like powers or potencies—and is not committed to any kind of natural necessity. But it is rather doubtful that this theory of the laws of nature saves their objective character, since "what regularities will end up being *laws* is based, at least partly, on epistemic criteria and, generally, on our subjective desideratum to organize our knowledge of the world in a deductive system."[12]

Another way to characterize the predicates suitable for lawlike statements is to assume they must pick from natural kinds. This objectifies the laws of nature at the price of relating them to the debate on the prospects of a theory of natural-kind predicates, which may be troublesome for someone who is not an essentialist and rejects hylomorphism. Psillos claims that "the least that is involved in the characterization of a *kind* of entities is that they are *like* each other in relevant degrees and respects. What respects of likeness are relevant to kind-membership? Here, the obvious answer would be: those respects in virtue of which entities have similar nomological and causal behaviour."[13] Despite being imprecise and ontologically wanting—the question of the nature and source of likeness shared by the members of the same natural kind remains unanswered—this way of reasoning is right in showing an intimate connection between the issue of what laws of nature are and the issue of what kinds are natural.

While analyzing RVC, Psillos also raises the problem of similarity that is crucial for the first of Hume's definitions of causation (all *C* placed in similar circumstances are followed by *E*). Following Leibniz (and going back to Heraclitus), he notices that nature never exactly repeats herself. He agrees with John Venn, who claims that reliance on similarity introduces a subjective element to the theory of causation. Finally, he analyzes RVC's thesis that causes are nomologically sufficient for the effects, a thesis that became the kernel of the deductive-nomological model of scientific explanation, advanced by Hempel and Oppenheim.[14]

1.4. Summary

One of the main arguments in favor of RVC brought by its followers is its falling within the empiricist worldview—that is, offering a theory of causal explanation which does not invoke metaphysical concepts of substance, substantial forms, or teleology. It has also been emphasized that RVC helped to legitimize social sciences. It provided a necessary background to challenge the claim that basic laws of nature are relatively few and that they only concern the subject matter of physics and chemistry. Inasmuch as the laws of nature take their truth from the way the world is (regularities of events in the world), we can find a separate and valid set of laws of nature in sociology, economics, psychology, and so on. These laws are no less real and basic than the laws of physics and chemistry. Consequently, RVC allowed its followers to keep the notion of laws of nature, abandoning the necessitarian account of physical laws and the idea of their dominance over all other scientific laws, which many thinkers treated as secondary consequences or implications of the laws of physics. Moreover, it has been argued that RVC opened our causal reflection on probabilistic descriptions and formulation of statistical laws.[15]

At the same time, however, I have listed a number of important problems challenging RVC and its two subcategories: INUSConVC and InfVC. Difficulties in specifying what makes a particular *C* to cause a particular *E* and in identifying the ground of regularity; the perception of causal relata in terms of events rather than entities; complications in stating what counts as a causal "condition"; the problem of generalization of causal claims based on regularities of the processes occurring in nature; challenges of overdetermination, undeterminative sufficiency, and directionality; the problem of limiting the number of state variables and forces that should be considered in formulating causal laws; and the difficulty of explaining the ultimate character of the laws of nature—these are all questions that are crucial for the plausibility of RVC.

It seems to me that many of these queries arise from rejecting a more robust ontology assumed by the theories of causation predominant in ancient and medieval philosophy and science. Thus, it may be possible that what proponents of RVC see as the major achievement of their position— that is, the dismissal of empirically not verifiable metaphysical concepts in explaining causal dependencies in nature—turns out to be a source of major problems challenging the theory of causation they themselves offer.

2. THE COUNTERFACTUAL VIEW OF CAUSATION (CVC)

Some of those who became discouraged with the problems of RVC decided to turn toward the other definition of causality in Hume's *Enquiry concerning Human Understanding*, in which he briefly states that the nature of a cause can be inferred from a simple rule of dependency, saying, *"If the first object had not been, the second never had existed."*[16] This suggests causality can be understood in terms of counterfactual dependency. Contemporary philosophy of language distinguishes between indicative conditionals (if p, then q) and subjunctive conditionals (if it were the case that p, then it would be the case that q). Counterfactuals (contrary-to-fact conditionals) belong to the class of subjunctive conditionals. They owe their name to the fact that they presuppose (rather than assert) the falsity of their antecedents. According to many neo-Humean philosophers (David Lewis, John L. Mackie, and Aidan Lyon in particular), causal relations are best described in terms of counterfactual dependencies.[17] Lewis points toward our intuition that "causation has something or other to do with counterfactuals," and says: "We think of a cause as something that makes a difference, and the difference it makes must be a difference from what would have happened without it. Had it been absent, its effects—some of them, at least, and usually all—would have been absent as well."[18]

Just like RVC, CVC raises metaphysical and methodological questions. Laurie Ann Paul asks whether CVC is to be analyzed as "a folk concept, a philosophical concept, a scientific concept, or something in between?"[19] If it is regarded as a conceptual analysis, is it descriptive or prescriptive? If it is not a conceptual analysis, it must have an ontological character, which departs significantly from Hume, for whom causation was a projection of a human impression of constant conjunction. A tendency to emphasize CVC's pragmatic character is not a solution either, as it simply gives up on the analysis of the hardest issue—namely, the objectivity and norm- and description-independency of the whole project.

Concerning its ontological commitments, CVC faces another fundamental challenge. Analyzing a situation in which C caused E, it assumes—by definition—that E did occur. It is then legitimate to ask, How can we determine what would have been if E did not occur? We cannot go back in time or replay the situation. Answering this question, Lewis argues in favor of the possible worlds modality, which allows us to speculate what a world that is just like our world, but in which E did not occur, would look like and how it would evolve. However, founding CVC on this kind of modal theory is problematic, as it seems to assume the possible ontological reality of countless parallel universes, the existence of which is not empirically verifiable. We thus find Edward Feser arguing that theories based on the possible worlds modality "are bound to be circular. . . . Their point is to explain modal notions like possibility, necessity, and impossibility in terms of possible worlds, but of course 'possible' is itself one of these modal concepts. Hence such theories presuppose precisely what they are supposed to explain."[20]

Trying to deal with this criticism of his position, Lewis formulates the postulate of modal realism—that is, a nominalist emphasis on all possible worlds being concrete particulars—things definable and understandable in straightforwardly nonmodal terms. Analyzing this claim in a broader context of the most recent debate on the nature of modality, we should classify it first—with respect to the *de dicto* aspect of modality—as a version of concretism. Emphasizing the importance of the proximity aspect of objects—which makes designations of "actual" and "merely possible" indexical (i.e., with respect to other things), and not absolute—this version of modality sees possible objects as real and existent (in parallel universes), even if they are not actual (i.e., they are not proximate to us). Consequently—with respect to the *de re* aspect of modality—concretism faces the problem of one and the same object existing in many possible worlds. Trying to resolve this difficulty, concretists reject transworld identity and embrace a counterpart theory, which says that different worlds represent a given object *A* not by involving *A* itself but by involving its counterpart (something very much like *A* but not identical to it). Transworld identity, on the other hand, is embraced by abstractionism. This ontological position treats possible worlds as abstract objects, and thus provides an alternative to Lewis's concretism, which—with respect to its acceptance of the counterpart theory—is accused of detachment of its *de re* modal facts from the things they are supposed to be about.

Moreover, even if Lewis's version of concretism offers a reductionist account of modal truths—cashed as nonmodal truths about existing parallel universes—its qualitative economy (i.e., freedom from assuming the reality of any fundamental modal truths) comes at a very high quantitative cost. For according to its principles, literally everything possible truly exists and/or happens (takes place). It has been observed that this version of modalism entails the problem of ethical fatalism. For if all that is necessary, possible, and impossible is settled once and for all, our choices cannot make any moral difference. In addition, concretism faces a number of further metaphysical challenges, the analysis of which goes beyond our interest here.[21]

Another challenge faced by CVC is that, like RVC—which derives general causal statements (laws) from the repeatable cases of concrete occurrences—CVC faces the problem of distinguishing between lawful and merely accidental generalizations of contrary-to-fact conditional dependencies. An explanation referring to the traditional method, which notes that lawful generalities support counterfactual claims, becomes unacceptably circular in the case of CVC. Losee thinks that the only way to establish the lawful status of generalizations under this view of causation is to show that each one of them is embedded in a theory—even if such theory is merely an approximation subject to ceteris paribus restrictions (e.g., Galileo's law of falling bodies or Kepler's laws of planetary motion). Understood as such, causal claims may be assessed in accordance with a rule such as the following:

"c causes e" if, and only if, in a universe just like our own in which c did not occur,
1. "not e" is true,
2. "if ~c then ~e" follows from a generalization, which generalization is
3. a lawful regularity, established by showing that
4. the generalization is embedded in a theory, which theory
5. is used currently by scientists.[22]

That this way of assessment of causal claims depends heavily on the currently applied scientific theories—which are subject to change—becomes obvious. This fact limits application of the method in numerous events, which cannot be linked to lawful regularities embedded in theories and yet are normally accepted as causal. Moreover, by referring to the possible worlds modality, the proposed way of assessing causal claims makes itself questionable for the same reasons I have mentioned above.

Apart from the fundamental problems concerning its ontological and methodological commitments, the acceptance of CVC leads to several crucial questions related to possible causal situations. First, it seems to be unable to deal with early, late, and simultaneous preemption, where two or more events compete to cause an effect. It is thus sensitive to extrinsic factors that function as possible causal backups. In case of an early preemption, one of the causes (C_1) not only causes E but also (at the same time) preempts another cause (C_2) which would otherwise have made E happen. Therefore, we cannot say that C_1 is a cause of E, for even if C_1 had not happened, E would have still taken place (caused by C_2). In case of a late preemption, C_1 preempts C_2 simply through its earlier occurrence in time. Again, had it not occurred, E would still have happened because of C_2. In case of a simultaneous preemption ("trumping"), both C_1 and C_2 are able to cause E, both exercise causal activity at time t, but only one of them (let us say C_1) is actually responsible for the occurrence of E (trumping the other potential cause). Once again, according to CVC, C_1 cannot be regarded as a cause of E. Had it not occurred, C_2 would still have caused E.[23]

Furthermore, CVC is challenged by the cases of over- and underdetermination. The former, similar to the case mentioned in the context of RVC, occurs when both C_1 and C_2 produce the same effect at the same time. Hence, none of them is a cause, in terms of CVC. Had C_1 not occurred, E would still have happened because of C_2, and vice versa. Finally, examples of underdetermination suggest cases of counterfactual dependence that are not causal. Jaegwon Kim gives an example of becoming an uncle. He defines it in terms of counterfactual dependence, saying, "If my sister had not given birth at t, I would not have become an uncle at t," and explains that although his becoming an uncle was determined by and dependent on the birth of the child, his becoming an uncle was not causally effected by that birth.[24]

Despite these difficulties, it has been noted that CVC has some important advantages over RVC. It helps avoid assigning causal character (C causes E) to joint effects of a common cause (C and E are both caused by C^*). It also seems to solve the problem of directionality in particular causal situations, in which the statements that C belongs to a minimal set of INUS conditions for E and that C is an effect of E are equally valid under RVC. Ned Hall mentions the ability of CVC to give a proper causal account in some cases of double prevention, recognized in many instances of physiological processes and everyday activities. These are situations in which C_1 is about to cause E, while C_2 is on the way to prevent it from happening, but before C_2 prevents C_1 from causing E, C_2 is itself prevented by C_3, in which case E actually happens. Consequently, we might say that C_3 is among causes that caused E, even if there is no process or fundamental regularity instantiated between C_3 and E (it thus cannot be described by RVC).[25] Moreover, it has been observed that CVC is applicable in cases of causation by omission (absence). Assuming that absences can be causes (I have expressed my doubt on this point in chapter 3), it seems that there can be causation without regular causal relations (e.g., the victim died because of a lack of air). Again, beyond the scope of RVC, such cases can be described as causal under CVC. At the same time, however, one might object that in defining the nature of causal influence Lewis restricted the scope of that influence to distinct events, and absences are not events. Here is what he himself says on this topic:

> Absences can be causes, as when an absence of food causes hunger; they can be effects, as when a vaccination prevents one from catching a disease; and they can be among the unactualized alterations of a cause or effect which figure in a pattern of influence.

Absences are not events. They are not anything: where an absence is, there is nothing relevant there at all. Absences are bogus entities. Yet the proposition that an absence occurs is not bogus. It is a perfectly good negative existential proposition. And it is by way of just such propositions, and only by way of such propositions, that absences enter into patterns of counterfactual dependence. Therefore, it is safe to say with the vulgar that there are such entities as absences, even though we know better.[26]

It thus remains questionable whether CVC can actually deal with negative propositions on the existential or ontological and not merely linguistic or explanatory level. It seems that it is capable of doing so only when one is treating absences not as bare privative causes, but as concrete actuality-dependent privations—that is, imperfections, deficiencies, gaps, and so on (see my comment on this point in the first two paragraphs of chapter 3, section 3.4).[27]

3. THE PROBABILITY VIEW OF CAUSATION (ProbVC)

Some philosophers notice that since modern quantum physics is incompatible with necessary and sufficient conditions, as well as InfVC, and because causal relations are usually accompanied by probabilistic dependencies (when C causes E, it raises E's probability), what we need to develop is a probability view of causation (ProbVC). According to its basic formulation, an event C is a probabilistic cause of an event E if, given that C has obtained, the probability of the occurrence of E has increased and is now higher than the probability of E if C had not obtained. That is $p(E/C) > p(E/-C)$.

A principal representative of this strategy, Patrick Suppes,[28] defines a prima-facie cause in terms of the basic formulation of ProbVC, to which he adds a time factor (C occurs earlier than E), and stipulates that causality is not entirely objective, as causal relations are relative to a conceptual framework and a particular conception of causal mechanism which is never complete. Although he intends to apply his account of ProbVC to both general and single-case relata, it becomes clear that the theory fails in the case of the latter. To give an example, it has been proven that smoking, in general, increases the probability of lung cancer. But in the case of a concrete patient with this type of cancer, what we are dealing with is not a raised probability but an actual development of a tumor. Suppose the patient smoked ten cigarettes/day for fifty years, quit it, lived for another five years without smoking, and then got a tumor. Suppose smoking increased the probability of patient's developing a lung cancer from 0.07 to 0.7. Now, there is still some probability (0.3) that the cancer that developed in a patient's body five years after the patient quit smoking was in fact caused by the effects of long-term exposure to radon (suppose our patient happens to be a retired miner, who worked underground for thirty years). Thus, even if the general statement that smoking raises the probability of lung cancer is true, we cannot say with an absolute certainty that smoking caused the development of a tumor in a concrete patient's body. This example shows that ProbVC is not sufficient to describe particular cases of causal dependencies.

In the face of this objection, it has been suggested that ProbVC should be modified and defined adequately in terms of individual events rather than their general types. The general formulation of the thus modified ProbVC argues that C is the cause of E if C and E are individual events (or facts) and the objective, single-case probability of E's occurrence is increased by the occurrence of C. Nevertheless, one might object that this kind of modification makes ProbVC vulnerable to another critical argument, which was raised already in reference to RVC and CVC. When defined in terms of individual (single-case) events, ProbVC seems to face the problem of forming lawful-kind generalizations based on individualized instances of probability-raising causal situations.

ProbVC is also vulnerable to the critical argument of Losee, who gives an example of being a bride or a groom participating in a marriage ceremony. He notes that according to ProbVC such activity increases the probability of being involved in a divorce, which leads to a counterintuitive conclusion that marriage is a cause of divorce. Taking into account that it is logically true that marriage is a necessary condition of divorce—just as having a lottery ticket is a logically necessary condition of winning a lottery prize—Losee states that ProbVC requires an exclusion of all logically necessary conditions from the ranks of putative causes. Providing an exhaustive list of such conditions may be extremely difficult, if not impossible.[29]

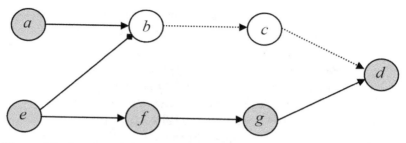

Figure 4.1. A neuron-firing network depicting Menzies's theoretical example of causes decreasing probability

Suppose that the probability that neuron *d* fires through the upper path from neuron *a* is high (0.8), the probability that the same neuron *d* fires through the lower path from neuron *e* is low (0.2), and the probability of neuron *b* being preempted by firing of the neuron *e* is moderate (0.4). Suppose further that in a concrete causal situation both *a* and *e* fire simultaneously, and *b* is preempted by *e*. In such a situation *d* fires nevertheless, and it is correct to say that neuron *e*'s firing caused neuron *d* to fire, despite the fact that by nature *e* actually lowers the probability of *d*'s firing.

Another critique of ProbVC comes from the realization that in many cases, an acknowledged cause of an event actually decreases, instead of increasing, the probability of its occurrence. If this is true, then probability raising ceases to be a necessary condition of causal dependencies. Instead of the cases in which $p(E/C) > p(E/\text{-}C)$, we are actually dealing with situations in which $p(E/C) < p(E/\text{-}C)$. Menzies offers a theoretical example of such situation, depicted in figure 4.1.[30]

Sosa and Tooley offer another related thought experiment, which becomes a strong logical objection to ProbVC, showing that causes do not always make their effects more apt to occur:

> Suppose, then, that there are two types of disease, *A* and *B*, satisfying the following conditions. First, it is a law that contracting disease *A* causes death with probability 0.1, and a law that contracting disease *B* causes death with probability 0.8. Secondly, it is a law that contracting disease *A* produces complete immunity to disease *B*, and a law that contracting disease *B* produces complete immunity to disease *A*. Thirdly, it is a causal law that in condition *C* an individual must contract either disease *A* or disease *B*. . . . Fourthly, individual *X* is in condition *C*, contracts disease *A*, and the latter causes his death.

Given that these conditions obtain, the question is what would have been the case if X, though being in condition C, had not contracted disease A, and the answer, it would seem, is given by the following counterfactual: if individual X had not contracted disease A, he would have contracted disease B. But if individual X would have contracted disease B if he had not contracted disease A, then his probability of dying had he not contracted disease A would have been 0.8, and so would have been higher than his probability of dying given that he had contracted disease A, since the latter is only 0.1. So, once again, it is not true that causes need make their effects more likely.[31]

These examples show that probability raising is not, in fact, a necessary condition of causal dependencies. But maybe it is a sufficient condition of their occurrence? Jonathan Schaffer's argument proves it is not. He presents a thought experiment that falls under the category of causal trumping (simultaneous preemption), which he classifies as causal overlapping. Suppose both Merlin (C_1) and Morgana (C_2) cast spells with a 0.5 probability of turning the prince into a frog (E), and suppose that the prince indeed turns into a frog. Even if we can regard both Merlin and Morgana as probability raisers of the prince turning into a frog, there may be the fact of the matter as to which spell actually did the causing. In such a case the other spell would be a direct probability-raising non-cause (insufficient to bring a causal change).[32]

The fact that probability raising is a neither necessary nor sufficient condition of causal relations seems to undermine ProbVC. But the theory faces—in my opinion—still more fundamental challenges, which arise from a thorough metaphysical analysis of its principles. Such an investigation makes us ask a number of key questions, which seem to remain unanswered by proponents of ProbVC. What is the ultimate nature of probability (chance)? Is it ontological or epistemological? Does it arise from ultimate indeterminism of events taking place in nature? Are causal relata population- or individual-level? Does ProbVC entail a subjective factor? Does it attempt to understand actual or merely potential causality? Do probabilities tell us about causal dependencies with or without any reference to physical knowledge?

I have tried to answer at least some of these queries in my investigation of Aristotle's teaching on chance in section 2.4 of the introduction.

My analysis of ProbVC offered here leads me to repeat once again that although the chance (probability) aspect of occurrences in nature has an ontological character, it should not be treated as causal *per se*, but merely *per accidens* (i.e., in reference to *per se* causes responsible for a given causal change). Hence, as much as I acknowledge the need to take probability into account in examining causal changes, I find dubious the project of developing a theory of causation based merely on probability raising.[33]

4. THE SINGULARITY VIEW OF CAUSATION (SVC)

Another view of causation in analytic tradition restricts it to a relation between concrete individual events, with no reference to general types. This was alluded to first by Elizabeth Anscombe, who addressed three claims typically associated with causation: (1) that causal relations are instances of exceptionless generalizations, (2) that causation needs to be identified with necessity, and (3) that causal relations presuppose the reality of some kind of law. She noticed that causal relations do not presuppose necessarily underlying laws (deterministic or probabilistic). Even if laws help us derive knowledge of effects from causes, laws do not point toward causes as sources of effects and are not relevant to causation qua causation.[34]

Anscombe's ideas were preceded by Curt J. Ducasse, who is even more explicit in his criticism of the analysis of causation in terms of laws and regularities between events or states of physical systems. He thinks that the idea of regularity or recurrence is totally irrelevant to causation, which must be defined in terms of a single case of causal dependency of events, directly observed. Suppose a ramp operated remotely by a railway employee did not close at the level crossing and that an accident followed. When asking what caused the accident, we are not asking about a constant conjunction of which this complex incident is a member. Rather, we want to know what the single difference in a normal operation of the ramp was. Thus, Ducasse insists that causal inquiry is an inquiry concerning single events and defines a cause (C) to be a particular change in the immediate environment in a particular situation that occurred just before the effect (E) in question.

Ducasse's theory of grounding causal ascriptions in reference to the immediate environment in particular causal situations becomes an attempt at reestablishing (contrary to Mill) the strong distinction between causes and conditions. This would make causation a triadic relationship between cause, effect, and the broader context of conditions of a given causal occurrence. Nevertheless, if we take Ducasse's definition literally, causal analysis requires still a specification of every change—that is, the motion of every molecule, atom, electron, and wave—in the immediate environment in a given $t\sim\Delta t$. Ducasse acknowledges that such interpretation may lead to counterintuitive causal ascriptions which say, for example, that "at the instant a brick strikes a window pane, the pane is struck, perhaps, by the air waves due to the song of a canary nearby."[35] Is then SVC, similar to InfVC, forced to limit causal description to a manageably small number of changes, at the price of offering a rather loose interpretation of a given causal situation? Moreover, is not the researcher's decision about which changes should be selected for examination vulnerable to the subjectivity and arbitrariness of his/her individual choice?

To answer these problems Ducasse introduces a distinction between two different meanings of cause. In the strict sense, it means "the fully concrete individual event which caused all the concrete detail of this breaking of this window." In the elliptical sense, it means "that which the cause of this breaking of this window has in common with the individual causes of certain other individual events of the same sort."[36] That the second definition—examining what has happened in other contexts—poses a problem becomes apparent. The regularity of causal events, which Ducasse opposes, sneaks back in with the requirement of the "commonality" of causes in events of the same sort. Hence, Psillos emphasizes that for SVC "(causal) laws are generalizations over causal facts and not (as RVC would have it) constitutive of causal facts."[37] He also refers to Donald Davidson, who notices that if Ducasse wants to consider a concrete causal event qua an instance of an event type, he needs to acknowledge the reality of general patterns under which events fall, which implicates regularity. Thus, Davidson would argue for a reconciliation between SVC and the Humean. Others point toward the first definition offered by Ducasse, asking whether the conditions of a causal event should be taken in a strong counterfactual sense, or merely in a weak sense of sufficiency.[38]

Summing up my analysis of SVC, I want to acknowledge that its proponents are right in their emphasis on the importance of the individual and singular aspects of causal relata for our investigation of cause/effect dependencies. At the same time, we have seen that Ducasse's attempt to develop a general theory of causation grounded merely in particulars, without paying enough attention to general causal assessments, encounters numerous difficulties. Defining causal relata as causal events brings Ducasse eventually back to a Humean RVC, the rejection of which motivated him to search for a new theory of causation. Paradoxically, his SVC faces the challenge of limiting causal description to possibly a small and manageable number of observable changes, which resembles the difficulty InfVC has with restricting the number of state variables in defining of the function F_k. We have seen that such restriction is necessary for identifying the recurrent states that F_k is supposed to describe. Both versions of practically limiting the number of events/variables taken into account lead to a certain imprecision, vagueness, and inadequacy of causal assertions offered under SVC and InfVC.

5. MANIPULABILITY-BASED VIEWS OF CAUSATION (M-basedVsC)

Another alternative proposed in the causation debate in analytic philosophy begins from our commonsense idea connecting causation with manipulation, telling us that if the relation between C and E has a causal (and not merely correlational) character, then any manipulation of C brings a relevant change in E. This observation provides us with another argument against RVC, especially in its version defined as InfVC. Those who support M-basedVsC begin with situations in which manipulation of one variable changes the other variable, but not vice versa. To give an example, the period of a simple pendulum can be approximated in accordance to the following formula:

$$T = 2\pi \sqrt{\frac{L}{g}},$$

where T = time, L = pendulum length, and g = gravity. This formula allows us to infer the pendulum's period when we know its length, as well as to infer its length when we know its period. While these two inferences are asymmetrical, our ability to actually change one of the two variables in question through manipulation of the other is not. Our changing (through manipulation) the pendulum's length changes its period, while our changing the pendulum's period does not change its length. It turns out that such practical manipulation helps us specify that it is the length of the pendulum that causes (has an influence on) its period, and not the other way around.

M-basedVsC, appealing for their ability to distinguish between causal and purely correlational claims, come in three versions defined as manipulation, agency (sometimes called instrumental), and interventionist views of causation (MVC, AVC, and IntVC).

5.1. The Manipulation View of Causation (MVC)

MVC was first developed by Georg Henrik von Wright, who thought that the concept of causation is secondary to the idea of human action. Hence, even if causal relations have a metaphysical status independently of our awareness of their occurrence, human knowledge of their reality seems to require in many cases a manipulability test. For, as we have seen, a mere observation of regularities in nature (under RVC) may make it hard for us to distinguish between lawful and merely accidental generalizations (see section 2 in this chapter, on CVC) or make us mistakenly qualify two events as cause (C) and effect (E) while they are, in fact, joint effects of some other cause (C^*). Emphasizing the power of appropriate manipulations in the process of distinguishing cases of mere correlation from causation, von Wright suggested that "what makes p a cause-factor relative of the effect-factor q is, I shall maintain, the fact that by *manipulating p*, i.e., by producing changes in it 'at will' as we say, we could bring about changes in q. This applies both to cause-factors which are sufficient and those which are necessary conditions of the corresponding effect-factor."[39]

Losee notes that despite the originality and practical usefulness of MVC, following its objectives suggests assigning causal status to functional laws that pass the manipulability test, such as those of Boyle, Ohm, or Snell. For instance, in the case of Snell's law—used to describe the relationship between the angles of incidence and refraction when one is referring to light or other waves passing through a boundary between two different isotropic media—manipulation of the angel of incidence changes the angle of refraction. Nevertheless, even though these and many other functional laws pass the manipulability test, philosophers of science commonly decline to assign causal status to them, realizing that laws of nature are descriptive rather than prescriptive. This fact obviously questions the explanatory value of MVC. In defense of von Wright's position, Losee refers to his distinction between "causal laws," which assert relations between successive states of physical systems, and "causal relationships," which do not assert such relations. But it is doubtful whether this distinction suffices to preclude assigning causal status to functional laws, which challenges the plausibility of MVC.[40]

5.2. The Agency View of Causation and the Action-Related View of Causation (AVC, ArelVC)

AVC, thought of as an improved version of MVC, was first introduced by Peter Menzies and Huw Price, who based their theory of formulating causal ascriptions on the connection between causation and free action. The category of free action can be understood both as a natural action that is unconstrained, unforced, and not deterministic and as an action due to voluntary choices of an agent. Menzies and Price seem to embrace the latter definition when they say that the reason we are able to distinguish causes from effects is that the former are always within the immediate control of free and conscious agents, while their influence on the latter is merely indirect. Thus, they declare that "the common idea to agency accounts of causation is that an event A is a cause of a distinct event B just in case bringing about the occurrence of A would be an effective means by which a free agent could bring about the occurrence of B."[41]

Interestingly, a similar definition of causation was proposed much earlier by Robin Collingwood, Douglas Gasking, and von Wright, who also understood causes as tools enabling us to manipulate nature:

> A cause is an event or state of things which it is in our power to produce or prevent, and by producing or preventing which we can produce or prevent that whose cause it is said to be.[42]

> A statement about the cause of something is very closely connected with a recipe for producing it or for preventing it.[43]

I now propose the following way of distinguishing between cause and effect by means of the notion of action: p is a cause relative to q, and q an effect relative to p, if and only if by doing p we could bring about q or by suppressing p we could remove q or prevent it from happening.[44]

But Menzies and Price think their proposition is more relevant. They claim that while earlier definitions assumed causal determinism, their theory presents an indeterministic approach, which incorporates the deterministic notion as its special or limiting case. Thus, they add to their definition that "A constitutes a means for achieving B just in case $P_A(B)$ is greater than $P_{-A}(B)$" (where P = probability).[45] Obviously, this statement brings them close to ProbVC. Acknowledging this fact, Menzies and Price claim that what makes their theory superior "is precisely its appeal to the notion of agency, by way of the notion of agent probability."[46]

One of the key motivations for developing this account of causation was an attempt to capture the practice of experimental science. For this reason, it has been well adopted by proponents of mechanistic explanations in biology. At the same time, however, AVC suffers from at least four major objections which seem to undermine its plausibility and explanatory power:

1. *Confusing metaphysics with epistemology.* AVC seems to suggest not only that we gain knowledge through agent manipulation and experiment but also that these methods become constitutive for causal interactions. Menzies and Price answer that the notion of agency enters causal explanation not as evidence for causal claims, but because the notion of causation, as an extrinsic notion, is by its very nature rooted in the idea of agent manipulation. Whether this explanation protects AVC from the fallacy of the causal version of verificationism, however, remains questionable.
2. *Unavoidable circularity.* The circularity of this view becomes evident in the light of Menzies and Price's definition, according to which A is a cause of B in the case that "bringing about" A would be an effective means by which a free agent could "bring about" B. Is not "bringing about" itself a causal notion? Is it not equivalent to "causing to occur"? It seems that this definition employs

as a part of an explanation the very concept it wants to explain. In an attempt to defuse this charge, Menzies and Price claim that our speaking of causation as "bringing about" becomes acceptable once we realize this is the outcome of successful instances of achieving our own ends due to acting in a certain way. They think such experience provides us with a direct, nonlinguistic acquaintance with the concept of bringing about an event, an acquaintance that is free from any prior acquisition of a causal notion. According to Menzies and Price, a reference to this experience makes our defining causation as "bringing about" free from the threat of circularity. But one might still object that having a grasp of the experience of agency which is independent of our grasp of the general concept of causation does not change the fact that our agency—that is, bringing about an effect—is always causal (even before we describe it in terms of causality). Hence, the argument presented by Menzies and Price does not change the fact that their AVC enables us to both define causation in terms of "bringing about" an effect and to describe "bringing about" an effect as an act of causation. If this is true, the charge of circularity still holds.

3. *The question about unmanipulative causes*—that is, causes in nature beyond the power of a human agent to manipulate. For example, we say that the friction between continental plates caused the 1989 San Francisco earthquake, even though we are not able to bring about such friction. In answer to this objection, Menzies and Price proposed a weakened form of AVC. It allows for classifying as causal those unmanipulative means-end relations which can be extrapolated to some analogous pairs of means-end-related events that allow for agent manipulation. Menzies and Price claim that this is possible "in virtue of certain basic intrinsic features of the situation involved, these features being essentially non-causal though not necessarily physical in character."[47] Thus, in the case of continental plates friction, we can think about artificial simulations of the movement of continental plates created by seismologists. However, deciding about the causal character of concrete occurrences based on approximations of this kind does not seem to qualify as a trustworthy method of procedure. For

what is the exact nature of the "intrinsic" but "non-causal" features in terms of which the friction of the continental plates resembles the laboratory simulation of tectonic plates movement? How can we make sure that that small-scale laboratory model will scale up? Can we assume our manipulable model resembles real phenomena taking place in macroscale events?

4. *Unacceptable anthropocentricity.* It seems that if AVC is correct, there can be no causal dependencies at times or places at which there are no agents. Here Menzies and Price state that—bearing in mind their answer to the first objection—we should understand that causal relations exist not only when agents have performed appropriate manipulations but in all situations in which "if a free agent were present and able, she could bring about the first event as a means to bringing about the second."[48] This claim seems to make AVC dependent on the possible worlds modality, which I find questionable here for the same reason I found it so above (section 2 of this chapter). Moreover, the reference to the possible human agent manipulation does not seem to solve the problem of anthropocentricity of AVC.

An original reformulation of AVC was suggested by Donald Gillies, who opted to replace it with an action-related view of causation (ArelVC).[49] He thought ArelVC is superior to AVC because it can describe both productive and avoidable actions. The latter are important in many contexts—for example, in medicine where an avoidance of an action A can prevent B from occurring. Gillies defines two types of avoidance action: (1) direct prevention of A from happening and causing B (sensitive to cases of overdetermination and trumping) and (2) alteration of the circumstances that usually accompany A causing B—that is, alteration of some ceteris paribus conditions crucial for a given causal occurrence (preemption).

Gillies looked for an answer to the question of unmanipulative causes, raised in reaction to AVC. He thought that although there is no way for a human agent to prevent them from happening, they can still be avoided. In the case of an earthquake, for example, one can refrain from going into areas on the boundary between continental plates. Thus, unlike AVC, ArelVC enables us to classify unmanipulative event sequences as causal, even if agents cannot intervene in their occurrence. However, does an avoidance action of the kind mentioned here change ceteribus paribus conditions under which earthquakes occur? And how can we decide which circumstances are relevant in a given causal situation? It seems that ArelVC does not provide a solid criterion of distinction between causally relevant and irrelevant avoidance actions.

5.3. The Interventionist View of Causation (IntVC)

IntVC is proposed as an alternative to AVC and ArelVC versions of MVC. Although it shares the same point of departure—that is, our recognition of the possible manipulability of C having an impact on its causing of E—it does not tie causation with human agency. Quite the contrary, IntVC states that manipulability helps us discover causal connections in purely natural processes, not involving human agency or intentionality at any point. These connections can count as interventions as long as they have the right causal and correlational characteristics. Thus, even when such manipulations are carried out by human beings, it is their causal character and not the fact they are carried out by human beings that matters for recognizing their true nature. Hence, states James Woodward,

> under this approach X will qualify as a (total) cause of Y as long as it is true that for some value of X... if X were to be changed to that value by a process having the right sort of causal characteristics, the value of Y would change. Obviously, this claim can be true even if human beings lack the power to manipulate X or even in a world in which human beings do not or could not exist. There is nothing in the interventionist version of a manipulability theory that commits us to the view that all causal claims are in some way dependent for their truth on the existence of human beings or involve a "projection" on to the world of our experience of agency.[50]

But what is the exact nature of an intervention, which is also described as "a 'surgical' change in A which is of such a character that if any change occurs in B, it occurs only as a result of its causal connection, if any, to A and not in any other way"?[51] Judea Pearl offers a mathematically sophisticated framework describing causal relations in complex cases.

Unlike Menzies and Price, whose AVC presents an indeterministic approach, Pearl argues that the quasi-deterministic conception of causality he wants to offer—often contrasted with the stochastic conception—enables us to define and analyze most of the causal entities we study. He builds functional causal models consisting of sets of equations of the form

$$x_i = f_i(pa_i, u_i), \qquad i = 1, \ldots, n,$$

where pa_i (connoting *parents*) stands for the set of variables judged to be immediate causes of x_i and where u_i represents errors (or disturbances) due to omitted factors.[52] Pearl depicts these models as network diagrams in which arrows stand for immediate causes. He pays attention to interventions that can be performed within causal networks and says that if one knows the antecedent probability estimates for the immediate causal relations in the network, one can estimate the quantitative changes that result from an intervention. Now, the simplest case of intervention amounts, according to Pearl, to "lifting X_i from the influence of the old functional mechanism $x_i = f_i(pa_i, u_i)$, and placing it under the influence of a new mechanism that sets the value x_i while keeping all other mechanisms unperturbed."[53]

Although such intervention determines x_i entirely, it is "surgical" in a sense that no other causal relationships in a given system are changed. This aspect is more apparent in the alternative description of an intervention (*I*), which does not refer to the relation between the variable intervened on and its effects, but defines *I* in terms of its influence on the variable *X* with respect to a second variable *Y* (where *X* and *Y* are expected to be related causally):

(M1) *I* must be the only cause of *X*; i.e., as with Pearl, the intervention must completely disrupt the causal relationship between *X* and its previous causes so that the value of *X* is set entirely by *I*,

(M2) *I* must not directly cause *Y* via a route that does not go through *X* as in the placebo example [in which a drug enhances recovery (*Y*) beyond the normal causal route (from *X* to *Y*)],

(M3) *I* should not itself be caused by any cause that affects *Y* via a route that does not go through *X*, and

(M4) *I* leaves the values taken by any causes of *Y* except those that are on the directed path from *I* to *X* to *Y* (should this exist) unchanged.[54]

The most natural way of defining causal effect within this framework is to characterize it as a difference made to the value of the variable *Y* by introducing (by an *I*) a change in the value of the variable *X*. Thus, causes—despite their ontological definition in terms of events, types of events, properties, facts, and so on—must be representable in IntVC as "variables" that can change or assume different values.

Woodward claims that the advantage of this sophisticated definition of causation is that it allows us to interpret as causal some specific and unusual cases, defined as

1. *Causation by omission* (e.g., a change under an intervention in whether the gardener waters the flower is associated with a change in the value of the variable of measuring whether the flower dies).
2. *Causation by disconnection* (e.g., a change under an intervention in whether a process of blood circulation has all parts sustaining its equilibrium is associated with a change in the value of the variable of measuring whether all other parts of the process are still connected).
3. *Double prevention*—that is, cases (discussed above, section 2) of removing a cause preventing another cause or of inhibiting an inhibitor (e.g., a change under an intervention in whether allolactose [an isomer formed of lactose; C_3] is present in the environment of *Escherichia coli* is associated with a change in the value of the variable of *E. coli*'s metabolizing lactose [*E*]—which happens through inactivation by allolactose [C_3] of the repressor protein [C_2] that binds to the operator region [C_1] of *E. coli*'s genome, preventing it from producing enzymes metabolizing lactose). See figure 4.2.

Moreover, proponents of IntVC find another advantage of this approach in enabling us to make sense of causal claims not only in contexts in which interventions do not in fact occur but also in cases in

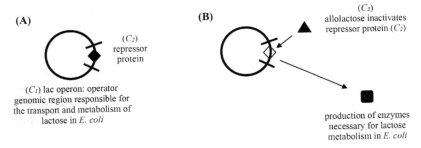

Figure 4.2. Lac operon in *Escherichia coli* as an example of double prevention (**A**) The expression of the gene that produces enzymes metabolizing lactose is prevented due to the repressor protein (C_2) inhibiting lac operon (C_1). (**B**) Double prevention. C_1 is not inhibited by C_2 due to the presence of allolactose (C_3), which prevents it. The whole situation effects in the production of enzymes necessary for lactose metabolism.

which they are causally impossible. Woodward claims it is possible "as long as we have some principled basis for answers to questions about what *would* happen to the value of some variable *if* an intervention were to occur on another variable."[55] He gives an example of the gravitational attraction of the moon causing the motion of the tides—a variable that cannot be altered (with respect to the tides) at present and may not be physically changeable by any physically possible process in the future. This strategy of dealing with unmanipulative causes seems to be more plausible than the one offered by AVC. Nevertheless, we need to notice that—similar to Woodward's basic definition of IntVC quoted in the opening of this section—his defense of the causal character of relations that cannot be intervened in or manipulated is based on subjunctive conditionals of the following type: if it were the case that *p*, then it would be the case that *q*. This fact makes the whole theory dependent on the possible worlds modality (similar to CVC and AVC), which seriously challenges its explanatory power.[56] What is more, similar to the more generally stated MVC, IntVC seems to face the problem of forming lawful-kind generalizations with a certain level of invariance, based on singular instances of interventions (*I*) changing the value of one variable (*X*), which changes—in turn—the value of another variable (*Y*).

Another challenge for IntVC comes from the fact that it is difficult, if not impossible, to apply it to fundamental physical theories that take account of the universe as a whole, in which cases "causality disappears because interventions disappear—the manipulator and the manipulated lose their distinction."[57] We can assume that causal claims either are literally true or do not apply at all to the context of fundamental physics. The latter, says Woodward—that is, "the view that fundamental physics is not a hospitable context for causation and that attempts to interpret fundamental physical theories in causal terms are unmotivated, misguided, and likely to breed confusion"—is "probably the dominant, although by no means universal, view among contemporary philosophers of physics."[58]

Consequently, he suggests we should agree that IntVC applies to what Pearl calls "small worlds"—that is, systems of medium-sized physical objects of the sort investigated by various special sciences. This claim leaves us, however, with a fundamental question of whether we can assume ontologically that at certain levels of complexity and description of the universe there are no cause/effect dependencies, based on the fact that our current physical and philosophical theories (IntVC in particular) do not "find" causes at those levels of description. Such a conclusion seems to be rather arbitrary and not justifiable.[59]

5.4. Summary

To sum up our analysis of IntVC and other versions of M-basedVsC we need to acknowledge that—similar to other theories discussed so far—they offer another important insight and provide another important criterion of distinguishing whether a given relationship between *A* and *B* is causal. These theories of causation gained popularity among social scientists, statisticians, and theorists of experimental design. Many natural scientists also seem to share an appreciation for making a connection between causation and manipulation in order to clarify the meaning of causal claims and understand their distinguished features. Moreover, we have seen that IntVC is able to provide a proper causal analysis of situations described as causation by omission, causation by disconnection, and double prevention.

At the same time, however, M-basedVsC were criticized by many philosophers. The main charges of being unilluminatingly circular, unacceptably anthropocentric (in case of AVC), dependent on possible worlds modality, and applicable only to "small worlds" (systems of medium size) that are available for manipulation impose some important limitations on their explanatory potential. I side with those who claim that even IntVC—which is regarded as the most advanced and allegedly free from anthropocentricity among other M-basedVsC—is still vulnerable to the charge of circularity, as the notion of intervention is itself causal in character. Many contemporary philosophers expect an acceptable theory of causation to be reductionist and based on noncausal, usually empiricist, terms such as regularity, spatiotemporal contiguity, and so on. This might be the reason why they question the consistency and usefulness of MVC, AVC, ArelVC, and IntVC.

6. THE PROCESS VIEW OF CAUSATION (ProcVC)

Critical of accounts of causation based on events, of the modal commitments of CVC, of the technical difficulties of ProbVC, and of causal theories appealing to agency (AVC), and influenced by Russell's idea of "causal lines" understood as trajectories of things through time (replacing the primitive notion of causation in the scientific view of the world), Wesley C. Salmon proposed his own theory of the causal world consisting in the nexus of processes and interactions.[60] He defines a process as anything with constancy of structure over time and says that causal processes can be distinguished from the pseudocausal ones by their capability of "transmitting a mark," where "mark" is explained as any local modification of a "characteristic" (signal, information, energy, etc.). He thinks that "causal processes constitute precisely the causal connections which Hume sought, but was unable to find."[61]

Salmon gives an example of a spotlight emitting white light and rotating in the center of a large, circular, darkened building. A pulse of light traveling from the spotlight to the wall is a causal process A. If we place a red glass (causal process B) in its way, the pulse of white light turns red and persists red until it reaches the wall. Placing the red glass (causal process B) transforms causal process A, which is now transmitting a new mark. Unlike the pulse, a spot of light moving around the wall as the spotlight rotates is a pseudo-process. If we put a red filter at some point on the wall, the spot will turn red when it reaches this particular point, but this modification (mark) will not be transmitted.[62]

Thus, according to Salmon, the terms "cause" and "effect" are properly predicated about processes. He defines causal processes as "the means by which structure and order are *propagated* or transmitted from one space-time region to other times and places," and causal interactions (temporally simultaneous intersections) as "the means by which *modifications* of structure (which are propagated by causal processes) are *produced*."[63] He claims that causal interactions are explicated in terms of interactive forks which are governed by basic laws of nature (see figure 4.3).

Apart from regular interactive forks, Salmon speaks about temporally asymmetrical conjunctive forks, which he thinks play a vital role in the production of structure and order (even if they are not closely tied to the laws of nature). They can be described as two or more processes, physically independent of one another, not interacting directly with each other, and yet correlated because they arise out of some special set of background conditions (Salmon refers his description to Reichenbach's "principle of the common cause"). Conjunctive forks can be discovered through noticing improbable coincidences that recur too frequently to be attributed to chance. The apparent lack of independence between them can be explained by reference to a common causal antecedent. To give an example, a similar response of cells in two unrelated human organisms (A and B) to an attack of the T-lymphotropic virus, which makes these cells cancerous, suggests a common causal antecedent. It might include (C) the production of γ-rays upon nuclear fission in the aftermath of an atomic bomb explosion, their passage through the air, and a modifying interaction with cells of a human body. In Reichenbach's terminology, C screens off (makes statistically irrelevant) A from B, and B from A (see figure 4.4).[64]

Salmon finds at least three main advantages of his project. First, he thinks it provides a strong argument to treat processes rather than Humean events as basic entities and causal relata. He believes that an analysis referring to processes—having much greater temporal duration and spatial extent (represented by lines in space-time diagrams)—represents the

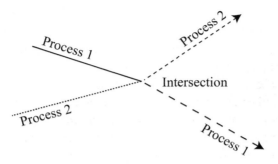

Figure 4.3. Interactive fork

In his explanation of interactive forks Salmon states:

> In everyday life, when we talk about cause-effect relations, we think typically (though not necessarily invariably) of situations in which one event (which we call the cause) is linked to another event (which we call the effect) by means of a causal process. Each of the two events which stands in this relation is an interaction between two (or more) intersecting processes. We say, for example, that the window was broken by boys playing baseball. In this situation, there is a collision of a bat with a ball (an interactive fork), the motion of the ball through space (a causal process), and a collision of the ball with the window (an interactive fork). (Salmon, "Causality: Production and Propagation," 168)

Earlier on in the same article Salmon develops a general account of the causal structure of the world:

> We live in a world which is full of processes (causal or pseudo), and these processes undergo frequent intersections with one another. Some of these intersections constitute causal interactions; others do not. If an intersection occurs which does not qualify as an interaction, we can draw no conclusion as to whether the processes involved are causal or pseudo. If two processes intersect in a manner which does qualify as a causal interaction, then we may conclude that both processes are causal, for each has been marked (i.e. modified) in the intersection with the other, and each process transmits the mark beyond the point of intersection. (Ibid., 165)

nature of reality and causation more adequately than one concentrated on events, which are relatively localized in space and time (symbolized by points in space-time diagrams). Second, Salmon thinks ProcVC answers the question of whether causes must precede their effect or whether causes and effects might occur simultaneously. He says propagation (transmission of a mark) involves lapse of time, while causal interactions exhibit the relation of simultaneity. Finally, concerning the question of whether statements about causal relations pertain to individual events or

Theories of Causation in Analytic Metaphysics 171

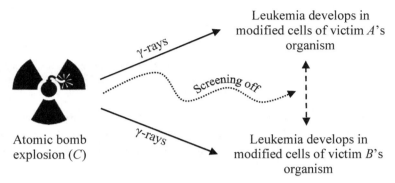

Figure 4.4. Conjunctive fork

classes of events, we find Salmon saying that, while each causal process is an individual entity (sustaining causal connection between an individual C and an individual E), we usually use statistical relations to describe interactive and conjunctive forks, making thus assertions which involve statistical realizations.[65]

At the same time, Salmon's version of ProcVC raises two fundamental problems. First, it raises a number of metaphysical questions concerning the ultimate nature of processes. How shall we understand classifying them as entities? The very term "process" seems to suggest (ontologically) that there is something that is in process. What is the ontological status of the "mark" or "structure" that is being transmitted in the process? What makes us classify process as a fundamental metaphysical building block of reality? Salmon's theory does not seem to offer satisfying answers to these questions. Secondly, the imprecise definition of causal process leads to further queries listed by Losee:

1. The extent of persistence of a mark through time required by the definition of a process is unclear. The motion of a positron emitted from a bombarded nucleus presumably qualifies as a process, even though its annihilation by an electron takes place a few seconds after emission. Does persistence of an orange for that same period of time qualify as a process as well?
2. The level of alteration within which a body (or wave) retains its self-identity is also uncertain. A billiard ball gaining electrons, a

falling snowflake with new molecules of H_2O organizing themselves on its surface in a hexagonal crystal structure of ice, and water vapor on its journey from my teacup along the desk all presumably count as processes. But the case of the growth of a sand dune may be controversial.

Because of these problems and because both mark-transmission and causal interaction tacitly involve counterfactuals (if no causal interaction had taken place, a modified mark would not have been transmitted), Phil Dowe suggested a replacement of Salmon's "mark-transmission" criterion of causal processes with the concept of "possession of conserved quantity." He notices that what qualifies as a conserved quantity is determined in accordance with current scientific theories, and he lists mass-energy, linear momentum, charge, and spin. He thus redefines causal process as "a world line of an object that possesses a conserved quantity," and causal interaction as "an intersection of world lines that involves exchange of a conserved quantity."[66] Because application of these definitions requires tracing concrete objects through time, Dowe accepts the fact that his theory presupposes that "identity-through-time" can correctly be, and should be, predicated of an object.

Answering Dowe, Salmon stated that his conserved quantity theory can improve without an explicit commitment to the self-identity of objects. He recommended identifying causal processes with the continuous transmission of a conserved quantity within a closed system, rather than the continuous possession of that quantity. Referring to Quine, Fair, and Aronson,[67] Salmon redefines causal process in terms of "an object that transmits a nonzero amount of a conserved quantity at each moment of its history (each spacetime point of its trajectory)," and causal interaction as "an intersection of world-lines that involves exchange of a conserved quantity."[68]

Dowe noticed also that Salmon's requirement of causal processes taking place within closed systems will in fact fit very few cases involving transfer of energy and linear momentum, as the majority of them are not isolated and include a loss of energy (e.g., through friction). He thus suggested softening the requirement and saying instead that "the CQ [conserved quantity] theory does not require that a causal process possesses a constant amount of the relevant quantity over the entire history of the process."[69] Moreover, defending his commitment to "identity-through-time" against Salmon's "transmission of conserved quantity Q at every space/time point of the process," Dowe notes that the latter position (without any assumptions of self-identity) makes it impossible for an observer to specify whether A and B interact with an exchange of Q, or just intersect without any interaction. Suppose "a stray neutrino passes through my body"—says Dowe—"Am I still myself, unaffected by the event, or am I now the thing that used to be the neutrino, having been radically transformed by the experience?"[70]

At the same time, Dowe is aware that ProcVC, both in his own version and in the one offered by Salmon, seems to be of little use in cases of omission and prevention, which are not subject to an interpretation in terms of interactions of causal processes that involve an exchange of conserved quantities. He thinks that an attempt to provide a causal interpretation of such cases under ProcVC requires a reference to counterfactual conditionals. In the situation of omission, instead of saying that "the father's inattention was the cause of the child's accident" (not interpretable in terms of an exchange of conserved quantities), we can state that even if the accident occurred, "it was possible for him [the father] to have prevented it" (his action would have been available to a conserved-quantities-type of interpretation). Thinking about prevention, Dowe suggests we can replace Ellis's example of "pulling down the windowshade [that] caused the room not to be light" with

1. a causal claim about the world saying that pulling down the windowshade caused light to be reflected from the shade (available to a conserved-quantities-type of interpretation but uninformative about prevention), and
2. a counterfactual conditional about a possible world stating that if light were not reflected from the shade, then the room would be light (not available to a conserved-quantities-type of interpretation but informative about prevention).[71]

Even if this hybrid reinterpretation of omissions and preventions may sound plausible, Losee notes that it is a bit artificial if one wants to be consequent in following the process definition of causation in terms of an actual exchange of a conserved quantity. For, strictly speaking, under ProcVC, "causation-by-omission" and "causation-by-prevention" are not instances of causation at all. Moreover, Losee mentions another problem of ProcVC, which I have addressed already in my evaluation of Salmon's theory. Though it provides an analysis of the structure of causal processes and their interactions, it does not define precisely the nature of causal relata (entities related as causes and effects). Losee acknowledges that Dowe—after David Armstrong and Ludwig Wittgenstein—defines the world as a collection of individual states of affairs, among which the most basic (atomic) states of affairs include the possession of properties by particulars and relations among particulars. He sees the constituent parts of states of affairs as events (changes in properties of objects at a time) and facts (objects having properties at a time or over a time period), which may serve as a foundation for a process-type of ontology. Nevertheless, an attempt at formulating its more elaborate version requires a much more thorough metaphysical analysis than the one offered by Salmon and Dowe.[72]

Finally, we need to remember that ProcVC is criticized for reducing the metaphysical account into a watered-down version of a physical theory—that is, an empirical analysis which may not be equivalent to any conceptual inquiry into the nature of causation. The same theory of causation—developed predominantly in the context of physical science—is further challenged by issues concerning causal dependencies in other branches of science, where analyses in terms of conserved quantities lead to a thoroughgoing reductionism.[73]

I believe that this exploration of ProcVC shows, once again, that as a very specialized theory—contributing to our better understanding and assessment of causal situations—it is limited and cannot serve as the ultimate explanation and definition of causal relata and their interdependence. This brings us to the final summary and evaluation of all theories of causation presented in this chapter.

7. SUMMARY

Although all six views of causation developed in analytic philosophy and discussed here contribute significantly to the contemporary causal debate, none has so far been able to win the stage and become predominant. None of them seems able to offer an exhaustive description and precise criteria of classification of causal relations and causal relata across different fields and methods of observation applied in scientific research. They leave open the question whether we should define causes and effects as particular objects, particular events, qualitative entities such as properties or universals, structured (complex) entities such as states of affairs or facts about reality, or processes. Moreover, all six concepts raise some crucial metaphysical doubts and questions which challenge their explanatory potential as philosophical theories. I have gathered and summarized main strengths and weaknesses of each one of them in table 4.1.

An awareness of difficulties of providing one satisfactory description of the nature of cause/effect dependencies in nature inspires at least three possible positions concerning the role of causality in scientific explanation:

1. *Anticausality position.* Some philosophers of science claim that while the notion of causality can be meaningful in the philosophical analysis of practical matters addressed by engineers, plumbers, surgeons, and so on, it is rather useless in hard science, especially in physics, where a grand unified theory is expected to be geometrical rather than causal. Others give up in their search for causal explanation, claiming that causation belongs to a group of concepts that remain "permanently resistant to definition and analysis because of the pivotal place they occupy in our conceptual scheme."[74] Still others give up classical causal definitions and turn toward a wide variety of formal methods for representing inferences about causal relationships, offered by statistics and computer science, which are labeled "causal modeling."[75]
2. *Limited scope position.* We cannot dismiss causal language from scientific explanation, but we should restrict it to experimental arrangements and research results, excluding any reference to causal concepts at a theoretical level. Philosophical analysis of causation should be thus limited to practical and methodological aspects of scientific procedures.
3. *Causal pluralism.* Accepting the indispensability of causal descriptions in scientific explanation, causal pluralists acknowledge that the apparently simple and univocal term "cause" is in fact masking an underlying diversity that can be captured only by various theories—each of them valid and contributing to our general description of changes in nature.

Table 4.1. Strengths and weaknesses of the six main views of causation in analytic metaphysics

Theory	Criteria for causal relatedness	Strengths	Weaknesses
RVC	regular co-occurrence (constant conjunction) of C and E	acceptable for empiricists	difficulty in identifying the ground of regularity; difficulty in specifying what makes a particular C to cause a particular E; problem of generalization of causal claims; treating accidental correlations as causal
INUSConVC	C as an Insufficient, Nonredundant part of a set of conditions that was itself Unnecessary, but Sufficient for E	allows to describe causal relations in many scientific contexts	difficulty in specifying what counts as a condition; problem of generalization of causal claims; problem of specifying directionality of causal changes; interpreting cases of underdeterminative sufficiency as causal; interpreting joint effects of common causes as causal; difficulty of specifying INUS conditions in cases of overdetermination
InfVC	functional relation (F_k) of variables, enabling prediction of future states of a system	allows to classify important scientific laws as causal relations	not applicable to all scientific laws; difficulty in specifying the number of state variables in formulating F_k
CVC	contrary-to-fact conditional dependency between C and E	helps avoid assigning causal character to joint effects of common causes; solves the problem of directionality in particular causal situations; applicable in some cases of causation by omission and double prevention (physiological processes and everyday activities)	unspecified character (is CVC a folk concept, a philosophical concept, a scientific concept, or something in between?); dependency on possible worlds modality may lead to circularity; the problem of distinguishing between lawful and merely accidental generalizations; unable to deal with cases of early, late, and simultaneous preemption and with over- and underdetermination

ProbVC	C raising probability of E $p(E/C) > p(E/-C)$	takes probability into account in examination of causal changes	applicable in general accounts, ProbVC fails in particular cases; difficulty in forming lawful-kind generalizations based on individualized instances of probability raising; requires an exclusion of all logically necessary conditions from the ranks of putative causes; in many cases, an acknowledged cause of an event actually decreases, instead of increasing the probability of its occurrence; difficulty in specifying the nature of probability (epistemological versus ontological)
SVC	single case of causal dependency between C and E, with C being a particular change in the immediate environment in a particular situation, a change that occurred just before E	emphasizes the importance of the individual and singular aspects of causal relata for investigation of cause/effect dependencies	difficulty in limiting description to a manageably small number of changes in the immediate environment of a singular causal change; the idea of regularity of causal events, which SVC opposes, sneaks back in with the requirement of the "commonality" of causes in events of the same sort
M-basedVsC (MVC)	appropriate manipulations on C effecting relevant changes in E	manipulability-based theories are helpful in distinguishing between lawful and merely accidental generalizations	suggests assigning causal status to functional laws that pass manipulability test, such as those of Boyle, Ohm, and Snell
AVC, AreIVC	a free agent bringing about the occurrence of C as an effective means by which he/she could bring about the occurrence of E	AreIVC enables us to describe both productive and avoidable actions	confusing metaphysics with epistemology (AVC may suggest agent interventions are constitutive for causal interactions); circularity (explaining causation in terms of "bringing about" a change); problem of unmanipulative causes; anthropocentricity

(*continued*)

Table 4.1. Strengths and weaknesses of the six main views of causation in analytic metaphysics (*continued*)

Theory	Criteria for causal relatedness	Strengths	Weaknesses
IntVC	for some value of C, if C were to be changed to that value by a process having the right sort of causal characteristics, the value of E would change	does not tie causation to human agency; useful in social sciences and statistics; applicable in cases of causation by omission, disconnection, and in double prevention; offers a strategy for dealing with unmanipulative causes	reference to possible worlds modality in dealing with unmanipulative causes; problem of forming lawful-kind generalizations with a certain level of invariance; difficult, if not impossible, to apply in cases related to fundamental physical theories that take into account the universe as a whole
ProcVC	causal process = the means by which structure and order (marks) are propagated in space and time or a world line of an object that possesses a conserved quantity causal interaction = the means by which modifications of structure (marks) are produced or an intersection of world lines that involves exchange of a conserved quantity	fitting with process-type ontologies; offers an answer to the question of whether causes must precede their effects, or whether causes and effects might occur simultaneously, and to the question of whether statements about causal relations pertain to individual events/processes or classes of events/processes; offers an explanation applicable in cases of omission and prevention with reference to counterfactuals	difficulty in specifying the ultimate nature of processes and the ontological status of the "mark" or "structure" that is being transmitted in the process; the extent of persistence of a mark through time is unclear; the level of alteration within which a body (or wave) retains its self-identity is uncertain; explanation of omission and prevention is a bit artificial (under ProcVC "causation-by-omission" and "causation-by-prevention" are, strictly speaking, not instances of causation at all); charged with reducing the metaphysical account into a watered-down version of a physical theory; challenged by issues concerning causal dependencies in branches of science other than physics

The last position, which seems to be the most popular one, comes in three variations. The first makes a general distinction between one type of causal relation that is an objective feature of the universe—satisfying an energy-transfer criterion and/or statistical relevance criteria—and the second type that consists of situations involving human conscious decisions, responsibility, and guilt, satisfying a counterfactual-conditional criterion. The second option distinguishes two types of causal description in scientific analysis, independently of phenomena of human consciousness and responsibility. On one level, a causal relation is a relation between event-tokens, related to changes that involve an energy transfer. On a second level, a causal relation is a relation between properties such that the realization of a property-cause (p-C) raises the probability of a property-effect (p-E) under maximally specific background conditions. Finally, the third option takes into account all six theories discussed in this chapter and their variations. It finds each one of them contributing significantly to our attempts at providing an adequate causal explanation.

My own view follows causal pluralism insofar as the latter acknowledges the role of various views of causation in the defense of the reality of causal dependencies in nature. At the same time, however, I want to emphasize that the six main theories explored so far do not cover wholly the complexity of the most recent reflection on causation in analytic metaphysics. Neither can they answer, in my opinion, some of the most puzzling and challenging metaphysical and ontological questions raised by the systems approach in biology. Though they struggle to overcome Hume's reductionist agenda—arguing in various ways in favor of the ontological character of causal relations—their inherent shortcoming is a tendency to look at causation only in terms of the efficient cause. I find this approach highly problematic, limiting, and unjustified. Moreover, this kind of attitude seems to prevent the participants in the contemporary causal debate from providing a universal view of causation—that is, a theory that would encompass all specific cases of causal dependencies that contemporary metaphysicians refer to, across scientific disciplines that investigate them.

In this context, and in reference to my analysis of EM and DC, which suggested expanding the typology of causes, we can now acknowledge and fully appreciate a unique character of one more analytic theory of causation—namely, the dispositional view of causation (DVC). This notion of causal dependencies is rooted in dispositional metaphysics and provides a ground for possible reintroduction of formal and final causation as defined in Aristotelian terms. I think that it also reopens the way to a unified theory of causation, the search for which has been abandoned in the recent philosophical debate. I deliberately did not list DVC with other theories investigated in this chapter. I think that its unique character, putting into question causal reductionism, places it beyond the mainstream of causal analysis in analytic metaphysics. It is not coincidental that the editors of *The Oxford Handbook of Causation* classified it among alternative, rather than standard, approaches to causation. Because of its metaphysical commitments, DVC may truly become a challenge to many contemporary theorists of causality. But the alternative that DVC offers should not be feared. Quite to the contrary, I hope to prove, in the remaining chapters of this book, that DVC brings an absolutely indispensable supplement to the most recent examination of the nature of causation—a supplement that cannot be ignored or easily dismissed.

CHAPTER 5

Dispositional Metaphysics and the Corresponding View of Causation

The departure point for the proponents of DVC is their recognition that the Humean ontology of "loose and separate" entities and/or events which are related only externally and contingently ("atoms in the void") is inadequate in light of contemporary science, which emphasizes the dynamic features of entities across the scale of magnitude, from organisms and macroscopic objects to elementary particles. While the acceptance of Humean premises led to the rejection of necessary connections between existences and the reduction of causality to constant conjunction, the new science opens the way to powers (dispositional) ontology, which "accepts necessary connections in nature, in which the causal interaction of a thing, in virtue of its properties, can be essential to it."[1]

Dispositionalists claim that, while all contemporary theories of causation might, to a certain degree, succeed in describing what is true when causation occurs, nevertheless, they are not about causality itself. In other words, dispositional metaphysicians may agree that regular occurrences and constant conjunctions, as well as counterfactual, probabilistic, interventionist, and process aspects of dependencies among inanimate and animate beings, in various settings and combinations, are real. But the very reason these all occur—they add—is that there are real causal connections in nature, to which contemporary theories fail to attend, or which they are hesitant to acknowledge. Dispositionalists, therefore, suggest replacing the metaphysics of distinct, discrete, passive, and unconnected entities and of contingently related causes and effects regulated by the laws of nature with a metaphysics of powers and their manifestations, which, although ontologically distinct, are necessarily connected. But what are dispositions and powers, and what are their manifestations?

1. DISPOSITIONS AND POWERS

We must begin with a terminological issue. Although the term "disposition" is used interchangeably with the term "power" by the majority of those who follow dispositional ontology, Alexander Bird claims the verbal difference marks an ontological distinction. He suggests that not every disposition (being fragile, soluble, etc.) is a power, understood in terms of a more robust essentialist ontology. Following Armstrong,[2] Bird argues that universals need not have dispositional essences to be characterized dispositionally. He is not willing to assign them natures/essences going beyond the trivial ontological assertions, such as self-identity. Therefore, notes Bird, a Humean like David Lewis may be willing to accept a dispositional analysis of a certain phenomenon—which he will find still reducible to a counterfactual explanation—while remaining skeptical about the existence of powers defined in terms of essentialist metaphysics.[3]

Nevertheless, since the majority of dispositionalists accept dispositions as an irreducible ontological category, characteristic of animate and inanimate things, we can ignore the terminological difference suggested by Bird for present purposes and use "dispositions" and "powers" interchangeably. Moreover, according to Molnar, both terms can be classified among a number of other interrelated concepts such as "capacity," "ability," "skill," "aptitude," "propensity," "tendency," "potential," "amplitude," and so on.[4] Mumford believes the "dispositional" is a genus that can accommodate as species various subclasses such as tendencies, capacities and incapacities, powers and forces, potentialities and propensities, abilities and liabilities.[5] Dispositionalists agree that powers should not be regarded as substance-like existents. However, they can be classified as properties—that is, properties of a unique character.[6] Bird defines powers as "properties with a certain kind of essence—an essence that can be characterized in dispositional terms."[7] Mumford suggests we should treat them as "a distinct and basic ontological category in their own right, irreducible to any other ontological category."[8] Dispositionality is for him and Anjum a primitive and unanalyzable modality, intermediate between pure possibility and necessity, which provides a metaphysical base for normativity (something ought to be the case but is not necessary), and intentionality (directness or aboutness, conscious and nonconscious). It can be understood as a "selection function" which picks up a limited number of possible outcomes that are manifestations toward which powers are directed.[9]

In another place Mumford emphasizes that powers are "actual, intrinsic states or properties rather than 'bare potentialities'" and that disposition ascriptions are in fact functional characterizations of properties—that is, such ascriptions characterize properties according to what effect the properties will produce in particular circumstances.[10] But this functional character of powers must not be equated with a conditional analysis and an empirical treatment of dispositions finding expression in the antirealism of verificationist and event philosophies. At the heart of the conditional analysis, introduced by Locke, we find a conviction that all we need for the ascription of a disposition is knowledge of whether a certain reaction occurs in certain conditions. Such analysis dispenses with the properties of objects and concentrates merely on their actual or possible behavior. It should be rejected, for although dispositions are related to conditionals, they are not equivalent, as they are something that lies behind what occurs and is verifiable.[11]

Because they are classified as properties, powers can be taken as universals which have their instantiations in substances and are decisive for their intrinsic features. As an example, we may think of fragility as the disposition residing in the crystalline structure of glass or pottery or in a protein crystal which can be obtained through a process of protein crystallization. While the glass is vulnerable to mechanical stress, the protein crystal is sensitive to temperature, pH, and other factors. Another example is solubility, understood as the disposition of ionic compounds such as sodium chloride (NaCl), which are soluble in water, or the disposition of petroleum jelly to dissolve in gasoline. Both examples (fragility and solubility) show the identity of powers across various and distinct instantiations. The fragility of glass or pottery is manifested due to external factors which differ from those that may affect the stability of protein crystals. In the case of solubility, we can notice that while ionic compounds dissolve in water because of the attraction between positive and negative charges, petroleum jelly dissolves in gasoline because they are both nonpolar hydrocarbons. Although the mechanisms of the two reactions differ, the reactions are still of the same type.

One more aspect of the metaphysical analysis of powers is the distinction between transitive and intransitive powers. The former manifest by changing or having an influence on something other than themselves in the environment, while the latter's exercise is simply their own activity (it can be still described in terms of change, but not as interaction). Also, some powers are always activated (e.g., the power of an electron field), while others manifest only under certain conditions (e.g., the power of a magnetic field—it manifests when an electric charge moves).[12]

2. GROUNDING OF POWERS

When arguing for the irreducibility of powers, Mumford obviously refers to the popular division of properties into dispositional and categorical, and the question of the ontological relation/dependence between the two—that is, whether they are reducible or eliminable with regard to each other. Attempting to specify the nature of dispositional properties, he states that disposition ascriptions are "ascriptions of properties that occupy a particular functional role as a matter of conceptual necessity and have particular shape or structure characterizations only a posteriori."[13] Categorical ascriptions, on the other hand, underlie dispositions (providing their base) and are "ascriptions of shapes and structures which have particular functional roles only a posteriori."[14] In other words, dispositional properties are responsible for things' modal characteristics, while categorical properties provide for their identity.

Bringing together classifications offered by Engelhard and Mumford,[15] we can name two major ontological positions among contemporary property theorists, each possessing specific subcategories (as depicted in figure 5.1):

1. *Ontological dualism*. It assumes that dispositional and nondispositional (categorical or occurrent) properties are two inherently

distinct kinds of qualities. This position may have two reductionist variations:
  a. *Dispositional reductionism*. It approves the existence of both dispositional and categorical properties, but only on some levels of description. At the end of the day the latter are reducible to the former.
  b. *Categorical reductionism*. It approves the existence of both dispositional and categorical properties, but only on some levels of description. At the end of the day, the former are reducible to the latter.
2. *Ontological monism*. It assumes the existence of only one kind of properties and can take one of four forms:
  a. *Pan-dispositionalism (dispositional eliminativism)*. It holds that all properties are dispositional (at least all sparse fundamental properties) and that, consequently, categorical properties do not exist.
  b. *Pan-categoricalism (categorical eliminativism)*. It states that all properties are categorical properties—that is, occurrences, states, episodes, shapes, structures, and molecular substructures, which are empirically verifiable. In other words, properties are instantiated only in events, which dispenses with the idea of properties corresponding to dispositional terms.
  c. *Neutral monism (dual aspect monism)*. It holds that there is ontologically only one kind of properties, which are neither dispositional nor categorical, and treats the dispositional/categorical distinction as merely linguistic. It says these are simply two aspects of, or ways of referring to, the same property.
  d. *Identity theory*. It states that the dispositional and the categorical are two modes of one and the same property, which is, thus, metaphysically simple. Those who follow this theory propose, as its model, a Necker cube, which can be seen now one way, now another, or the duck-rabbit figure. Therefore, unlike in neutral monism, identity theory finds properties being both categorical and dispositional.

Among those who approve the ontological dualism of dispositional and categorical properties (1), Molnar, Ellis, and Lowe try to avoid any

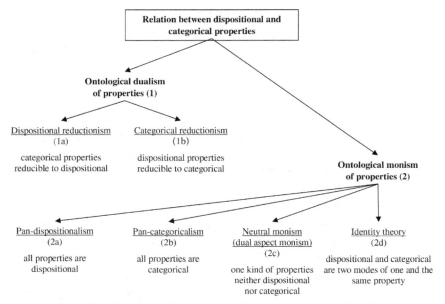

Figure 5.1. Classification of the major positions among dispositionalists concerning the relation between dispositional and categorical properties

kind of reductionism. Molnar treats distinctively the reality of powers and nonpowers, which he defines as S-properties responsible for symmetry operations. He lists among S-properties spatial location, temporal location, spatial orientation, and so on and claims they have causal relevance; for how objects are located in space and time and how they are oriented makes a difference for what effects they have on each other.[16] Brian Ellis follows the same reasoning and presents an even more extended list of categorical properties, which are recognizable in terms of spatiotemporal relations. He includes length, orientation, distance, time, shape, size, speed, hardness, direction, and angular separation, saying this list can be continued. Like Molnar, he states that without categorical properties, causal powers would be nowhere, no-when, directionless, and lacking identity. He ascribes a causal role to categorical properties without reducing them to causal powers, based on the fact that the latter have dimensions described by the former.[17]

Lowe concurs with Molnar in disliking "the common practice of contrasting the dispositional with the *categorical*, partly because this invites—if, indeed, it doesn't simply arise from—the questionable assumption that what is 'dispositional' is merely *hypothetical* and so, in the idiom of possible worlds, concerns what is (categorically) the case in non-actual possible worlds."[18] Arguing in favor of the reality of both types of properties, he suggests replacing the dispositional/categorical distinction with a dispositional/occurrent distinction—for example, contrasting an object's disposition to dissolve in water (grounded in the characterization of a substantial kind by a property) with its actual dissolving on some occasion (grounded in the characterization of an individual substance by a property instance).[19]

Unlike Molnar, Ellis, and Lowe, Armstrong supports categorical reductionism (1b) when he claims that every dispositional property entails a categorical basis to which it is identical. He conceives dispositional properties in terms of conditional statements in the following form: if a certain stimulus is present to a thing that has the property in question, a certain manifestation will follow. He agrees that conditionals of this type must have "truthmakers," but he argues that these are not powers or potencies. He sees them as categorical properties of things together with laws of nature, rather than dispositional properties.[20] Dispositional reductionism (1a) seems to be a merely theoretical option, as it is difficult to find a representative of this position.

Among property monists (2), Martin and Heil support the identity theory (2d),[21] whereas Quine can be regarded as a categorical eliminativist (2b), as he claims that if there is no categorical basis that supports a putative disposition, that disposition term can and should be eliminated from the vocabulary.[22] Mumford seems to have changed his opinion on this matter over the years. While in *Dispositions* he claims that his modest form of realism causes him to embrace a neutral monism (2c: "monist because dispositional and categorical tokens can be identical, neutral because it refrains from the classification of reality as either 'really' categorical or 'really' dispositional"),[23] in his later works he states that all properties are dispositional and should be regarded as powers or clusters of powers.[24] The latter position obviously classifies as pandispositionalism (2a), which is also supported by Bird, Popper, Mellor, and Shoemaker.[25]

This variety of opinions concerning dispositional and categorical properties reflects a hot debate over this issue in current analytic metaphysics, a debate that, from my perspective, has a decisive significance for the entire enterprise. The bottom-line question, reflecting Bird's distinction between powers and dispositions, mentioned in the first section of the present chapter, is this: Can we speak about dispositions in terms of counterfactual dependencies, and without acknowledging the reality of powers, understood as irreducible ontological features of living and nonliving things? Approaching this problem from the Aristotelian point of view, I am less sympathetic toward eliminativist pan-categoricalism (2b) and the version of property dualism that reduces dispositional to categorical properties (1b). Because I want to argue in favor of the reality and ontological character of powers, I am closer to positions defending irreducibility of dispositions. Among these positions, I favor the genuine (nonreductive) property dualism (1, but not 1a). Because I think that dispositions need grounding, I find property dualism more metaphysically sound than monistic propositions of neutral monism (2c), which sees the dispositional/categorical distinction as merely linguistic, identity theory (2d), which defines it in terms of modes of the one type of property, and eliminativist pan-dispositionalism (2a), which questions the reality of categorical properties.

But is the position embracing property dualism metaphysically relevant at all? Does it not place us among substantial dualists of the Cartesian school, replacing a passive *res extensa* with categorical properties and an active *res cogitans* with dispositional properties? Does it not end up with the old Cartesian-type dilemma of the relation between categorical and dispositional properties? I find this challenge real and crucial for the proponents of dispositional metaphysics. The best way to avoid this dualistic charge—in my opinion—is to interpret the categorical/dispositional distinction in terms of the Aristotelian theory of act and potency, referring it to one and the same substance. I will say more about this strategy in the next chapter. In the meantime, we need to ask about manifestations, their character and relation to dispositions.

3. MANIFESTATIONS

The variety of powers ranges from the most basic physical properties (e.g., charge, mass, force) to the very sophisticated, such as human agency.[26] What is emphasized by dispositionalists is that the anti-Humean ontology of powers assumes that each power is essentially, or necessarily, related to its manifestations of a specific kind. However, powers and their manifestations are distinct existences, and even if not manifested, powers are real. This necessary relation of powers and their manifestations, even if they are not existent, can be understood either in terms of subjunctive conditionals of the form "if the power had been appropriately tested then the manifestation would occur" or in terms of natural directedness of powers toward their manifestations.[27]

Looking once more at the examples of dispositions listed above, we can deduce that the shattering of glass, the destabilization of protein crystals, the dissolution of sodium chloride in water, and the dissolution of petroleum jelly in gasoline are all examples of manifestations of powers instantiated in various substances. But what is the metaphysical nature of manifestations? They may be regarded merely as instantiations of new properties (pan-dispositionalism). But it seems more plausible to define them as concrete events that are effects of realizations of dispositions, or as contributions to these effects when they have a complex structure. Many thinkers believe a single power has only one manifestation-type, while Jennifer McKitrick and Heil claim it can have different effects (some of which are instantiations of nondispositional properties).[28] Toby Handfield claims manifestations are causal processes that form natural kinds. Each manifestation kind has at least two constituent properties (e.g., the property of being a rock and being glass are constituent for a natural kind of shattering of glass by a rock).[29] His suggestion brings us to the topic of causation.

4. CAUSATION

Powers ontology provides a metaphysical framework for a theory of causation which describes each event as an effect of powers manifesting themselves in a causal process. It treats causation as a metaphysically real type. In Molnar's view, causation is a manifestation of reciprocal powers.[30] Ellis clarifies that causal power is "a disposition to engage in a certain kind of process: a causal process."[31] His analysis is further developed by Mumford and Anjum, who state that "according to causal dispositionalism, causation involves an irreducible dispositional modality."[32] They question Hume's two-event model of causation and his insistence on temporal priority of causes before effects as the only possible way of defending the asymmetric direction of cause/effect dependencies. The fact that dispositionalism explains the ontological priority of causes by reference to their dispositions enables Mumford and Anjum to suggest that causation may involve simultaneity of causes and effects without any time gap between them and may include reciprocity, as in the case of two books leaning against each other or a kitchen magnet remaining attached to the front of the refrigerator.[33] Causation is about the unfolding of a process in which one thing affects/turns into another at the same time, whether instantly or over an extended time, often in a transparent and perceptible way. It comes in temporally extended wholes. Consequently, they claim that "an ontology in which particulars and events endure through processes is more suitable for dispositionalism. Processes are seen as dynamic in the sense that change is undergone throughout the process, which means it is to be found in any part of it, and it thus cannot be broken down into a string of changeless parts."[34]

This notion of causation has several advantages over concurrent theories. These advantages can be summarized in the following ten points:

1. DVC is not merely a metaphysical antidote to the Humean program of reducing causation to constant conjunctions of separate causes and effects in the mind of an observer. It also becomes a viable alternative to both neo-Humean views of causation (RVC and CVC) in justifying the generalization of causal claims. Because dispositions are decisive about intrinsic features of entities that possess them, we can make general claims about their causal activity and reactivity, even if they are not manifested. Our judgment does not depend on an unspecified number of observations of regularities or counterfactual conditional dependencies between events in nature.[35]

2. Moreover, DVC offers an intuitive solution to the difficulty of RVC and to the inability of CVC to distinguish between causes and causal conditions. With respect to the first, RVC, a clear distinction between kind-specific dispositions of action and reaction, on the one hand, and the circumstances of their manifestations, on the other, enables DVC to avoid the complexity and deficiency of the explanation offered in INUSConVC (which

is a version of RVC). Concerning the other neo-Humean theory of causation, it seems that CVC finds it difficult to distinguish between causes and conditions, as it moves us to think, for example, that the Big Bang is the cause of everything (had there been no Big Bang, nothing would have happened), or that our birth is the cause of our death (we cannot die if we are not born). The dispositional approach links causation to the essence of a nonliving thing or an organism, showing thus, respectively, that the Big Bang is just a condition corresponding to the causality which is an outcome of an intrinsic disposition of each being, and that our birth is a condition relevant to our disposition to die, but not a direct cause of our death.[36]

3. One great advantage of the dispositional view of causation is that it avoids both pure necessity and pure contingency. Because manifestations of powers occur only in certain circumstances in which they are stimulated, enabled, or released, and provided they encounter no impediments, the necessity of causal interactions is suppositional. To give an example, baby sea turtles are necessarily caused by their sea turtle parents, but only on the supposition that the temperature of the sand where they lay their eggs is within a certain range (73–91°F) and that the area harbors no turtle-egg predators, such as crabs, raccoons, birds, or coyotes. At the same time, baby sea turtles can be caused only by their sea turtle parents, and the connection between them is not purely contingent in the Humean sense, where every birth of a baby sea turtle is understood as an event which is self-contained and not related to other, similar occurrences in any other way than just in our mind. Followers of the powers view of causality will argue for the necessary character of the relation between the power of sea turtle parents to cause baby sea turtles and the manifestation of that power in the form of actual baby sea turtles.[37]

4. Another gain of the dispositional theory is that it respects both singularist and plural approaches to causality. That is, it is attentive to singular causal claims like "aspirin relieves headaches"[38] and to the cases of polygenic causation, where more than one cause is responsible for an effect (e.g., the winning of the final

game of the men's World Championship in Volleyball in 2018 by Poland was an effect caused by all players of the team).[39] Molnar explains that the idea of the polygeny of events was introduced by Dupré, who, in turn, takes it from genetics, which acknowledges that many genes typically contribute to the production of one trait.[40] Molnar notes not only that events are polygenic but also that powers, conversely, are pleiotropic and flexible and can contribute to many different effects. Consequently, he states that as the immediate consequence of recognizing polygeny and pleiotropy, we must distinguish between effects and manifestations, remembering that "manifestations are isomorphic with powers because each power gets its identity from its manifestation. Effects, that is, occurrences that have causes, are not isomorphic with the exercise of powers, considered distributively."[41]

5. Advantages of the dispositional approach mentioned in points 2 and 3 contribute to its further explanatory success. The awareness of the polygeny of causal situations and pleiotropy of causal powers not only favors the acknowledgment of the temporal simultaneity of causal relations (mentioned already above). It also enables DVC to avoid problems of early, late, and simultaneous preemption, as well as over- and underdetermination, which remain a serious challenge and an unresolved problem for the neo-Humean RVC and CVC.

6. The acknowledgment of the suppositional character of causal necessity in dispositional metaphysics enables it to deal with probabilistic cases in which causation is chancy yet probabilistically constrained. This aspect of DVC becomes a fitting alternative to ProbVC. Recognizing the importance of the probability aspect of causal occurrences, it does not try to ground causation as such on probabilistic dependencies, which—as we have seen in chapter 4, section 3—raises metaphysical objections.[42]

7. Paying attention to concrete individual cases of causal dependencies and yet being able to abstract from them to form general rules of causation, DVC avoids the evident limitations of SVC and the anthropocentricity of AVC (and other M-based VsC). At the same time, it avoids the other extreme of CVC and its dependency on the possible worlds modality, which assumes the

ontological reality of countless alternative worlds, the existence of which is not empirically verifiable.[43]

8. The ability of DVC to account for plural causal situations offers a solution to the puzzling and counterintuitive observation that sometimes causes are absences (e.g., my not watering the flower seems to be the cause of its death).[44] Because most of the cases of genuine causation involve manifestations of many dispositions, where an absence of action is invoked, we do not mean that it has a causal power, but rather that an absence of a particular power's manifestation changes the resultant causal effect which is brought by manifestation of other dispositions, responsible for the causal work. Hence, absences can be classified as enabling conditions, rather than causes *sensu stricto*.[45]

9. The same, plural view of causation enables us also to develop a more holistic understanding of causal processes, paying attention to context sensitivity and the complexity of single causal events, which usually require contribution of many powers to cross a threshold at which the effect is achieved despite the powers that dispose away from it.

10. Finally, DVC proves helpful in determining the character of the laws of nature. While theorists of causation agree that natural laws are descriptive rather than prescriptive, the question of what they really describe remains. Since, according to Humean metaphysics, laws of nature are merely generalizations of perceptions of constant conjunctions, they can be treated instrumentally rather than realistically, as useful tools for making predictions and developing technologies. Contrary to the Humean position, the dispositional approach shows that the laws of nature describe reciprocity and causal relations between real and distinct objects, relations that are realizations of their powers.[46]

5. MODELING CAUSES AS VECTORS IN QUALITY SPACE

Because causation is related to motion and change, some theorists within analytic metaphysics develop graphic models, trying to depict and explain causal dependencies, and specify directions of causal processes

(I have analyzed an example of such modeling in section 3 of chapter 4). Interestingly, the growing interest in DVC is supported by development of a new causal modeling theory proposed by Mumford and Anjum. I find it very helpful in understanding the originality of DVC, described in the previous section.

To begin, Mumford and Anjum rightly note, "Since the influential work of Lewis on causation (Lewis 1973), there has been a dominant convention of representing particular causal situations in the form of neuron diagrams."[47] Unfortunately, neuron diagrams (see figure 5.2) are conducive to a Humean ontology and thus impose important presuppositions which are foreign to other ontologies, including the ontology of powers and manifestations.

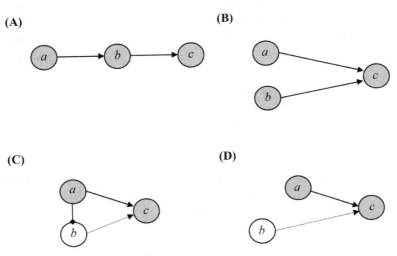

Figure 5.2. Neuron causal diagrams

(A) A classical diagram depicting the contingent relation between distinct events. *a* causes *b*, which in turns causes *c*. (B) An example of overdetermination. Both *a* and *b* cause *c*. Even if not *a*, *c* will occur because of *b*. Therefore, we would be wrong to say that if not *a*, then not *c*. (C) An example of early preemption. We could not say that if not *a*, *c* would not happen, because of *b*, which would still do the causal work. And yet *a* is a cause of *c* because it preempted *b*. (D) An example of late preemption. *a* causes *c* before *b* does. The problem is again that *a* is a cause of *c*, but had it not occurred, *c* would still have happened because of *b*. **B, C,** and **D** show the main objections to the neo-Humean counterfactual view of causation (CVC) (see chapter 4, section 2; Paul, "Counterfactual Theories," 171–82).

The major fallacies of neuron diagrams are these:

1. They assume that causation is a contingent relation between two distinct and discrete events, which is problematic in the context of the simultaneous and processual view of causes and effects in powers metaphysics, its holistic approach to causation, and its awareness of complexity and the context sensitivity of causal occurrences.
2. Causal dependencies are depicted as "all or nothing" cases. When the cause occurs and there is no inhibitor, the effect must follow. Stimulatory action guarantees its occurrence. This model cannot give account of situations where there is only a certain probability of an effect to occur or where an effect occurs only to some degree, depending on the strength of its cause.
3. Similarly, neuron diagrams are insufficient to illustrate cases in which effects are inhibited to some degree or in which the probability of an effect's occurrence is lowered but not reduced to nothing.

After specifying major shortcomings of neuron diagrams, Mumford and Anjum offer a new way of modeling causes as vectors in a quality space, the notion of which was first proposed in 1986 by Lawrence Lombard. In what follows (figures 5.3, 5.4, and 5.5) I present a series of diagrams which explain their approach to the modeling of causes.[48]

In their analysis Mumford and Anjum deal with three major problems concerning composition of powers.[49] First, it has been argued that introduction of the resultant power makes compositional powers impotent. Second, the same theory may lead to causal overdetermination. If compositional powers do causal work, it may look problematic to introduce a resultant power and ascribe to it causal activity as well. Finally, vector addition can be regarded as a merely mental operation. One may therefore question whether it truly reflects nature. Answering all three objections, Mumford and Anjum argue, "In a dispositionalist theory of causation, commitment to the component powers is central. It is they that drive causal transactions. If the dispositionalist were forced to choose between component and resultant powers, if there really were a danger of overdetermination, it would make more sense to be an anti-realist about

Figure 5.3. Vector modeling of causes

The figure above contains a single vector plotted on a space with a one-dimensional quality; the vector shows a disposition toward X. The straight vertical line in the middle represents the starting point. Each causal vector diagram shows a causal situation at a particular moment, specifying what, at that moment, is disposed to be caused. We must note, however, that a moment does not necessarily mean an instant or an unextended temporal slice. The metaphysics of powers suggests a view of the world as full of dynamic particulars, which are in a constant flux that often cannot be captured at an instant. The key point of one-dimensional causal vector diagrams is presentation of one causal situation related to powers that are causally relevant to one property dimension and one subject of such change, which is typically a particular thing or process. The advantage of using vectors is that they indicate both a direction (the way the arrowhead is pointing) and an intensity of causes (length of the vector). Although in most causal situations we encounter many powers disposing toward certain effects, stimulating/enforcing/inhibiting one another, there might be cases of "lonely" powers as depicted here. According to the Copenhagen interpretation of quantum mechanics, radioactive decay can be regarded as an example of such a lonely power manifesting itself with no stimulus.

the resultants. Perhaps they have just instrumental value, as summaries of all the distinct powers that are brought to bear on the situation."[50]

Mumford and Anjum seem to claim that multiple-powers causal vector diagrams enable us to pay attention not only to efficient causes but also to manifestations of other powers engaged in causal processes.[51] However, they do not specify the character of these powers. In my opinion, we cannot construct a vector model which could bring together various types of causal vectors—referring to causes in the Aristotelian sense—into one quality space, enabling mathematical operations on them and specifying a resultant causal vector which would show the direction of the causal process. But is such a project possible within a multidimensional quality space? Examples of such modeling are depicted in figure 5.5. Mumford and Anjum note that operations on causal vectors

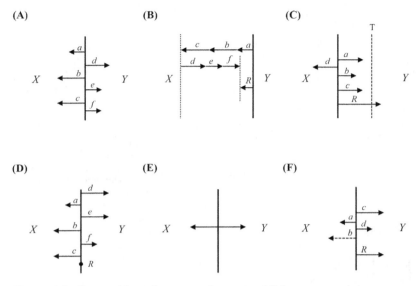

Figure 5.4. Composition of causes and vectors addition

Causal processes are usually complex resultants of many powers, some of which dispose toward a certain outcome (*X*), while others tend to prevent it (disposing toward *Y*). **(A)** Multiple powers engaged in a causal process. **(B)** Vector diagrams enable an addition of powers. *a*, *b*, and *c* dispose toward *X*, while *d*, *e*, and *f* dispose toward *Y*. *R* is a resultant vector depicting the final direction of the causal process. **(C)** Vector diagrams help us depict the notion of causal thresholds, which must be crossed to achieve certain causal effects. Causes *a*, *b*, and *c* together result in *R*, which crosses *T* and causes *Y*, despite *d*, which causes toward *X*. **(D)** A special case of vector addition. Dispositions toward *X* and toward *Y* balance out perfectly. Resultant vector *R* is directed neither toward *X* nor toward *Y*. Mumford and Anjum call it a zero resultant vector. We should note that all four diagrams (**A**, **B**, **C**, and **D**) make it possible to describe situations in which causal effects occur only to a certain degree, according to the strength of causes, as well as to describe cases of causal inhibition. Diagram **D** is of a special importance as it depicts states of causal equilibrium, in which the causes are present and active while nothing is happening outwardly. **(E)** Probabilistic causation is a special case which may appear to be vector addition, while in fact it is plotted as a single double-headed vector, disposing partly toward *X* and partly toward *Y*. **(F)** Vector diagrams account for cases of "causation by absence." The absence of *b* changes the resultant vector *R* from "zero" value to disposing toward *Y*.

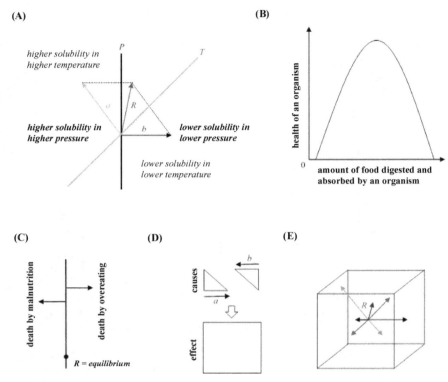

Figure 5.5. Multidimensional causal vector diagrams

(A) an example of the two-dimensional vector diagram illustrating solubility of chloride in water. It shows a dependence of the efficiency of the reaction on two factors: pressure and temperature. Two long axes (P and T) show the starting points for powers disposing, respectively, toward higher or lower solubility in reference to the pressure and temperature. a disposes toward higher solubility because of the high temperature, while b disposes toward lower solubility because of the low pressure. The resultant vector R shows the actual efficiency of the causal process of the solubility of chloride in water. **(B)** a simple two-dimensional causal vector modeling can be challenged by nonlinear composition of causes, as in any example of overdose cases (e.g., overeating). Mumford and Anjum argue that at least some cases of nonlinear causation can be described in a standard vector model, to which we need to apply the notion of an equilibrium **(C)**. However, they acknowledge that there might be cases of emergent powers where the resultant is composed by a nonlinear function which is not describable in a vector diagram. These are the situations depicted in **(D)**, in which compositional powers (symbolized by triangles) produce an effect (square) which is not equivalent to their simple mathematical sum. **(E)** a three-dimensional causal vector diagram. Vectors go from a central point in six directions. The resultant vector R shows the direction of the process.

can be performed in principle on any number of dimensions (quality spaces), though the mathematics of such cases might be complicated. One might suggest that a four-dimensional diagram would be suitable to depict Aristotle's four types of causation (material, formal, final, and efficient) and show their joint influence on a causal process. But the idea of bringing into one model all four classical dimensions of causal relations and the puzzling reality of chance and fortune, with their physical and metaphysical characteristics, is rather questionable, as some of these causal and quasi-causal factors (as in case of chance and fortune) are classified as strictly metaphysical principles (categories) and thus cannot be empirically observed, experimentally manipulated, and quantitatively represented. Nevertheless, this does not diminish the value and usefulness of the modeling theory presented by Mumford and Anjum. I believe the model they offer is very helpful in understanding the metaphysics of powers and shows the advantages of DVC over concurrent theories offered in analytic metaphysics. What is more, even if vector modeling of the Aristotelian four causes in a four-dimensional quality space may not be possible, it does not mean we can identify no common points between Aristotelian and dispositional views of causation. Quite to the contrary, certain similarities of the two theories become apparent. In the following chapter, I will analyze both convergences and divergences concerning these two notions of causality.

CHAPTER 6

From Powers to Forms and Teleology

New Aristotelianism

Anyone familiar with the philosophy of Aristotle cannot but see that the powers/dispositions and manifestations ontology resembles his theory of potency and act. Hence, Cartwright and Pemberton say that Aristotelian powers "are part of the basic ontology of nature—at least as nature is pictured through the lens of modern science."[1] The distinction between potentiality and actuality is one of the most fundamental distinctions in Aristotle's philosophy, influencing his metaphysics of substance and his theory of causation. It is one of the basic characteristics of being ("that which is"), both differentiated from and intrinsically related to the meaning of being, specified by substance and other categories (quantity, quality, etc.). The theory of potency and act strongly emphasizes the dynamic aspect of Aristotelian ontology, which often goes unnoticed by contemporary critics of substance metaphysics.[2] Hence, to evaluate the link between the classical Greek and the contemporary analytic notion of powers, we need to analyze first the topic of potency and act in Aristotle.

1. ARISTOTLE ON POTENCY AND ACT

The theory distinguishing between potency and act has its origin in Aristotle's position on the topic of change versus permanence, in which he avoids the two extreme positions of Heraclitus, holding that being is endless becoming, and Parmenides, denying the reality of change (see introduction, section 1). This juxtaposition of the two extremes is paralleled by a similar opposition between singularity and multiplicity. Heraclitus denied singularity and did not see any principle unifying various objects of our experience. There can, for example, be many individual round objects, but no one thing, roundness, that they all instantiate. Parmenides denied multiplicity, rejecting the idea that there can be more than one being. The genius of Aristotle's metaphysical realism consists in his avoiding both extremes and developing an intermediate position, showing that "Being-in-potency is ... a middle ground between being-in-act on the one hand, and sheer nothingness or non-being on the other. And change is not a matter of being arising from non-being, but rather of being-in-act arising from being-in-potency. It is the *actualization of a potential*—of something previously non-actual but still *real*."[3]

1.1. In Defense of Potency

Aristotle lists the Greek term δύναμις ("potency," "strength," "power," "ability," "capacity," or "faculty") among thirty basic metaphysical categories of his philosophical lexicon in the fifth book of the *Metaphysics*. He defines it as "the source, in general, of change or movement [ἀρχή μεταβολῆς ἢ κινήσεως, *archē metabolēs ē kinēseōs*] in another thing or in the same thing *qua* other, and also the source of a thing's being moved by another thing or by itself *qua* other."[4] This basic definition proposed by Aristotle introduces already an important distinction between active and passive potency. The former means a capacity to bring about an effect in another thing (e.g., fire burning a wooden log) or in the same thing as if it were the other ("*qua* other"—e.g., a doctor healing herself). The latter means a capacity to be affected by another thing (e.g., a wooden log's being burned by fire), or by the same thing as if it were the other ("*qua* other"—e.g., a doctor's being healed by her own action). It is important to note that Aristotle defines δύναμις in terms of a general category of "change" (μεταβολή, *metabolē*) and does not limit it to physical "motion" (κίνησις, *kinēsis*). In the opening paragraph of book 9 of the *Metaphysics* he states even more explicitly, "Potency and actuality [δύναμις καὶ ἡ ἐνέργεια, *dynamis kai hē energeia*] extend beyond the cases that involve a reference to motion."⁵ Moreover, he also distinguishes between nonrational and rational active powers, showing thus that potency is truly one of the most general characteristics, applicable to all material beings: "Since some such originative sources are present in soulless things, and others in things possessed of soul, and in soul, and in the rational part of the soul, clearly some potencies will be non-rational and some will be accompanied by a rational formula."⁶

Aristotle notes that nonrational and rational active potencies differ in three ways. Potency characteristic of a nonrational being is (1) capable of one effect, which (2) is the same in the patient as in the agent that actualizes it (e.g., fire burning a wooden log) and (3) is produced in accordance with suppositional necessity (within proper conditions, and in the absence of impediments, the effect will necessarily occur). Potency characteristic of a rational being, on the other hand, is (1) able to produce contrary effects (e.g., a doctor can heal or kill), which (2) may differ qualitatively and ontologically in the patient in comparison with the agent (e.g., an idea of a house in the mind of a house builder and an actual house) and (3) require the desire or will of an agent to be instantiated.⁷

In the third chapter of book 9 of the *Metaphysics* Aristotle defends the reality of potencies against the actualism of the Megarian school of philosophy, which was founded by Euclid of Megara and flourished in the fourth century BC. No matter how uncontroversial our intuitive recognition of the existence of potentialities and abilities is, enabling animate and inanimate things to change, to do different things, and to be engaged in processes other than those in which they are participating at present, Aristotle notes, "There are some who say, as the Megaric school does, that a thing 'can' act only when it is acting, and when it is not acting it 'cannot' act, e.g. that he who is not building cannot build, but only he who is building, when he is building; and so in all other cases."⁸

Although he finds absurd the Megarian claim, questioning the reality of abilities and potentialities of animate and inanimate entities even when they are not actualized or perceived, Aristotle considers it worthy of refutation by an argument. He presents three reasons why actualism should be rejected:

1. *The* Technē *Argument.* It defends the reality of active potencies typical of rational beings, such as house building, even when they

are not exercised. Since acquiring *technē* is extended in time, the likelihood is slight that a person who stops building loses it instantaneously or regains it immediately whenever she begins to build again. Our experience points toward stability and persistence of abilities understood as active potencies of rational beings. Similar examples can be given in defense of passive potencies of rational beings—for example, an ability to learn.
2. *The Perception Argument.* It defends the reality of passive powers typical of nonrational beings and is based on the example of perceptivity. The likelihood is very slight that things lose their properties, such as being cold or hot or sweet, when they are not perceived. Again, our experience seems to prove the stability of such properties and of the passive potencies of things that make them real and perceivable, even when they are not observed or tested. Similar argument can be given in defense of active potencies of nonrational beings—for example, the ability of fire to heat.
3. *The Immobility Argument.* Based on the logical consequences of the Megarian position, it shows that actualism eliminates all movement and change. For if something lacks potency (δύναμις), it is incapable (ἀδύνατος, *adynatos*) in that respect. Thus, we cannot say it "is" or "will be" with regard to the property in question. Hence, "that which stands will always stand, and that which sits will always sit, since if it is sitting it will not get up; for that which, as we are told, cannot get up will be incapable [ἀδύνατος] of getting up."[9]

Witt finds the third argument most convincing, as it is most general, not limited to active or passive potencies, and applicable to "nature" understood as the inner source of change of a unified substance.[10] Moreover, unlike the other two arguments, it accuses the Megarians not only of arguing in favor of an implausible theory of potency but of supporting a theory which has counterintuitive and absurd consequences, making impossible understanding of movement and change, which is strategic for the enterprise of sciences, ethics, politics, and crafts.[11]

Aristotle's refutation of Megarian actuality and defense of the real character of potentiality gives him a reason to deny the equivalence of possibility and potentiality. For, as notes Irwin, "Purely external changes can make something impossible for a subject without changing the subject itself, and some possibilities may be open for a subject without corresponding to any appropriate permanent state."[12] For example, an external change of the humidity of the air may make impossible the combustion of a match at its striking, without changing its intrinsic potentiality of combustion. The same match can become a part of a matchstick model of a house, which does not correspond directly to its substantial form, but can be listed among its accidental features. Hence, we can conclude that possibility is both unnecessary and insufficient for potentiality, which helps us realize that the latter has a deeper ground—that is, it is intrinsically related to actuality. Indeed, as we have already seen in my analysis of primary matter and substantial form (see section 2.1 of the introduction), the account of Aristotelian ontology shows that there is no potentiality without actuality, just as there is no actuality without potentiality.[13]

1.2. The Relation between Potency and Act

Speaking of the relation between potentiality and actuality in Aristotle, we must acknowledge, first, that the distinction between them is for him real, and not merely logical or formal. In other words, it is not simply intellectual (the view of Suárez), nor is it a distinction between "formalities" (the view of Scotus). To give an example of the latter position, intermediate between real and logical distinction, we may think about a human's rationality and animality. On the one hand, says Scotus, there is no real distinction between them (a rational animal is in reality one thing, not two). On the other hand, while the animality of a human is like the animality of nonhuman animals, the animality of a dog is different from human rationality, which distinguishes the two qualities in a human being. Hence, we can view human nature either in terms of its rationality or its animality. Scotus speaks here about the difference between the formality of animality and the formality of rationality. The problem of his position is that it eventually falls into either logical or real distinction.[14]

Leaving aside these medieval considerations, which Aristotle would probably find somewhat confusing, we must emphasize once again that for him the distinction between potency and act is definitely real, and thus ontological, although it does not entail separability. At the same time, thinking about the relation between these two aspects of beings, Aristotle repeatedly stresses the superiority of actuality over potentiality in several crucial respects:

1. Actuality is prior to potentiality in formula (definition), since we know a concrete potency (e.g., to build or to see) only in terms of the actuality toward which it is directed (the actual activity of building or seeing).
2. Actuality is prior to potentiality in time. First, "the actual which is identical in species though not in number with a potentially existing thing is prior to it."[15] Second, "from the potentially existing the actually existing is always produced by an actually existing thing—for example, man from man, musician by musician."[16]
3. Actuality is prior to potentiality in substantiality (ontologically). First, "the things that are posterior in becoming are prior in form and in substantiality (e.g., man is prior to boy and human being to seed; for the one already has its form, and the other has not)."[17] Second, "eternal things are prior in substance to perishable things, and no eternal thing exists potentially."[18]

One more aspect of the relation between potency and act is expressed in Aristotle's refusal to describe actualization as a mere alteration. Such description leaves out an essential feature of the process which requires a reference to potentiality that is being actualized.

1.3. From Potency and Act to Causes, Essences, and Teleology

The real distinction between potency and act has a strategic meaning for the whole metaphysics of Aristotle.[19] Since it makes change possible, it turns out to be crucial for the theory of causation. This becomes evident once we realize that to characterize any causal relation or occurrence, it is not enough to describe an actual event or process of change. We must explain why the change is possible at all, what decides about its reality, and why the efficacy of its agent cause is limited to bringing (producing) particular outcomes. Here the theory of potency and act tells us that change occurs always due to a realization of active potencies to bring a change and of passive potencies to receive it. The reality of a change is guaranteed by a real distinction and transition from potency to act, while the efficacy of the change's agent cause is limited primarily by the character of potentialities (active and passive) realized in the change's occurrence (apart from limits imposed by circumstances and possible impediments). The whole theory thus emphasizes the reality of causal powers, as distinct from their exercise and the vehicles by which they operate.

At the same time, however, the theory of potency and act refers us to the possessors of active and passive potencies. For Aristotle, unlike for Hume, causes are not events but concrete substances—that is, non-living as well as living beings involved in causal processes. A process (often characterized in terms of counterfactual conditionals) describes a change which is brought about through a realization of active (powers) and passive potencies characteristic of beings. This assertion makes us realize there needs to be something about the nature of animate and inanimate things that is the principle of their potentiality (a possibility to change), and something about their nature that is the principle of actuality (a possibility to bring a change). As we have seen in the introduction, Aristotle defines these principles as primary matter and substantial form. He states that they are characteristic features of both living and nonliving beings, constituting their essences (natures). Primary matter is the principle of passive potentiality, while substantial form determines actuality, realized and manifested in actual and perceivable characteristics of a nonliving being or an organism, as well as in its not yet realized (not yet manifested) active powers and dispositions.

At this point we must emphasize once again that Aristotle's realism about forms and essences is a moderate one. He finds a middle ground between two extremes. The first, essentially nominalist position assumes natures or essences have individuality *per se*, which denies the reality of universality in the world. The other, rooted in the Platonic theory of forms, assumes natures or essences have universality *per se*, which denies the reality of individuality and particularity in the world. Aristotle's realism is free from commitment to the metaphysical and epistemological baggage of Platonism. Even if forms are for him universal by an intellectual abstraction, the product of this mental activity is always rooted and instantiated in a mind-independent reality of concrete beings.

Defined along these lines, Aristotelian hylomorphism becomes a relevant and convincing response to the Humean "bundle" or "trope" theories of substance, which define its nature as a collection of properties (accidents) or in reference to causal processes that a given substance can enter. As I said before (chapter 2, section 2), the challenges of specifying the principle of unification of tropes, the possibility of accidental changes, the identity of indiscernibles, and the principle of individuation seem to make the bundle theory less appealing than Aristotle's ontology assuming the reality of substantial and accidental forms. For the latter provides a plausible and coherent explanation of the principles of unity and complexity (composition) as well as the principles of generalization and individuation and of stability and change. The Humean accusation that the classical theory of hylomorphism is based on an unknown and unverifiable concept of a featureless substratum of accidents—described by Locke as "something I know not what"[20]—is wrongheaded. Such accusation should be aimed, rather, at the contemporary ontological bare substratum theory in analytic metaphysics (again, revisit section 2 of chapter 2). For a moderate realist like Aristotle, accidents always presuppose the existence of a substantial form, which is in turn always realized in particular beings, characterized by specific accidents.

Finally, Aristotle's theory of act and potency is related to his idea of a natural directedness toward an end in natural objects. Such a tendency is possible provided there is a real and objective potency in beings that can be actualized. As I emphasized in the introduction (section 2.2), the classical notion of teleology is not limited to the realm of conscious and living beings, researched by human and biological sciences. It extends to inorganic systems as well, insofar as they are cyclical and show tendencies (potencies) toward certain end states (actualities).

Thus, we can see how the theory of potency and act influences and grounds all fundamental principles of Aristotle's metaphysics, including his notion of material and formal, as well as efficient and final causation. In this context, we may reasonably ask to what extent the contemporary analytic theory of powers/dispositions and their manifestations brings about a revival of Aristotelian philosophy of nature.

2. POWERS METAPHYSICS AS NEW ARISTOTELIANISM

My analysis of Aristotle's notion of potency and act demonstrates clearly why many followers of his philosophy find contemporary analytic metaphysics intriguing and inspiring. The rediscovery of powers as states distinct from their manifestations (events/processes) leads many contemporary metaphysicians, in turn, to acknowledge the neo-Aristotelian flavor of their analyses, placing them in stark contrast to the standard Humean alternatives concerning causation. Being aware of certain differences, questions, and skepticism coming from some theorists on both sides of the conversation, I will now point toward some strategic points of convergence and divergence between the two theories.

2.1. Potentiality | Actuality ‖ Disposition | Manifestation

We have seen that the differentiation between potency and act underlies all basic principles of Aristotle's metaphysics. Similar is the importance and role of the distinction between dispositions and manifestations in analytic metaphysics. First and foremost, the distinction allows for the reality of change and causation, making better sense of the analytic method applied in natural sciences than Humean approaches can do. The ontology of powers helps us realize that regularities, fundamental for a Humean view of causation, are in fact artifacts of what Cartwright and Pemberton call "nomological machines" and define as "sufficiently stable arrangements of components and capacities or powers that can, under suitable circumstances, give rise to causal regularities."[21] Powers can thus be regarded as equivalent to Aristotelian potencies, enabling causal relations to occur and calling into question the relevance of Humean regularity theory. They show that "the problem with Humean analyses is not so much that regularity and counterfactual dependence are not real aspects of causation, but that they are the *consequences of* causal relationships rather than being *constitutive of* causal relationships. What *is* constitutive is what can only be captured in the language of powers and their manifestations."[22]

However, even if the relation between potency and act is in some way relevant to that between dispositions and their manifestations, analytic ontology of powers seems to be insufficiently nuanced at some crucial points, in comparison with the principles of Aristotelian metaphysics. Here, we have to remember that Aristotle's theory of potency and act was rediscovered, adapted, and further developed by Scholastic philosophers, who introduced some other important metaphysical distinctions, derived logically from the first distinction of potentiality and actuality (see appendix 1). One of their suggestions seems to be very helpful for our better understanding and critical evaluation of the project of dispositional metaphysics. In their analysis, scholastics distinguished between logical and real potencies. Their proposal helps us differentiate unrealized *possibilia*, which are merely objects of thought (e.g., a satyr), from causal powers, which are real potencies (active and passive), grounded in concrete objects. Understood this way, powers—although beyond the category of mere *possibilia*—are still both actual and potential: actual, because they are real properties of concrete entities or dynamical systems; potential, because they can (although need not be) manifested. Aristotle himself introduces in *De an.* 2.1.412a22–23 a helpful distinction when he says that "the word actuality has two senses corresponding respectively to the possession of knowledge and the actual exercise of knowledge."[23] Applied to powers, Aristotle's distinction enables us to classify them as being actual in the first meaning, and not actual (potential) in the second meaning of the category of actuality. In other words, powers are as actual and real as my knowledge of, let's say, the Pythagorean theorem. At the same time, as long as/while they are not manifested, powers are like my knowledge of the Pythagorean theorem at times when I do not actively think about it or use it to solve concrete geometrical problems.

It appears that powers metaphysicians share the same understanding of the nature of dispositional properties. Nevertheless, the lack of reference to the categories introduced by Aristotle and his followers makes their account much less transparent and lucid. While emphasizing the actuality (reality) of powers ascriptions, they seem to lose their potentiality aspect, and vice versa:

> Pure powers . . . are thought of by Armstrong as mere potencies: potential rather than actual. . . . [But] the realist about dispositions or causal powers will accept such powers to be real enough. . . . They are certainly assumed as actual in their own right, whether or not they are manifested.[24]

What is not actual cannot be a cause or any part of a cause. Merely possible events are not actual, and that makes them causally impotent. This suffices to show that powers are not to be equated with mere possibilities. . . . The thought that powers are not actual properties is a mistake.[25]

Dispositions are actual though their manifestations may not be. It is a common but elementary confusion to think of unmanifesting dispositions as unactualised *possibilia*; though that may characterize unmanifested manifestations.[26]

Moreover, the Aristotelian distinction between active and passive potencies further nuances the theory of powers. It helps us realize that, although in the standard example of sodium chloride's (NaCl) power of solubility in water (H_2O) both substances are engaged in the causal process, the powers they manifest are different in character. In a solid NaCl crystal each Na^+ ion is surrounded by six Cl^- ions, while each Cl^- ion is surrounded by six Na^+ ions. When a solid crystal of NaCl is put into water, Na^+ and Cl^- ions are constantly moving away from the surface of the crystal due to the attraction of H_2O molecules and are being pulled back by attractions to oppositely charged ions in the crystal. Na^+ ions are attracted by negatively charged O^- ions of H_2O molecules and pulled back by Cl^- ions of the NaCl crystal structure, while Cl^- ions are attracted by positively charged H^+ ions of H_2O molecules and pulled back by Na^+ ions of the NaCl crystal structure. Water molecules collide with Na^+ and Cl^- ions, pushing them farther away from the solid NaCl, disrupting their attraction to oppositely charged ions on its surface, and stabilizing Na^+ ions by attractions to O^- ions and stabilizing Cl^- ions by attractions to H^+ ions of H_2O molecules surrounding them (see figure 6.1).

We can thus clearly see that water shows an active power (potency) to dissolve, manifested in H_2O molecules bombarding the NaCl crystal structure and pulling Na^+ and Cl^- ions away from its surface. An NaCl crystal, on the other hand, shows a passive power (potency) to be dissolved in H_2O, due to the positive and negative charges of its constituent Na^+ and Cl^- ions. Its power to dissolve is manifested in proper conditions (the favorable electrostatic interactions with H_2O molecules and entropy gradient). This important distinction of active and passive powers

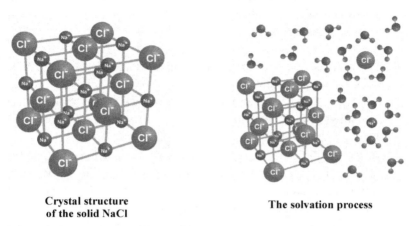

Figure 6.1. The process of NaCl dissolving in water
Source: (author) Dawid Socha

is somehow overlooked in analytic powers metaphysics. We will see that this neglect has some serious consequences for that entire project. For the distinction in question not only helps us realize that causal dependencies often engage manifestation of different powers on the side of agent and patient, but also asks a question about the source of active and passive potencies. It thus alludes to primary matter and substantial form, opening the way back to classical Aristotelian hylomorphism.

The question concerning the metaphysical status of dispositions and kinds of potency is closely related to the distinction between dispositional and categorical properties, which was the object of our analysis in the previous chapter (section 2). I have said the Aristotelian position supports the genuine (nonreductive) dualism of properties. For, although it defends the irreducibility and ontological character of dispositions, it does not question the reality of categorical properties, treating them as merely linguistically different, or conflating them with dispositional properties as two different modes of a generic type of property. Relating the dispositional/categorical to the potency/act distinction, we can also defend property dualism from the charge of substance dualism (of the

Cartesian type). Just as we cannot have potency without act or act without potency (in the material world), we cannot have dispositional properties without categorical, and vice versa. At the same time, however, both aspects of being refer to one and the same substance. Thus, the theory escapes the charge of those who support dispositional or categorical reductionism, eliminativism of either type of properties, neutral monism, or identity theory, claiming that any kind of dualist position about properties violates substantial (ontological) monism.[27]

We should also note that both categorical monism and categorical eliminativism (pan-categoricalism), reducing or questioning altogether the reality of dispositions (powers), resemble Megarian actualism, criticized by Aristotle in his defense of potency. This new version of actualism is one of the tacit assumptions of the Humean view of causation, still predominant in many scientific circles, but criticized by analytic metaphysicians arguing in favor of the reality of powers and their manifestations.

2.2. Dispositionalism and Teleology

My analysis of the most important analogies and convergences between the distinctions of potency and act on the one hand and disposition and manifestation on the other prepared the ground for a deeper investigation of Aristotelian aspects of analytic metaphysics of causation. We saw in the previous chapter how powers metaphysics helps avoid both pure necessity and pure contingency of causes. I have emphasized its great advantage in dealing with singularist and pluralist approaches to causation, especially in reference to the polygenic and pleiotropic character of causal powers. I have discussed the challenge it poses to the Humean event model of causation and the temporal priority of causes, arguing for the simultaneity of causes and effects in time and the possible reciprocity between them. I have shown the explanatory value of the modeling theory depicting powers as vectors in quality space, which enables us to capture and analyze more precisely the complexity of actual causal situations. Although all these advantages of the powers view of causation are implicitly related to Aristotle's distinction between potency and act, showing thus the relevance of the Aristotelian heritage in contemporary analytic metaphysics, the time has come to consider the revival of the two core aspects of Aristotelian metaphysics of powers—namely, teleology and hylomorphism (essentialism). I will begin with the former.

Teleology comes back in contemporary analytic metaphysics with the recognition of dispositions "pointing" or being "directed" toward their characteristic manifestations. Hence Molnar speaks about "physical intentionality," and Heil about "natural intentionality," while Place describes dispositions as "intentional states" pointing beyond themselves:

> Mental states and the physical powers that determine the behaviour of their bearers share a number of traits that suggest intentionality. It is arguable that their similarity is sufficiently strong, in respect of the central criteria of intentionality, to create a case for *physical intentionality* (PI) as a concept that is in fundamental respects analogous to *mental intentionality* (MI). . . . I accept the intentionality of the mental, and go on to argue that something very much like intentionality is a pervasive and ineliminable feature of the physical world.[28]

> Dispositions are of or for particular kinds of manifestation with particular kinds of disposition partner. Dispositions preserve the mark of intentionality in being of or for particular kinds of manifestation with particular kinds of non-existent—possible, but non-actual—objects. This is not mysterious or spooky; it is a feature of dispositions possessed by rocks, or blades of grass, or quarks. My suggestion is that we make use of the "natural intentionality" afforded by dispositions in making sense of the kinds of intentionality we find in the minds of intelligent agents.[29]

> The intentionality of dispositional properties whereby, in Armstrong's words, they "point beyond themselves to what does not exist" is, in Hume's phrase, the very "cement of the universe" without which there would be no causation, no change, no time, no space, no universe.[30]

All three philosophers quoted here refer to the four criteria of intentionality of the mental formulated by Franz Brentano:[31]

1. An intentional state or property is directed to something beyond itself, toward the "intentional object." For example, the thought "the cat is on the mat" is directed toward a concrete state of affairs.
2. The intentional object can be existent or nonexistent. For example, one can have the thought "the cat is on the mat" even if no cat is sitting on the mat.
3. The intentional object can be indeterminate (fuzziness of an object). For example, one can think of a cat sitting on the mat without thinking of its color, size and weight.
4. Non-truth-functionality and referential opacity of intentional states is possible (not all intentional objects are truth bearers). For example, if the name of the cat is "Felix," having the thought "the cat is on the mat" does not follow that one has in mind Felix sitting on the mat.

All three thinkers argue that intentionality is an ineliminable feature of not only the mental but also the physical world. Molnar finds parallels between Brentano's criteria of mental intentionality and his idea of physical intentionality, the latter of which can be summarized in the following four points:

1. Powers are directed toward their manifestations, which are something "outside themselves."
2. The manifestation toward which the power is directed need never in fact exist, which does not question the reality of the power.
3. A power can have an indeterminate object (fuzziness of an object), as in the case of indeterminate timing of a given radium atom's decay.
4. Power ascriptions can exhibit referential opacity. For example, the power of the sun to accelerate the ripening process of an apple and make it turn red does not entail that the sun has the power to turn this particular apple the color of Post Office pillar boxes (to use Molnar's example of redness).[32]

However, points 3 and 4 were criticized by Bird, who argues (contra 3) that the timing indeterminacy of a radioactive decay is not quite the same as a vagueness of thought, because "the half-life of a radioactive nucleus is perfectly precise, as is the probability of its decaying within a given time interval."[33] When he considers referential opacity (point 4), Bird notes that Molnar's argument fails because ripe apples being red does not mean ripe apples are the color of Post Office pillar boxes, unless we take their color as a rigid designator of redness. But if we do so, powers' referential opacity disappears, as from "ripe apples are red" it will follow that "the sun has a power to make this particular apple the color of Post Office pillar boxes."[34]

Nevertheless, the first two aspects of "physical intentionality" are plausible and offer a strong defense of the directedness of powers. What is at the core of analytic metaphysicians' argumentation is, therefore, the fact of a stripped-down notion of finality, characteristic of inanimate objects as well as nonsentient, sentient, and conscious forms of living organisms. Oderberg emphasizes that this notion brings back the idea of natural (inorganic and organic) teleology, which "fell by the wayside under the anti-Aristotelian assaults of empiricism and materialism and has not yet recovered."[35] I side with Feser, who claims that the Aristotelian terminology of teleology and final causation is less vulnerable to the objection of panpsychism, which can be raised against the notion of physical intentionality. Molnar responds to this charge, claiming that consciousness, rather than intentionality, is the characteristic of the mental.[36] Despite these terminological difficulties, however, we can certainly conclude that dispositional metaphysics does bring an important revival of the classical notion of teleology and that it is neo-Aristotelian in this respect.[37]

2.3. Dispositionalism and Hylomorphism (Essentialism)

Dispositionalism's straightforward emphasis on teleology undoubtedly speaks in favor of its antireductionist character and its remaining in stark contrast with contemporary reductive and eliminative materialism. Although the reductionist version of modern atomism is willing to accept the reality of complex structures, it still strives to describe them in terms of the mechanical combination and recombination of the basic physical constituents. Using the typical phrase "nothing more than . . . ," it leads to the reduction of mental to biological, to chemical, and finally to physical properties. The eliminativist position simply rejects the reality of higher phenomena, stating boldly, "There are just fermions and bosons and combinations of them."[38]

While the retrieval of teleology in contemporary analytic metaphysics of powers and dispositions seriously challenges the materialist position, the question remains as to dispositionalism's understanding of the very nature of nonliving and living beings, as well as the dynamical systems they enter. More precisely, what is at stake is the question of whether powers metaphysics is in fact a hylomorphic position. The answer to this question is rather complex and requires, first, an introduction to the most recent versions of hylomorphism, as the Aristotelian reference to the composition of matter and form finds a number of commentators (both advocates and critics) among contemporary ontologists trained in the analytic tradition. As we will see, a closer analysis of these positions shows both some similarities and differences between them and Aristotle's original theory presented in the introduction (sections 2.1, 2.3, and 3). One might argue that approaching the same questions of unity, stability, and change from a different perspective, the proponents of the analytic versions of hylomorphism never pretended to be, or aimed at, bringing about a restoration the original Aristotelian ontology of material beings. Nevertheless, my inquiry shall prove that the reinterpretations of hylomorphism they offer may put them at risk of falling into contemporary nonreductionist or reductionist materialism.[39]

In order to better understand analytic versions of hylomorphism I will refer once again to an important classification of different levels of substantial unity proposed by Thomas Aquinas in his *Commentary on Aristotle's Metaphysics*, and mentioned briefly in section 3 of the introduction.[40] He lists several types of formal causation, saying:

1. In Aristotle's examples of the goblet made of silver or statue made of bronze, the term "form" is used in a general and somehow unqualified sense.[41]
2. But sometimes many things are brought together to constitute one thing. This can happen in three ways, which are associated with three respective kinds of formal causation:
    a. things can be united merely by their arrangement (*secundum ordinem*), as men in the army or houses in a city (USO = unity *secundum ordinem*);

b. things can be united by contact and bond (*secundum contactum et colligationem*), which is evident in the example of a house and its parts (USCC = unity *secundum contactum et colligationem*);

c. and sometimes an alteration of the component parts (*alteratio componentium*) is added to (b) above, which occurs in the case of a compound (*mixtione*) and includes a substantial change of components on the way of becoming one substance (UAC = unity by *alteratio componentium*).

This classification enables us to distinguish various levels of formal causality proper for different levels of complexity of material entities. At the same time, it helps us realize that, although form can be regarded as a unifying principle of material entities (USO and USCC), it is only in the case of UAC that the relevant parts of a unity (which are now virtual—see below, section 2.3.5) become literally one substance and are able to enter into a complex network of dynamic relations, decisive for and expressing its character and nature. Acknowledgment of this fact will prove useful in my assessment of different versions of hylomorphism in analytic tradition.

2.3.1. Mereological and Structural Versions of Hylomorphism

One of the new versions of Aristotle's basic ontology was offered by Kathrin Koslicki and is grounded on her reinterpretation of *Meta.* 7.17.1041b11–33. In this passage—closely related to my analysis of form in section 2.1 of the introduction—Aristotle states that the oneness of a whole composed of parts (and thus differentiated from a mere heap) requires an explanation in terms of "something else" that is defined neither as another part—that is, an element added to all other elements— nor as a mere composition of elements, since both options would lead to an infinite regress. He suggests that one interprets this "something else" in terms of substance (οὐσία, *ousia*—probably in the more relative sense of a "substance of something"), a primary cause (reason) of being (αἴτιον πρῶτον τοῦ εἶναι, *aition prōton tou einai*), a kind of nature (φύσις) "which is not an element but a principle" (ἀρχή). Although Koslicki agrees that such a principle must belong to a separate ontological category (distinct from the material parts of a whole) and can be best classified as form, she argues that form can be still treated as a part of what she calls "the matter/form compound." Knowing that Aristotle himself does not speak of matter and form as parts of a "compound," she, nevertheless, states, "A convincing case can . . . be made for such a mereological construal of Aristotle's hylomorphism."[42] Further development of this proposal leads Koslicki to suggest that the formal part of the whole plays a special unifying role. For it is precisely in relation to the formal part that other parts turn out to be mereologically complex. She says that this unique part must be indivisible and possess its unity in a primitive and underived manner, and she coins for it the name "mereological atom." On another occasion, Koslicki adds a structural aspect to her version of hylomorphism. She attributes the unity of a substance to its material constituents' "satisfying structural constraints"—that is, occupying their proper positions in a given structure, which she identifies with substantial form.[43]

A closer look at Koslicki's version of hylomorphism shows that it is heterodox in relation to Aristotle's original thought. For if he calls form a "part" of a compound, he seems to treat it as a "part" in a metaphysical sense, remembering that its proper correlate is prime matter, which can also be called a metaphysical "part," in distinction from physical mereological parts of a whole. In other words, while Koslicki might be on the right track in acknowledging the reality of Aquinas's categories of USO and USCC, she seems to be missing the deepest level of substantial unity, defined as UAC.

A slightly different, semimereological, version of hylomorphism was proposed by Mark Johnston, who defined it as "the idea that each complex item admits of a real definition, or statement of its essence, in terms of its matter, understood as parts or components, and its form, understood as a principle of unity."[44] Although he seems to understand matter mereologically in terms of physical parts, Johnston differentiates his view from the one presented by Koslicki when he states that the form playing a unifying role "cannot be a *mere* part alongside the other parts" of the whole.[45] He adds that form as a principle of unity can be either static—that is, the holding of parts as they are—or dynamic—that is, allowing or requiring that the parts it holds vary over time. Giving as an example of the latter an organism, he defines its form in terms of a "complex structure of biochemical reactions."[46] This enables us to classify Johnston's version of hylomorphism as both semimereological (with respect to his definition of matter) and semistructural (with respect to his definition of form), which obviously situates it beyond the classical hylomorphic orthodoxy. Similar to Koslicki's position, it seems to acknowledge the reality of USO and USCC, but not of UAC.

Another version of mereological and structural hylomorphism was suggested by Kit Fine. Critical of explanations of complex objects in terms of a simple mereological sum, aggregate, or compound, Fine offers his own version of part-whole theory, which he calls "the theory of embodiment" and classifies as falling "within the hylomorphic tradition of Aristotle."[47] Although he supports "the idea that there is both a formal and material aspect to most material things," Fine does not understand them in strictly Aristotelian terms.[48] He seems to treat matter as physical stuff and explains form in terms of the two kinds of embodiment. The first, rigid embodiment, refers to concrete types of parts required for a given whole, which accounts for its mereological structure at a time (e.g., an engine, a chassis, etc., as indispensable parts of there being a car at a particular point in time). The second, variable embodiment, refers to parts of a given whole that can be replaced, and this embodiment accounts for its variation over time. Thus, the car can be described as a variable embodiment /E/ whose principle E picks out various rigid embodiments. Fine concludes by saying: "I believe that it was the failure of Aristotle to distinguish between these two different roles for form that principally accounts for the inadequacy of his hylomorphic conception of material things, which, in other broad respects, is so close to our own."[49]

I find his thesis misguided, for at least two reasons. First, we have learned about the Aristotelian-Thomistic theory of virtual presence (see introduction, section 3), which provides the necessary ground for an explanation of the exchange of parts in a substantial whole. I will say more about it below in section 2.3.5 of the present chapter. Second, Fine's mereological and structural version of hylomorphism, following Aquinas's concepts of USO and USCC, does not take the crucial step of defining unity of compounds as UAC, which includes substantial change of parts, where the continuity of the process of transition from many physical parts (substances) to one new substance is grounded in the metaphysical principle of potentiality—that is, primary matter. This fact makes the statement about the proximity of his view to the one offered by

Aristotle questionable. Moreover, apart from these arguments, it seems that according to Fine's theory any relation holding among any plurality of entities qualifies as a distinct rigid embodiment, which proliferates his universe with a vast number of "ontological monsters," many of which share the same material components at at least one point in time.

One more version of neo-hylomorphism defined along the lines of mereology and the structuralist approach to reality comes from William Jaworski, who states explicitly that "hylomorphism is about structure" and adds that "structure (or organization, form, arrangement, order, or configuration) is a basic ontological and explanatory principle."[50] Based on the reference to dispositionalism and its observation "that the activities of structured individuals involve coordinated manifestations of the powers of their parts," Jaworski defines substantial forms as "individual-making structures."[51] He classifies his ontological position as nonreductive naturalism, which he understands as the conjunction of a broadly Quinean thesis that we are committed to all the entities postulated by our best descriptions and explanations of reality, with a broad empiricism maintaining that our best descriptions and explanations of reality derive from empirical analysis of natural science. He distinguishes his view from NP and emergentism. Nevertheless, his insistence on structure being "un-controversially part of the physical world" makes his position, in my opinion, vulnerable to materialist reductionism.[52]

Similar to the thought of Koslicki, Johnston, and Fine, Jaworski's structural version of hylomorphism is based on a mereological understanding of matter in terms of physical parts (embracing USO and USCC, but not UAC). Unlike other neo-hylomorphists discussed so far, however, Jaworski is aware that his view of substantial form departs considerably from the classical one, which makes him acknowledge openly and willingly that he cannot vouch for its similarity to the ontology of Aristotle and Aquinas. But does his own structural version of hylomorphism have the same explanatory power as that of Aristotle? When pressed by the questions of what gives an origin to structures and what the source of the novel powers of a higher-level dynamic system is—the ultimate queries within the limits of his own rhetoric that seem to be successfully addressed in the Aristotelian-Thomistic tradition—Jaworski states, "On [my version of] the hylomorphic view, structure itself, like material, is a basic principle, one that stands in need of no further explanation. . . .

Asking why either structure or materials exist . . . comes close to asking why the universe as a whole exists, or why there is something instead of nothing."[53] I find this statement hardly satisfactory from the standpoint of both philosophy and natural science, as it clearly does not answer their primary concern with the reality and explanation of the phenomena of stability and change in nature.

The last mereological version of hylomorphism that needs to be mentioned here was proposed by Robert Koons. Critical of the views of Fine, Johnston, and Koslicki, Koons strives to redefine hylomorphism in terms of the relation between emergent wholes and their material parts. He is looking for a theory that will overcome the dilemma of assuming both that the whole is more than its parts (against nonreductionist materialism) and that it is still composed of smaller material elements (against substance dualism).[54] Trying to find a proper candidate, Koons discusses the following six theories:

1. *Aristotelian Parts-Nihilism*: Emergent substances have no actual parts at all. He rejects this view on the supposition that Aristotle's theory assumes the reality of material (mereological) substrates that endure through any substantial change.
2. *Reverse Mereological Essentialism*: Existence of parts depends on their being parts of the whole of a given kind. This view can be rejected for the same reason as option 1.
3. *Downward Sustenance*: Persistence and operation of the whole cause the persistence of parts. The view falls into substance dualism.
4. *Upward Sustenance*: Persistence and operation of the parts cause the persistence of the whole, the substantial form of which is seen as a process. Taken by itself, this view falls into nonreductionist materialism.
5. *Upward Power Migration*: Some or all causal powers migrate from parts to the whole. The theory assumes each elementary particle has certain primary material powers (grounding the primary causal powers of a whole) and two sets of secondary powers, one corresponding to its existence as a separate substance, and the other corresponding to its status of being a part of a whole (and grounded by the whole).

6. *Teleological Subordination*: Causal powers of the parts are teleologically ordered to some end pertaining to the whole. This view can be charged with promoting holistic epiphenomenalism (i.e., the position wherein there are no real ontologically independent causal powers of the whole).

After analyzing these six propositions, Koons suggests a solution which combines options 4, 5, and 6 into a single account which he calls the sustaining instruments theory: "On this account, the persistence of the whole is grounded in the ongoing cooperation of the parts, and the active and passive powers of the parts are grounded in corresponding primary powers of the whole. In addition, the whole acts through the parts, as teleologically subordinate instruments."[55]

Similar to other neo-hylomorphists discussed in this section, Koons fails to recognize the Aristotelian notion of primary matter, with the result that he defines the material aspect of hylomorphism in terms of mereologically understood elementary particles. Consequently, in light of classical Aristotelianism, his argument against options 1 and 2 does not stand. For, as we have seen, what persists through each and every substantial change on Aristotle's account is primary matter and not mereological physical substrates. As we will see in section 2.3.5 below, according to Aquinas such physical substrates, defined as proximate or secondary matter, can indeed enter and be released from compounds, but both changes in question are substantial changes in which the substantial form of a given substrate is first corrupted and then retrieved.

2.3.2. Incomplete Entities Version of Hylomorphism

Another contemporary commentator on hylomorphism, Edward Jonathan Lowe, acknowledges that Aristotle's "combination" of matter and form rules out thinking of them as "*parts* of the substance, at least in the normal sense of 'part,'" and that "a true individual substance never has other individual substances as parts," which differentiates it from aggregates, heaps, and bundles.[56] At the same time, however, he attributes to the hylomorphic position the explanation of the matter and form of an individual substance as "'incomplete' entities, completed by each other in their union in that substance," which can be defined as a "combination of items neither of which can exist independently of the other in just such combination."[57]

That Lowe's idea of the mutual saturation of incomplete entities in hylomorphic composition acknowledges the reality of USO and USCC, but not of UAC, becomes clear in an analysis of the examples he gives. Although he mentions the category of "prime matter," he sees the process of the coming to be of an individual house as engaging immediate matter—that is, "some bricks, mortar, and timber, . . . organized in a certain distinctive way"—due to the form of the house. Similarly, "an individual horse has as its immediate matter some flesh, blood, and bones, which are organized in a certain distinctive way fit to sustain a certain kind of life, that of a herbivorous quadruped."[58] These two examples show, first, that Lowe's analysis fails to capture the classical Aristotelian-Thomistic distinction between accidental form—that is, a union of secondary (proximate or immediate) material parts by contact and bond, as in the case of a house—and substantial form—that is, a union effecting in an alteration of the component parts, which includes actualization of primary matter by a new substantial form, as in the example of a horse (see introduction, section 3). Second, defining the relation between matter and form in terms of "organization" does not follow the original version of hylomorphism, which analyzes it in causal terms and in reference to the distinction between potentiality and actuality.

In his final example of the coming into being of a hydrogen atom, Lowe seems to deal a final blow to hylomorphic ontology. Although he claims to be "perfectly happy to describe the case of the newly created hydrogen atom in terms of 'combination' and 'constitution,' and indeed in terms of 'form,'" he does not find such description meaningful, since "the only things that do any 'combining' are *the proton and the electron*, when the former captures the latter and the latter occupies an orbital around the former," and "the only things that *constitute* the atom are, again, the proton and the electron, which are its *parts*, in the perfectly familiar sense of 'part.'" To this he adds that "in the newly created hydrogen atom, the proton remains exactly what it was before, just *a proton*, and the electron remains just *an electron*." Finally, the fact that fundamental particle physicists "don't nowadays speak of protons and electrons as having, or being composed of, *matter*—although they might happily speak of them as being 'packets of energy' and certainly as possessing *mass*"—makes Lowe state that he has "no serious need for the hylomorphist's category of *matter*," defined as an "'incomplete' constituent of the atom."[59]

I find this argument problematic in at least three ways. First, hylomorphic actualization of primary matter by substantial form cannot be rightly described in terms of "combination" or "constitution." Second, a thesis that the constitution (coming into being) of a hydrogen atom is simply an effect of the combination of one proton and one electron, both of which nevertheless remain unchanged, proves to be wrong on account of the fact that an electron in a new hydrogen atom is now able to form, for example, a covalent bond—a feature that it did not possess before becoming "a part" of the new substance. We may say that the electron does this "on behalf" of the hydrogen atom as virtually present in it, or—more accurately—that the hydrogen atom possesses the power of forming a covalent bond "through" an electron virtually present in it.[60] Finally, Lowe's dismissal of the hylomorphic category of matter—based on the discharging of the concept of physical matter in the domain of quantum physics—may make sense in terms of his own metaphysical definition of matter as an incomplete constituent of the atom. Yet, as we already know, such a mereologically oriented definition has little to do with the orthodox hylomorphic understanding of primary matter as the principle of potentiality, which protects it from an easy refutation on scientific terms.

2.3.3. Mixed Version of Hylomorphism

Gordon Barnes offered an alternative proposition of hylomorphic ontology, which seems to bring together the mereological and the incomplete entities views, discussed in the previous sections. In his article "The Paradoxes of Hylomorphism" he offers, first, an answer to some basic lines of criticism of the hylomorphic position. Addressing the charge of substantial forms being both abstract and concrete, he defines abstractness as being outside space and time and concreteness as having determinate location in time and space and says that, in the light of these definitions, the charge against hylomorphism does not stand. For hylomorphic ontology assumes that substantial forms are determinately located in space and time, and thus sees them as immanent universals.

Moving next to the question of forms being both universal and particular, Barnes notes this paradox goes all the way back to Aristotle.

On the one hand, we find him saying we can provide a definition of a substantial form, while emphasizing—on numerous occasions—that only universals are definable: "the formula is of the universal," "definition is of the universal and of the form."[61] This clearly supports the view of those who see forms as universals. On the other hand, however, Aristotle acknowledges that substantial form must be predicated about "this something" (τόδε τι, *tode ti*) and that "thisness [in plural τόδε ταῖς, *tode tais*] belongs only to [concrete] substances"—that is, particulars.[62] Does it mean that form can have two, distinct, modes of existence? asks Barnes. He answers: "The hylomorphist should say that abstraction is the power to apprehend a real universal, which is exemplified by a substance in virtue of the fact that it has its own, particular substantial form. Thus, the substantial form itself is a particular, not a universal."[63]

But this answer, says Barnes, only raises another objection. If substantial form is a particular, what sort of particular is it? Is it an entity, a property instance, an event, or a process? One might argue that substantial form as "actualization of potentiality" is a basic principle and that "to ask for some account of it in terms of other concepts is to misunderstand the order of conceptual priority in the Aristotelian system."[64] However, Barnes does not find this strategy helpful and says that classification of substantial form as "principle" raises the question: What sort of being is a principle? He rejects an explanation that a principle is not a being (for him it would mean it is nothing) and suggests that hylomorphists should accept Francis Suárez's view that substantial form is an "incomplete substance," in need of a complementary entity, matter, in order to compose a complete entity.[65] That this classifies Barnes's position as a type of the incomplete entities version of hylomorphism becomes obvious.

Arguing in defense of both the real distinction between the form and the matter of an individual substance and the real distinction between the form and the substance itself, Barnes refers to our commonsense and prephilosophical intuition that even the most radical changes in nature involve some sort of continuity. He finds an explanation for this fact in the reality of matter, which underlies the changes and exists in two kinds: substance-independent matter (subatomic particles, whose identity does not depend on their composing a complex substance—e.g., a human being), and substance-dependent matter (e.g., a human heart, lungs, kidneys, etc., which are essentially specified by the substantial form of a human body). Barnes states that because only the former kind of matter persists through substantial change, we can argue in favor of its real distinction from the form of an individual substance. This theory of matter, devoid of any references to the concept of primary matter that persists through any substantial change, subscribes to modern and contemporary atomism. Thus, it can be classified, together with definitions given by Koslicki, Johnston, and Lowe, as mereological—that is, accepting USO and USCC, but not UAC.[66]

Moreover, Barnes's definition of form as "incomplete substance" and his lack of reference to the concept of primary matter are not the only ways in which he departs from the classical Aristotelian version of hylomorphism. Continuing his reflection, Barnes wrongly suggests that the view in question assumes that "potentiality and actuality are complementary *entities*, naturally suited to *compose* [*naturally combine into*] a single, individual substance."[67] He also rejects the Aristotelian-Thomistic doctrine of "virtual presence" (presence by power) of elements in wholes actualized by new substantial forms. Referring once again to his example of human beings, he concludes, saying:

> The matter that composes a human being is not a mere potentiality.... [It] must have a plurality of substantial forms, of which the form of a human being is only one.... So I am suggesting that one and the same "parcel" of basic matter can be in-formed by two distinct substantial forms and thereby compose two distinct substances.... If there is an intrinsic property that is instantiated in the space occupied by some basic particles of matter, and if the instantiation of this property is not a logical consequence of the properties and relations of these basic particles of matter, then the basic particles of matter coincide with something that is really an intrinsic unity and not merely the aggregate of the basic particles that coincide with it.[68]

Anyone familiar with the classical version of hylomorphism realizes how little of it is left in the theory offered by Barnes. His suggestion that the substantial form of the whole coincides with the substantial forms of its parts because more than one form can inform the same mereological-type "parcel" of basic matter has little to do with Aristotle's concept of substantial form thought of as a causal principle of actuality in-forming primary matter, which is the principle of pure potentiality.[69]

2.3.4. Re-identification of the Parts Version of Hylomorphism

Another version of hylomorphism, remaining closer to Aristotle's original idea, has been defended by Anna Marmodoro. In her article "Aristotle's Hylomorphism without Reconditioning," in reference to Scaltsas, she states that "being unified into a whole re-identifies the parts in a way they cannot be when apart from the whole. The parts are re-identified according to the unifying principle of the whole, the substantial form. Once re-identified, they have no distinctness in the substance; they exist in it holistically."[70]

Alluding to *Meta.* 7.16.1041b5–15, Marmodoro adds that "partitioning a substance generates parts that are *not* parts of the substance," and defines substantial form as a principle or "*an operation* on the elements of a substance, stripping them of their distinctness, rather than being an item in the ontology." She thinks "the unification of elements is achieved through their *re-identification* in terms of the role allotted to them by the substantial form," and states that "re-identification, rather than combination or relationality, is Aristotle's solution to hylomorphic unity." Finally, she notes that "substantial forms demarcate the joints of nature; entities are unitary if they fall between these joints, that is, if they are *one* by the principle of a substantial form. This does not presuppose the *simplicity* of a form, but only its *fundamentality in nature*, for 'measuring' being."[71]

Defining substantial form as a principle or an operation recalls Aristotle's idea of form as causal principle of actuality. Moreover, an emphasis on the fact that partitioning a particular substance generates parts that are not parts of the substance in question proves that Marmodoro has a better grasp of the classical concept of form than other contemporary hylomorphists and might be close to accepting not only USO and USCC, but also UAC. At the same time, however, she is not entirely clear in her understanding of the material component of the hylomorphic union. Although she ascribes to Aristotle the idea of "matter *as being potentially* the form,"[72] she does not define it as primary substratum. Nor does she mention that the "re-identification" of material parts (components) in a substantial change happens due to the potentiality of the primary matter which underlies them and becomes properly disposed to receive a new form.

Consequently, Marmodoro's description of hylomorphism in terms of re-identification of parts due to substantial form may be charged with being semimereological, as she follows the example of other contemporary philosophers who speak about hylomorphism as unifying physical (mereologically describable) "elements of a substance."

2.3.5. Virtual Presence and Neo-Hylomorphism

It seems that the main limitation and shortcoming shared by all versions of neo-hylomorphism analyzed so far lies in their lack of reference to Aristotle's idea of primary matter understood as the ultimate principle of potentiality and substantial form defined as the principle of actuality. Although not obvious, difficult to comprehend, and exegetically challenging, this basic ontological distinction grounds Aristotle's original and innovative metaphysical system and should not be ignored by anyone who wishes to follow its main objectives. And yet we could see that this chief rule of hylomorphism, crucial for acknowledging the reality of UAC, seems to be absent in various propositions of its contemporary reinterpretations.

However, this accusation may be challenged on the basis of a reference to the theory of the virtual presence of parts (elements) in compounds, which I briefly introduced in section 3 of the introduction. One could argue that at least some of the mereologically and structurally oriented accounts of hylomorphism discussed above are defendable once we realize that what their proponents have in mind when they speak of matter is proximate (secondary) matter—that is, elementary physical substrates virtually present in the wholes of composite (mixed) entities (bodies), in-formed by a higher form of a compound. Such an assumption might suggest that at least some neo-hylomorphists accept Aristotle's idea of primary matter and substantial form in principle, even if they do not mention it directly.

To answer to this challenge, we have to ask, first, about the proper interpretation of the Aristotelian-Thomistic concept of virtual presence in light of the most recent developments in natural science. I already mentioned in chapter 3, section 3.5, footnote 95, that Aristotle listed four elements: fire, air, water, and earth. We have also seen him defining an element as "the primary component immanent in a thing and indivisible in kind into other kinds."[73] The key aspect of this assertion is its emphasis on the fact that elements are organized in such a way that dividing each one of them does not give us parts belonging to different natural kinds. In other words, elements are indivisible according to form. Note that this definition is not atomistic, since, unlike Democritean atoms, Aristotelian elements can be divided. Moreover, in contrast to Democritus's claim that all atoms are of the same and unchangeable kind (nature), both Aristotle and Aquinas stressed the fact that—being "composed" of primary matter and substantial form—elements are subject to change into one another in the process of substantial change. Trying to explain the way in which they are present and remain in complex entities, Aquinas—in reference to Aristotle's work *De generatione et corruptione*—states that the elements remain in their powers and qualities, but without their substantial forms, which—nonetheless—are not entirely corrupted away in the process of substantial change effecting in the origin of a compound. At the same time, substantial forms of these elements are retrievable in the process of the corruption of a given compound.[74]

What becomes apparent to us today is that the four ancient elements can hardly be regarded as "primary components." We know that fire is not an element, but a process of rapid oxidation of a given material in the exothermic chemical reaction of combustion; air, a mixture of gases; earth, a mixture of minerals, organic matter, gases, liquids, and countless living organisms; water, a chemical compound made of the elements hydrogen and oxygen. Moreover, each chemical element seems to be further divisible in a way that results in obtaining parts that belong to different natural kinds. Every atom can be "divided" into protons, neutrons, and electrons; and protons and neutrons, into quarks. Are quarks divisible into more-elementary particles? What about all other elementary particles, including not only different types of quarks and leptons but also those particles that are not matter particles but, instead, transmit the electromagnetic (photons), gravitational (gravitons), weak (W^+, W^-, and Z^0 bosons), and strong (gluons) nuclear force, or are quantum excitations of the Higgs field (Higgs bosons)—are they all ultimate elements? What is their intrinsic nature? Should we classify them as substances or rather as mere "outcroppings" of local energy? Can they exist as separate entities? Can they be further divided into particles that are still more elementary? What are the qualities they possess that can be altered in the formation of compounds?

These questions do not find easy answers. It has been pointed out, for example, that quarks under the current laws of nature are, in fact, never free—that is, they cannot exist separately but only as constituents of hadrons (mesons and baryons). If this is the case, how can we know the active and passive qualities of quarks, or the way they act upon and are acted upon when entering into a compound? Is the process of retrieving them from hadrons limited merely by our present lack of means to produce the proper level of heat and energy? Moreover, both electrons and quarks are usually classified as point particles, which may suggest they are without dimensions. Does the fact that they are physical—and not mathematical—points, with a radius, mass (defined as the amount of "stuff" in a physical object), a rest energy, an electric charge, a color charge (in case of quarks only), a spin, and a baryon number (in case of quarks only) assigned to them enough to classify these particles as substances?

Some thinkers are willing to classify at least some elementary particles (especially quarks and leptons) as substances.[75] Their proposition may sound plausible based on the assumption that every manifestation of actuality in nature—and these elementary particles are described by physicists as being actual—is due to a certain substantial form in-forming primary matter. Nevertheless, I tend to be more hesitant in assigning to elementary particles described in quantum physics a substantial status in a straightforward way, due to the persistent uncertainty concerning their qualities, nature, constant alterations when they are not being measured, and the possibility of retrieving some of them from more complex compounds and analyzing them in their "free" state (e.g., retrieving up and down quarks from protons).

Approaching these questions from a philosophical point of view, I leave it up to the physicists to specify the most basic "primary components" that can be classified as physical objects. I also remain careful about an oversimplified attempt to relate the "fluctuations in the quantum vacuum" (or the "field quanta") model of elementary particles to Aristotle's metaphysical theory of hylomorphism. At the same time, depending on the physicists' decision concerning basic physical elements, I assign to these entities the principles of primary matter and substantial form. I claim that as such they can enter compounds and remain virtually present in them, with their powers retained yet (possibly) altered and substantial forms not entirely corrupted away and retrievable in the processes of corruption of these "mixed" (composite) bodies or in the reclaiming of given elements from complex substances, which nevertheless "keep" their substantial form.[76] While on the physical, chemical, and biochemical level of observation a given primary component or a more complex entity such as an atom, molecule, or chemical compound can be perfectly traceable in a composite being, this fact does not prevent or invalidate a philosophical reflection stating that the properties of that primary component are, nonetheless, altered and its substantial form corrupted away upon entering the whole, as it now "belongs" to (is a "part" of) a new being, characterized by a new substantial form.[77]

I already provided in section 2.3.2 of this chapter an example of an electron entering the compound of a hydrogen atom. I spoke about the alteration of its qualities with an important qualification, stating that the qualities in question are, in fact, the qualities of a new hydrogen atom, predicated in reference to, or due to, an electron present in that atom virtually (by power).[78] Another possible example is the case of sodium cations (Na^+) gathered around ion channels in a nerve cell membrane, described as participating in the production of action potential, and traceable with basic analytical tools of empirical science. On the one hand, we may say that these sodium cations have a new power of contributing to the occurrence of an action potential—power that cannot be ascribed to them when separated from a living organism. Moreover, their usual activity and reactivity—although retained—seems to be suppressed in various ways on account of their becoming a part of the same organism. On the other hand, we must not forget that the phenomenon of action potential is, *sensu stricto*, a property (quality, power) of an organism, which is predicated in reference to sodium cations "subsumed" in substantial change and now present in it virtually (by power). I believe that these and countless other conceivable examples prove the plausibility of the Aristotelian-Thomistic metaphysical theory of virtual presence in the context of contemporary science.

Going back to the primary interest of this section, I can now address the possible suggestion that my charging a number of the most recent proponents of different versions of neo-hylomorphism with ignoring the Aristotelian distinction between primary matter and substantial form may be wrongheaded. Is it possible that what underlies and inspires their accounts of hylomorphism is in fact an acceptance of the contemporary interpretation of the theory of virtual presence? Is it plausible to claim that what they classify as mereological and structural parts of complex wholes, organized by substantial forms, are chunks and bits of proximate (secondary) matter? Is it justifiable to claim that a reference to proximate (secondary) matter entails their implicit acceptance of the fundamental part of Aristotle's original theory of hylomorphism—namely, his idea of the composition of substances of primary matter and substantial form?

Although someone might find such a scenario both plausible and defendable, we need to realize that—apart from Barnes, who explicitly rejects the theory of virtual presence—none of the thinkers described in my analysis and evaluation of different versions of contemporary neo-hylomorphism directly refers to its main objectives. It is true that both Johnston and Fine can be regarded as loosely referring to virtual presence when they speak, respectively, about the dynamic aspects of form allowing the variation of the parts it holds over time and about the "variable embodiment" of parts within a given whole that are subject to replacement in time. Moreover, Koons's suggestion concerning the migration of some or even all powers from parts to the whole and/or their teleological subordination to an end pertaining to the whole might be understood as an allusion to virtual presence as well. Finally, Marmodoro's idea of the re-identification of parts entering into a compound may be read and interpreted in reference to both the corruption of their substantial forms and the alteration of their properties on the course of the process of such change.[79]

While noticing these references and loose allusions, however, we cannot ignore the fact that none of the thinkers mentioned here embraces the Aristotelian-Thomistic terminology of substantial change and the reality of UAC, which entails (1) corruption of substantial forms, (2) alteration of powers of more basic parts (primary components) at their entering more complex compounds, and (3) the retrievability of those forms after they're corrupted or experience a given change in which a complex whole alters accidentally but persists as one and the same substance. A lack of a more explicit reference to both the concept of virtual presence itself and the language in which it was originally defined and commented on throughout the centuries prevents us—in my opinion—from arguing that contemporary neo-hylomorphists are using it as the ground and inspiration for their theories. Consequently, the assumption that they implicitly accept the fundamental aspect of Aristotle's original idea, describing the composition of any given substance in terms of primary matter and substantial form, is all the more disputable and doubtful.[80]

2.3.6. Powers Version of Hylomorphism

My critical analysis of different versions of hylomorphism in analytic metaphysics discussed so far brings us to one more original theory, which was proposed by Michael Rea.[81] Critical of classical hylomorphism's "(i) commitment to the universal-particular distinction; (ii) commitment to a primitive or problematic notion of inherence or constituency; [and] (iii) inability to identify viable candidates for matter and form in nature, or to characterize them in terms of primitives widely regarded to be intelligible," Rea embraces the ontology of powers, in its particular version of dispositional monism or pan-dispositionalism (see point 2a in figure 5.1 in chapter 5).[82] He states that everything hylomorphists want to explain can be characterized in terms of the concepts of "location" and "power," in a way that avoids all three drawbacks of the classical Aristotelian theory. He thus lists the four premises (P) and the three central theses (T) of his position:

(P1) there is no universal-particular distinction,
(P2) properties are powers,
(P3) powers can be located in space-time, and
(P4) objects can be reduced to or identified with powers;

therefore,

(T1) natures are powers (i.e., the natures of substances are fundamental powers),
(T2) the natures of composite objects unite other powers (in particular, the powers that are the natures of their parts), and
(T3) natures can enter into compounds with "individuators" and play the role of form.[83]

Rea further develops his version of powers ontology by noting that although powers are, *sensu stricto*, neither universals nor particulars, they are in a certain respect like universals (they can be present in multiple regions of space-time), while in another respect they resemble particulars (they enter into causal relations). Next, he distinguishes between fundamental powers (substance natures)—which are natural properties, grounding nonnatural powers, and irreducible to other powers—and powers that function as principles of unity. Comparing the function of the latter to EM, Rea defines his own version of hylomorphism in terms of the unification of relevant powers by one uniting power. In such cases, "the natures of material objects play the role of form, and they enter into compounds with things or stuffs that play the role of matter. On one common way of understanding the roles of form and matter, forms are constituents that are shared among objects of the same kind, whereas matter is what individuates objects of a kind."[84]

Speaking of the principle of individuation, Rea rejects the classical Aristotelian-Thomistic view, which says it is matter that provides for numerical unity (while form is responsible for specific unity).[85] He defines "individuators" for simple objects as regions in space-time occupied by powers (i.e., forms located in these regions), while in the case of complex material things he sees them as regions in space-time that are occupied by "natures of certain kinds of objects standing in certain kinds of relations." To illustrate his version of hylomorphism, Rea gives an example of a human organism and states,

> The manifestation of humanity in a region depends causally upon the cooperative manifestation of the natures of the simple parts of the human organism. Not just any sort of cooperative manifestation will do, however. Take all of the simple parts of a human and force-fit them into a one-quart cylindrical container and you will not have a human organism, even if, at that time, the natures of the erstwhile parts of the human are engaged in some sort of cooperative manifestation. Thus, the presence of humanity in a region depends upon a particular sort of cooperative manifestation of the natures of the relevant parts.[86]

Following the lines of criticism of Rea's position suggested by Marmodoro, we can say, first, that the dismissal of the universal/particular distinction in his ontology makes it difficult for him to give an account of resemblance—that is, similarity in nature. Although we learn from Rea that similar powers can be located in different regions of space-time, he does not explain what makes these powers similar. But without an explanation of similarity, any reference to "objects of the same kind" in the definition of hylomorphism remains ungrounded. Moreover, when saying, "In the straightforward senses of 'in,' nothing in a hydrogen atom looks like a kind property,"[87] Rea shows his mereological and physicalist approach to reality. He forgets that "it is not through physical division of an object that one can isolate the form from it, but through *division by abstraction*. The sense in which the universal form is 'in' a particular is that it can be abstracted *from* the particular."[88]

Second, concerning emergent uniting powers (P's), Rea does not seem to explain what makes them to be over and above the manifestation of powers that they supposedly unify. Rea's example of the manifestation of humanity seems to define the role of uniting powers as the structural and functional organization of other powers on which they depend. However, does the dependence of uniting (formal) powers on the manifestation of other powers explain how this dependence organizes the latter structurally and functionally? "Calling it [i.e., the dependence] a 'cooperative manifestation' gives it a name, but not an account that will deliver what Aristotle's substantial forms do, metaphysically, in substances. Rea describes the items in his ontology by negation, telling us what they do not do, but not how they can do what they are expected to do in his system."[89] Moreover, the fact that the uniting power P can be described as a structural and functional organization of lower-level powers and the fact that Rea defines the material principle in terms of "things or stuffs" lead me to classify his project within the category of structural and mereological hylomorphism, ready to acknowledge the reality of USO and USCC, but not of UAC.

Finally, in his rejection of Aristotelian concepts of primary matter and substantial form, Rea seems to make the common mistake of trying to bring physics and metaphysics into one and the same level of inquiry. He speaks about "the looming danger of disconnecting our metaphysics of material objects from empirical reality," and asks, "Where in physics, or chemistry, or biology do we find something answering to the description 'something in a material object that actualizes its potential to be a dog [or a hydrogen atom, or a sodium chloride molecule]'?"[90] This question suggests a methodological mistake. For metaphysics, by definition, does not deal with those aspects of reality that are available or amenable to a scientific description and definition. Quite the contrary, finding in the latter its point of departure, metaphysics offers a meta-physical explanation. Such explanation—remaining complementary to scientific analysis—introduces principles that go beyond the methodology and language of physics, chemistry, and biology.[91]

But does this mean that dispositional metaphysics is altogether foreign to the Aristotelian version of hylomorphism? After all, Rea deliberately grounds his version of hylomorphism on powers ontology to distance himself from the classical metaphysical concepts of act and potency,[92] universals and particulars, and instantiation—all of them crucial for Aristotle's theory of hylomorphic union. To address this question, we need to remember that Rea follows a particular version of dispositional ontology—namely, dispositional monism (or pan-dispositionalism), which is controversial as it lacks the foundation for the grounding of powers. I already mentioned in section 2 of chapter 5 and in section 2.1 of the present chapter that the genuine (nonreductive) categorical/dispositional property dualism version of dispositional ontology (see point 1 in figure 5.1 in chapter 5) opens the way back to Aristotle's theory of act and potency. In what follows, I will try to show that the same metaphysical position may enable its followers to reembrace Aristotelian hylomorphism.

2.3.7. Dispositionalism and Aristotelian Hylomorphism (Essentialism)

Going back to my analysis of dispositional metaphysics, we need to realize that, on the one hand, powers metaphysicians—with their insistence on certain dispositions and their manifestations being strategic and decisive for the identity of animate and inanimate entities—seem to remain in radical contrast to the contemporary antiessentialist and antihylomorphist positions of Quine, Popper, or Wittgenstein.[93] We saw dispositionalists defining dispositions as genuine, mind-independent, and nonrepeatably particular "tropes" or "unit properties" (Molnar), intrinsic states or properties of particular entities (Mumford), or "properties with a certain kind of essence" (Bird). Their emphasis on the necessary relation of each power to its specific kind of manifestations has also a strong essentialist flavor, along with the emphasis of some dispositionalists on the reality of both dispositional and—grounding them—categorical properties.

On the other hand, however, we do not find dispositionalists subscribing openly and unanimously to the classical version of hylomorphism as defined by Aristotle. It is true that Cartwright and Ellis recognize the reality of natural kinds (natures) in physics and chemistry:

> Modern experimental physics looks at the world under precisely controlled or highly contrived circumstance; and in the best of cases, one look is enough. That, I claim, is just how one looks for natures.[94]

> Every distinct type of chemical substance would appear to be an example of a natural kind, since the known kinds of chemical substances all exist independently of human knowledge and understanding, and the distinctions between them are all real and absolute.[95]

But are they willing to extend essentialism to biological kinds as well? And what is their exact definition of natures/essences/natural kinds in the context of the analytic metaphysics of powers? The answer to the second question is complex. On the one hand, powers metaphysicians may be seen as accepting the Humean "bundle" or "trope" theory of substance, defining its nature in terms of a collection or cluster of properties (here powers and/or categorical properties) and processes that a given substance enters (here manifestations of dispositions). This brings them close to the theories of natural kinds proposed by the new mechanical philosophy, which I analyzed in the introduction (section 6). These theories develop out of Boyd's "homeostatic property cluster" view, offered as a third way between essentialism and nominalism.[96] According to this theory, a natural kind is defined as (1) a cluster of properties that concur regularly and (2) a similarity-generating mechanism that explains why these properties have a tendency to concur. It thus sees natural kinds as property clusters explained by mechanisms. To some extent, this view is close to the one offered by Rea in his powers version of hylomorphism.[97]

On the other hand, dispositionalists emphasize that the range of powers characteristic for a given being is strictly defined and constrained. They seem to approve the reality of a more robust principle of unity, which goes beyond the mechanistic explanation of the similarity-generating constitution of the properties. They speak about "essential" and "intrinsic" properties and powers, which suggests they approve the reality of essences: "Dispositions are not relations to actual or possible manifestations, however. Objects possess dispositions by virtue of possessing particular intrinsic properties. The nature of these properties ensures that they will yield manifestations of particular sorts with reciprocal disposition partners of particular sorts."[98]

And yet dispositionalist analyses referring (both indirectly and directly) to essences do not quite follow Aristotelian hylomorphism, as they do not refer to primary matter and substantial form. Nor do they subscribe to Kripke's version of essentialism based on the notion of a "rigid designator" and possible worlds modality[99] (even if they may remain sympathetic to his idea of "internal structures" or Putnam's theory of "hidden structures" of things).[100] Powers metaphysicians also do not seem to subscribe to the versions of hylomorphism I have so far examined (the mereological, structural, incomplete entities, mixed, re-identification of parts, and Rea's powers versions). What is, then, their ultimate position on hylomorphic ontology?

As I face the difficulty of specifying the most fundamental and ultimate metaphysical principles of powers metaphysics, I want to argue that those among dispositionalists who make a clear and undeniable reference to the classical theory of potency and act, as well as emphasize the kind-specific and intrinsic character of powers, situate themselves closer to Aristotelian hylomorphism and essentialism than to the Humean "bundle" theory of substance and different versions of contemporary neo-hylomorphism, analyzed in previous sections. Nevertheless, as I have said, the absence of an emphasis on the distinction between passive and active potencies, the lack of a thorough analysis of the powers metaphysics in terms of the essentialist and natural-kinds-approving ontology, and the departure from the language of primary matter and substantial form seem to prevent dispositionalists from rediscovering the classical Aristotelian hylomorphism. That is why I find highly relevant the following critical remarks by Oderberg:

> The contemporary debate about dispositions and powers has so far shown little interest in the distinction between active and passive power, and hence it is no surprise that hylomorphism has played no role in the debate.[101]

Unfortunately, ... talk of essence (despite the work of Ellis) is still in short supply among dispositionalists. Yet without essence we cannot explain *why* a thing has the powers it does. Essentialism tells us that a thing has the powers it does because of the kind of thing it is. Its essence bestows on it a range of powers, none of which is ever exhaustively manifested; ... otherwise, the object would lose all potentiality and be in a state of pure actuality, which is impossible.[102]

Referring to Aristotle's ideas of primary matter and substantial form, Oderberg adds that the passive qualities of a nonliving thing's or an organism's powers point toward its prime matter, while its active qualities point toward its form.[103] He argues from the point of view of classical Aristotelian hylomorphism, which he puts into conversation with analytic metaphysics and natural science. His position differs significantly from all versions of neo-hylomorphism discussed above in that he grounds it on the distinction between potency and act and defines it in terms of substantial form and primary matter. He explains the former as an irreducibly holistic "principle of specificity of any thing, that by which it is what it is." He sees it as a principle remaining beyond shape and actualizing primary matter, which should be understood as "pure passive potentiality ... wholly receptive ... conceptually beyond sensible ... or secondary, or proximate matter ... conserved throughout substantial change." He also distinguishes between substantial and accidental changes and advocates in favor of the reality of the theory of virtual presence.[104] Concluding his analysis of powers ontology, Oderberg says, "Only the essentialist—i.e. the believer in unitary essences explained in terms of substantial forms—can explain the necessary relationships between certain kinds of power, namely the ones possessed essentially. The unitary essence explains the powers: the powers *flow* from it, to repeat the metaphor. Hylemorphism explains what needs explaining."[105]

This statement brings us back to my original question about the dispositionalist view of hylomorphism. My investigation of the many nuances of the possible relation between these two ontologies shows the complexity and difficulty of the debate on this issue in contemporary metaphysics. Nevertheless, I want to argue that, acknowledging the differences between powers metaphysics and classical Aristotelian ontology and philosophy of nature, one can still conclude that of all metaphysical theories concerning the nature of nonliving and living beings in analytic philosophy, powers metaphysics remains the closest to the orthodox notion of hylomorphism and essentialism. Therefore, despite some reservations, I claim that in this respect—in addition to its bringing a revival of teleology—dispositionalism can indeed be regarded as a neo-Aristotelian type of ontology.

2.4. Event Ontologies

Before closing this chapter, I must say a word about some other important aspects of dispositionalism and its possible relation to classical Aristotelianism. First, we need to consider powers metaphysics in relation to so-called event ontologies, which strive to emphasize, and define reality in terms of, causal processes rather than substances. Here powers metaphysics seems to be standing close to the classical Aristotelian position, acknowledging the importance of process accounts for the metaphysical description of reality, but without questioning the reality of substances. Christopher Shields reminds us that it is an outcome of Aristotle's grounding of his theory of efficient causation on the distinction between potency and act. He suggests a framework definition of efficient causation: "x_1 is an efficient cause of $x_2 =_{def}$ (i) x_1 and x_2 are processes (instantaneous or extended in time, and terminus-directed); (ii) x_1 and x_2 are reciprocally paired, categorically and nomologically; (iii) x_1 is the actualization of x_2 (in the sphere of mobile or immobile); and (iv) x_1 and x_2 are necessarily co-extensive (yet not identical and not extensionally discrete)."[106]

Describing beings in terms of manifestations (actualizations) of their dispositions (potencies), powers metaphysics recognizes the role of the process approach, which alludes to Aristotle's notion of actualization of potentiality in the course of substantial and accidental changes in nature. At the same time, powers metaphysicians repeatedly emphasize that dispositions are real and decisive about the nature (essence) of beings, even when not manifested, which shows that actualization of potentialities taking place in causal processes must be grounded in the concrete natures of their participants. This position, unlike what is usually thought by many contemporary philosophers of science and metaphysicians, is by no means proven false by relativity and quantum mechanics, since they (by themselves) do not entail anything about substances. And if they do, they do not get rid of substance but just relocate it. Relativity sees the entire space-time continuum as a single four-dimensional substance, while quantum mechanics tends to replace substance with bundles of events, which are treated as substances building up into ordinary-size objects.

We need to realize that since the neo-Humean event ontologies are basically recapitulations of the Heraclitean position, assuming that all is flux and becoming, they face similar problems and objections. Their most important flaw is an inability to explain the unity and stability of objects of our experience. The emphasis on the processual aspect of reality begs a question of what undergoes the process, followed by a difficulty in specifying whether something survives it. In this context, the moderate position of neo-Aristotelian powers metaphysics proves more adequate as it is able to capture the stability and ground of identity of natures of animate and inanimate things, as well as their changeability in the causal processes they enter, realizing their active and passive potencies. The explanatory power of this metaphysics becomes a suitable base for contemporary scientific realism and the theory of the laws of nature, which I will discuss shortly in the remaining two sections of the present chapter.

2.5. Scientific Realism

Scientific realism, a view embraced today by the vast majority of philosophers of science, assumes that our best scientific theories correctly describe both observable and unobservable, mind-independent aspects of reality (as opposed to being merely useful instruments for making predictions about future behavior of things and dynamical systems). It is proposed as an alternative to a thoroughly pessimistic induction, which assumes that, since most of the past scientific theories proved to be false, most of our present-day theories will also prove false and be replaced in the future. This approach is criticized because it resembles Kant's disbelief in our access to things as they really are (noumena).

But scientific realism, as Anjan Chakravartty notes, traditionally does not come as a unified position. Realists are selective and divided into two broadly defined camps of entity realism and structural realism. Entity realism states that under certain conditions, which enable us to exploit (manipulate) entities causally in various ways, we have a good reason to believe that the entities described by our scientific theories really exist. Structural realism, by contrast, suggests that insofar as scientific theories offer plausible descriptions of the world, they do not inform us about the true nature of things or their unobservable parts, but tell us about the structure of relations that they enter. The disparity between these two kinds of scientific realism becomes apparent. Entity realism advocates realism about concrete entities described by scientific theories, while it remains antirealist concerning the descriptions of relations between these entities. Conversely, structural realism accepts realism with respect to certain structures, promoting antirealism concerning knowledge of the nature of entities building up these structures.[107]

Chakravartty argues that the stark opposition between entity and structural realism, highly detrimental to the position of scientific realism about theories and models, can be resolved in reference to powers metaphysics, which synthesizes the best insights of the two options. We have seen that its realism about dispositions necessarily relates them to the concrete entities that have them. It allows us to classify dispositionalism as a neo-Aristotelian position, which shows that masses, charges, accelerations, volumes, temperatures, and so on point toward dispositions to behave in a certain way in proper circumstances—dispositions that are grounded in beings and available for manipulation and experiments. At the same time, the dispositional account of properties analyzed by sciences recognizes their relational character and emphasizes that structures should be defined in terms of relations. It sees the tendency of scientific theories to describe the behavior of things in terms of their relations, usually expressed in the form of mathematical equations, which by definition relate variables (their values depend on the magnitudes of the properties of the things in question). Dispositional realism follows the same path of reasoning metaphysically, offering an account of structure by means of relations understood as manifestations of dispositional properties. In the conclusion of his argument Chakravartty writes:

> Dispositional thinking synthesizes the very notion of a causal property of scientific interest with that of the structure of natural phenomena. Knowledge of one entails knowledge of the other. Particularly in connection with many of the unobservable entities with which scientific realism is most concerned, it is only by means of a knowledge of their relations that one is able to interact with them

causally, and thus, the epistemic warrant that entity realists identify with certain forms of causal knowledge is evidently parasitic on structural knowledge. Lacking the relevant structural knowledge, it would be simply impossible to design and perform experiments to detect and manipulate such entities.[108]

2.6. Laws of Nature

The other area of philosophy of science that can benefit from the neo-Aristotelian character of powers metaphysics is the theory linking scientific realism to natural kinds and laws of nature. We have seen that realism about dispositional properties points toward the realism of concrete entities, taken separately or defined in terms of their relational position in structures. The properties in question can thus be regarded as typical of the natural kinds to which those entities belong. What many exemplars of one natural kind share in common is a common nature (essence), expressed in similar dispositions and manifestations. Laws of nature, in turn, can be understood in terms of behavioral regularities that follow upon manifestations of the causal powers things possess by virtue of belonging to the same natural kind. Thus, we can assert with Oderberg: "*The laws of nature are the laws of natures.* For natures just are abstract essences in concrete operation. Nature is the collection of all the natures of things. So to say the laws are *of nature* is to say that they are *of the natures* of things."[109]

In other words, our theory of laws of nature no longer depends on regularities or universal generalizations (a regulatory account). Nor is it defined in terms of necessary connections between properties or in terms of entities which ground them, such as universals (a nomological view). The former is challenged by a large number of regularities which seem to be accidental. The latter assumes laws of nature are external to entities which they relate, and it attributes to those laws a unique ontological and causal character that allows them to "make" things behave the way they do. Both dispositionalism and the classical Aristotelian hylomorphism suggest treating laws as "ontologically derived entities . . . and not as a fundamental category in the ontology." Moreover, "one might even suggest that there are no causal laws as such, except insofar as they are derived from the behavior of causally powerful properties."[110]

The link between laws of nature and natures of concrete beings, belonging to natural kinds, provides one of the most important arguments in favor of the necessary character of the laws of nature. It is important to notice that their necessity is metaphysical precisely because it derives from the essences of the inanimate objects and organisms representing natural kinds. Because laws of nature are abstracted from the behavior of real entities, their necessary character is usually perceived in terms of an absolute necessity. However, we must not forget that the necessity of causal dependencies among actual objects is always suppositional.

This view differs considerably from the one supported by neo-Humean metaphysicians. Because they see as fundamental truths only the categorical features distributed across time and space, they seem to treat laws of nature as axioms of the best scientific theory of the world (characterized by the best combination of simplicity, accuracy, and generality), without grounding them ontologically. This strategy makes laws of nature depend on our own practices and preferences, which are both subjective and conventional.[111]

Summing up, despite remaining difficulties and despite dispositionalists' reluctance to embrace hylomorphic ontology, my comparative analysis of powers metaphysics and Aristotle's classical philosophy of nature, offered in this chapter, shows many points of convergence between the two theories. It reveals (1) the resemblance of the distinction of disposition and manifestation to that of potency and act, (2) the influence of that distinction on the philosophy of causation, and (3) the ontological position of powers metaphysics on the nature of reality, scientific realism, and the theories of the laws of nature and natural kinds. I have argued that even if analytic metaphysics of powers does not pay enough attention to the distinction of active and passive powers or to the hylomorphic distinction of primary matter and substantial form, it still remains fairly close to classical essentialism and thus can be classified as neo-Aristotelian. Since its attractiveness lies in the fact that it is developed in the context of the contemporary science, the next and the final step of my project will be to apply it to both the classical DC-based EM and Deacon's theories of EM.

CHAPTER 7

Dispositional Metaphysics, Downward Causation, and Dynamical Depth

In the course of the investigation pursued in the first part of my project I analyzed the most prominent nonreductionist propositions offered by contemporary analytic metaphysics. I have listed and evaluated five versions of NP, based on (1) Davidson's "anomality of the mental," arguing in favor of the irreducibility of teleology to efficient causation, (2) Putnam's "multiple realizability," arguing in favor of the primacy of form over matter, (3) nomological SUP, arguing in favor of the real existence of higher properties, (4) Shoemaker's distinction between manifest and latent properties, and (5) DC. I have spent considerable time discussing the last of these positions, which is considered the sine qua non of the ontological (strong) version of the classical notion of EM. I have suggested that the problem of all these approaches lies in their ultimate atomist-cum-physicalist assumption that the level of basic particles and efficient causation is metaphysically fundamental and "more real" than higher-level phenomena and their description.

In light of these findings I have suggested a return to a more robust metaphysics and the broader notion of causation offered by Aristotle. I have investigated and evaluated an original notion of EM, alternative to the classical one offered by Deacon, in which he argues in favor of the retrieval of Aristotle's formal and final causes. My critical evaluation of his project emphasized both its strengths and weaknesses, showing that the dynamical depth model of EM needs to be grounded in a more thorough metaphysics.

Trying to situate the whole debate on EM in the context of the most recent philosophical theories of causation developed in the analytic tradition, I have investigated and evaluated six main contemporary views of the nature of causality: RVC (including INUSConVC and InfVC), CVC, ProbVC, SVC, M-basedVsC (including MVC, AVC, ArelVC, and IntVC), and ProcVC. What speaks in their favor is that they strive to overcome the Humean reductionist agenda and to argue in favor of the ontological character of causal dependencies. Also, each one of them does capture some important metaphysical aspects of causal dependencies in nature. At the same time, however, none of the six theories is ultimately convincing, as they all raise some serious metaphysical questions. Moreover, it seems to me that the major shortcoming of these concepts of causation is their inherent tendency to look at causal relations only in terms of the efficient type of causation and to describe the phenomena that accompany these relations, rather than their intrinsic character and nature. In answer to these difficulties, I have introduced and analyzed the dispositional view of causation, which is rooted in the metaphysics of dispositions/powers and their manifestations and does not limit causal relationships to physical interactions. Having answered the question concerning the Aristotelian legacy of dispositional view, I must now apply dispositional metaphysics and the view of causation it offers to the theory of EM—first to the top-down (DC-based) concept of EM, then to Deacon's dynamical depth model of EM. My goal is to describe EM and DC in causal terms, free from Kim's charge of causal overdetermination. Therefore, I envision it as a plausible alternative to the reinterpretations of DC in noncausal terms, described in section 6 of chapter 2.

1. DISPOSITIONALISM AND DOWNWARD CAUSATION

I said above that one of the main problems challenging the classical version of the contemporary (top-down) emergentism is the difficulty in specifying the exact nature of DC, exercised by the whole on its constituent parts—a sine qua non condition of strong (ontological) EM. One of the main arguments against classical emergentism, brought by Kim, shows that when defined in terms of efficient (physical) causation, DC seems to be epiphenomenal and reducible to efficient causes operating on lower levels of complexity (see chapter 2, section 5). However, this argument is valid only within the Humean ontology, which sees all entities in nature as loose, separate, and related only externally and contingently, and defines causation in terms of regularity of causal events, or in terms of counterfactuals. The acceptance of Humean metaphysics clearly makes the defense of the irreducibility of DC impossible.

But the case of dispositionalism is entirely different. Applied to the theory of EM, it sees DC as a specific kind of manifestation of a new and unique set of dispositions, which are irreducible and nonrepeatably particular, defined as intrinsic properties "with a certain kind of essence." I want to argue that understood this way, DC has several complementary aspects contributing to its metaphysical character and nature. Their analysis sends us back to Aristotle and his robust theory of causation. I contend that revisiting his thought is relevant and justified in the context of dispositionalism, which is classified as a new version of Aristotelianism, developed within the framework of the analytic metaphysics. I will now analyze the main aspects of DC from the perspective of the new and the old Aristotelian metaphysics and theory of causation.

1.1. Downward Causation and Hylomorphism

I have said that dispositionalism sees DC as a kind-specific manifestation of unique dispositions, which are irreducible and nonrepeatably particular. They can be defined as intrinsic properties "with a certain kind of essence." The use of essentialist language in this explanation is significant. It can be referred to the Aristotelian concept of formal causation, which I think should be regarded as the first and the fundamental aspect of DC.[1] Because it is irreducible to efficient causation and decisive about the intrinsic nature of each entity, it becomes a powerful argument in defense of the irreducible character of DC, as well as a viable explanation of its nature.

As I have noted in the introduction (section 2.1), the formal type of causation is inextricably related to primary matter in Aristotelian metaphysics. These two principles form the first and the most basic ontological composition underlying the reality of all entities. I am aware that dispositionalism is not simply a version or an expression of hylomorphism and that there is no easy transition from one position to the other. Nevertheless, we have seen that dispositionalism does become an advocate of the reality of essences. I argue that those who want to embrace essentialism should find the concept of hylomorphic composition not only acceptable but also metaphysically superior to other theories proposed as the foundation of essences. It provides a solid base that not only unifies all properties of an entity but also becomes a source of their qualitative characteristics (see chapter 6 section 1.3). It thus solves the problem of (1) Locke's substratum theory of particulars, as well as (2) Humean "bundle" or "trope" theory of substance and, related to it, (3) Boyd's "homeostatic property cluster" view of natural kinds, which all lack an exact explanation of the principle unifying properties typical of the particular nature/kind/essence/substance. Unlike these theories—referring to (1) bare substratum-grounding properties, (2) a special relation tying all the attributes in a bundle, or (3) a similarity-generating mechanism—hylomorphism stands as an explanation which is neither mechanical nor devoid of reference to any qualitative features. Quite to the contrary, as a source of identity and novel properties, intrinsic for complex entities and systems, Aristotle's hylomorphic composition has an irreducible and qualitative character, which goes beyond physical and mathematical description. It thus needs to be distinguished from the structural neo-hylomorphism proposed by some analytic philosophers, which defines form as a structure, and matter as the materials being structured—a position which falls, in my opinion, into metaphysical reductionism (see chapter 6, section 2.3).

Applying Aristotle's hylomorphism to a top-down (DC-based) version of EM enables those who embrace it to solve the ambiguity of defining emergents in terms of properties. It suggests characterizing them in reference to new substantial forms of more complex/higher entities (e.g., chemical compounds and substances) and organisms, and new accidental forms of more complex dynamical systems engaging particular entities (e.g., large conglomerates of molecules of H_2O showing surface tension or forming an eddy).[2] The novelty of properties characteristic of emergent entities or dynamical systems finds its ground in the novelty of the substantial and accidental forms proper to them. For what happens in the case of each and every ontological emergent transition is either an ontological change arising from the actuality of a new substantial form in-forming primary matter—which is properly disposed to receive it by having been in-formed by the substantial forms of lower-level entities—or a manifestation (occurrence) of a new accidental form that involves a large quantity of substances and is a function of their natures. Thus, reference to the formal aspect of DC makes us acknowledge that emergent properties describe and reveal, rather than decide about, the emergent character of given entities or dynamical systems. This emergent quality can be assigned to them due to their natures (in case of entities) or the natures of multiple substances organized in a way that brings a manifestation of a new and unique accidental form.

1.2. Downward Causation and Efficient Cause

Among properties characteristic of emergent entities and dynamical systems (both living and nonliving), we find those that express and actualize their abilities to act and react in a particular way, in response to internal and external factors, in specific circumstances. This ability to act and react is described in Aristotelian philosophy in terms of efficient causation. Thus, we can say that the new substantial forms of emergent entities/organisms and dynamical systems are decisive about the nature of their efficacy.

This observation becomes crucial for the understanding of the nature of DC. Because the new type of activity and reactivity proper to an emergent entity is inseparably related to its very nature, efficient causation becomes another indispensable explanatory aspect of DC. However, my position differs significantly from the one supported by causal reductionists, who strive to explain DC in terms of efficient causation only. I find the argument about the relation between DC and efficient causation plausible only with two important qualifications. First, as we have already noticed, efficient cause analyzed philosophically in reference to the Aristotelian fourfold notion of causation does not stand on its own, without a reference to the substantial form and accidental features of an entity that shows it. This fact relates the efficient aspect of DC necessarily to hylomorphic analysis of emergent entities and dynamical systems, even if it is otherwise a subject of scientific analysis and mathematical description. Second, we have to remember that within the

Aristotelian theory of causation, efficient causality has broader meaning than the one defined within the domain of the natural sciences. We have seen that in his basic definition of efficient causation Aristotle goes beyond physical efficacy and extends it to mental activity (in the introduction, section 2.2, see his example of a man giving advice). Interestingly, it is precisely causal activity of the mind that became one of the most common examples of DC in contemporary emergentism, which pays attention to causes that originate changes in the behavior of emergent wholes but are hardly measurable quantitatively and hardly describable in the mathematical language of natural sciences.

Translating this argumentation into the language of dispositionalism, we can state that even if DC has an efficient character and is a subject of scientific analysis and mathematical description, its irreducible nature is kept and defensible within dispositional metaphysics, because of its grounding of all cases of activity and reactivity in kind-specific powers, characteristic for an intrinsic nature (essence) of a given entity (organism/system). Moreover, my analysis of dispositionalism showed the diversity and flexibility of powers ascriptions among its followers. We saw Mumford classifying the following as dispositions: fragility, belief, bravery, thermostats, and divisibility by 2 (see chapter 5, note 26). If we want to analyze the efficient aspect of the manifestations of these dispositions, it becomes clear that our definition of efficient causation—similarly to the one offered by Aristotle—must go beyond purely physical exchange of matter and energy, which also provides another argument in favor of the irreducibility of emergents and DC they exercise.

1.3. Downward Causation and Teleology

Finally, the teleological character of dispositions "pointing" or being "directed" toward kind-specific manifestations, described as their "physical" or "natural intentionality," becomes one more argument in favor of the irreducible character of DC, and the third crucial aspect explaining its nature.

We have seen Deacon's emphasis on the importance of intrinsic teleology at the higher levels of organization of matter and his defining it as a crucial factor in distinguishing living from nonliving dynamical systems, enabling us to attribute to the former a quality of a primitive "self." I will go back to his argumentation in the second part of this chapter. But even in reference to the classical version of the theory of EM, we can analyze DC in terms of the new types of teleology—that is, tendencies to realize new potencies characteristic of a given emergent entity/organism or dynamic system. Because these tendencies are crucial for the future development of an organism (or future states of a nonliving entity), final cause becomes an indispensable and irreducible aspect of DC.

1.4. The True Nature of Downward Causation

My analysis of teleology, together with my reflection on the hylomorphic aspects and efficient causation proper for emergents, shows the interrelatedness of various types of causality which fall under the general term DC in contemporary philosophical reflection on biological complexity. Armed with the neo-Aristotelian and the classical Aristotelian philosophy of nature and metaphysics, I was able to specify the exact character of this new type of causation, without falling into reductionism or reinterpreting this causation in noncausal terms. For unlike physically/mechanically interpreted efficient causation characteristic of higher levels of complexity, which is reducible to lower-level interactions and amenable to mathematical and quantitative description, formal and final types of causation—proper for a given order of complexity and qualitative in nature—are always irreducible. One cannot, for instance, reduce the form and teleology proper for a living organism to the form and teleology of elementary particles building it. Even efficient cause, in its classical Aristotelian interpretation, is not always describable in terms of physical exchange of matter and energy. This fact allows us to ascribe to it an irreducible character as well.

I claim that this redefinition of DC in terms of new causal dispositions that require explanation in terms of formal causation (with reference to essentialism and hylomorphism), final causation, and efficient cases of action and reaction distinctive for emergent entities and dynamical systems both saves its irreducibility and explains its meaning. It answers the criticism coming from those philosophers who find the concept of DC—as it is defined in writings of its proponents—inconsistent, spooky, and undefined.

At the same time, I want to emphasize that the analysis of DC in terms of Aristotelian four causes presented here does not make the very term "DC" redundant nor spurious. The distinctiveness and explanatory power of this analysis is preserved in its emphasis on and expression of the interrelatedness of all four types of causation listed by Aristotle. Because their interdependence is characteristic and unique for each particular emergent complex entity/system/organism, my use of the term "DC" is still legitimate as an expression of this uniqueness and particularity.

The interrelatedness of causes was noticed already by Aristotle. We have seen him saying in *Phys.* 2.7.198a25–27 that "the last three [the form, the mover, that for the sake of which] often coincide; for the 'what' and 'that for the sake of which' are one, while the primary source of motion is the same in species as these." Commenting on this aspect of Aristotle's theory of causation, Aquinas emphasizes not only the hylomorphic relation between material and formal causes but also the interdependency of efficient and final causality. The efficient cause is for him the cause of the final because the former begins motion toward the latter, while the final cause is the cause of the efficient as it is the reason for the latter's activity.[3] Going still further, Aquinas points toward the relation between formal and final causes by saying: "The form and the end coincide in the same thing," and "it must belong to the natural philosophy to consider the form not only insofar as it is form but also insofar as it is the end."[4] On numerous occasions we can find him saying, following Aristotle, about the relations between three causes: "Notice, also, that three causes can coincide in one thing, namely, the form, the end and the efficient cause."[5]

At the same time, however, although both Aristotle and Aquinas suggest a reference to all four causes in explaining natural phenomena, Aquinas notices that we often have difficulty distinguishing and naming each of them in particular occurrences.[6] This thought of Aquinas expresses the concern and difficulties encountered by those who try to specify and distinguish causes in particular examples of causal interactions engaging entities/systems/organisms classified as emergent. In this context, the term "DC" seems to be justified and useful as expressing the unique interdependency of causes in emergents.

1.5. Downward Causation and Probability

Before moving to the second part of this chapter, I must refer to the probability aspect of causal dependencies in emergent entities/systems/organisms. As we have learned from Aristotle (see introduction, section

2.4), chance occurrences have an ontological character, and thus belong to the fabric of the universe. Their reality contradicts deterministic views of nature and makes us treat the necessity of all necessary occurrences as suppositional, rather than absolute. It is, therefore, not surprising that systems biology—using complex mathematical techniques in high-throughput data collecting and interpreting—needs to take into account, in the algorithmic setup of its instruments, the probability factor. As the quantified measure of the likelihood of occurrence of a given event, probability theory proves very helpful in the scientific description of the underlying mechanics and regularities of complex systems. At the same time, we must not forget that an axiomatic mathematical formalization of chance in probability theory can only "tame" it, not eliminate it. Therefore, it remains a real—and yet not causal—aspect of cause-effect dependencies observed in the universe.

This fact brings us to a deeper metaphysical question about the importance of chance in our description of the causality of emergents. Shall we regard it as one more crucial aspect of DC, indispensable for the characteristics of its nature? Should chance be added to the list of causes intrinsically interrelated in providing for the identity of emergent entities/systems/organisms?

To answer this question, we should recall once again Aristotle's teaching on chance, which I briefly discussed in the introduction. Following his position, I acknowledge both the reality and the ontological character of chance events in nature. At the same time, however, I want to emphasize that chance as such is not a cause of anything. I regard it, after Aristotle, as a *per accidens* cause, which has to be related to *per se* causes engaged in a given causal situation. As we have seen, both old and new Aristotelianism help us specify the *per se* causes in question as intrinsically interrelated and interdependent: formal cause (with reference to essentialism and hylomorphism), final cause, and efficient cause. Dispositional metaphysics sees these causes as both sources and manifestations of concrete, kind-specific, and irreducible dispositions. The top-down version of the theory of EM suggests interpreting them as different aspects of DC, exercised by emergents.

I want to emphasize that the reality of *per accidens* "causality" of chance can be traced and analyzed only in reference to *per se* causation of both simple and complex (emergent), nonliving and living entities/

systems. Moreover, our ability to "tame" chance and to present it in the form of the mathematical formalization of probability theory is possible precisely due to the reference to the *per se* causes engaged in causal situations classified as chance occurrences. Without such reference, chance remains an ultimate mystery, penetrating each and every level of the complexity of the universe.

Consequently—following the argumentation of the advocates of DVC (see chapter 5, section 4)—I want to recognize the importance of the chance factor and probability theory in analyzing causal dependencies, without trying to make of them a foundation of the theory of causation as such. Moreover, because chance can be classified as cause only *per accidens*, I do not recommend regarding it as an additional core aspect of DC. Rather, I suggest treating it as an important feature of many causal situations, which needs to be taken into account in both speculative investigations and practical applications of causal theories.

This remark closes my reinterpretation of the classical, DC-based concept of EM in the context of the old and the new Aristotelianism and brings us to the last part of my project, in which I will share my constructive proposal of applying the same metaphysics in Deacon's dynamical depth model of EM.

2. DISPOSITIONALISM AND THE DYNAMICAL DEPTH MODEL OF EMERGENCE

I closed chapter 3, completing the first part of my project, with a clear evaluation of the ontological aspects of Deacon's theory of EM. I said that although his redefinition of the four Aristotelian causes and his theory of the causation of absences is very original and thought provoking, it needs to be more carefully analyzed and specified at the level of the metaphysical principles on which it is founded. I can summarize the main ontological aspects of Deacon's theory of dynamical depth and my doubts and questions concerning their explanatory power in the following six points:

1. *The description of EM in terms of the reversal of orthograde and contragrade changes, generating new constraints decisive*

for emergent transitions. The theory begs the question concerning the source of both spontaneous orthograde and forced contragrade processes. Deacon repeatedly emphasizes their importance in his description of all three levels of rising complexity in his model of dynamical depth: from the spontaneous tendency of higher-order linear thermodynamic phenomena to eliminate constraints, thus increasing global entropy and reaching the state of equilibrium in homeodynamics; through the spontaneous character of the regularizing tendency of various classes of phenomena to become more organized and globally constrained over time in morphodynamics; to the reciprocal coupling between morphodynamic processes of autocatalysis and self-assembly in autogenesis, developing spontaneously toward the "higher attractor" of a teleodynamic target state. As I remarked in chapter 3 (section 4), the spontaneity of these processes cannot be explained by referring to the laws of thermodynamics and entropy gradients, as they themselves are merely descriptive and not prescriptive in their nature. This is one of the most significant metaphysical challenges for Deacon's project.

2. *The thesis about the emergent character of teleology.* Deacon claims teleology emerges on the highest level of complexity in his model of dynamical depth, from mere physico-dynamic processes classified as a contemporary notion of Aristotle's efficient causation. His theory leaves unanswered the question concerning the natural goal-directedness of processes observed at the lower levels of his dynamical depth.

3. *The tendency to explain form in terms of the geometric properties of phase-space (Aristotle's μορφή = "shape").* This approach distances Deacon from the classical definition of form understood as a principle of unity and identity in animate and inanimate things, defining their natures, as well as their potencies and powers (Aristotle's ὁ λόγος τοῦ τί ἦν εἶναι = "the statement of the essence"). Deacon finds this classical notion of form homuncular. At the same time, his own redefinition of the formal principle seems to follow the structural hylomorphism of Koslicki and Jaworski (see chapter 6, section 2.3.1), which is metaphysically reductionist.

4. *The theory assigning causal powers to absences.* My long debate with Deacon on this topic, which I summarized in the section 3.4 of chapter 3, does not bring a final answer to the question of the exact nature of this causation. I find Deacon struggling between the claim that absences in fact do not do any causal (efficient) work and the position attributing to them an ability of "allowing," "enabling," "channeling," or "opening up for" an energy release. Aware of the causal character of this terminology, he strives—together with Cashman—to provide a more precise definition of the causation of absences. He sees it not in terms of physical (efficient) causation or the primary notion of formal causation in Aristotle (ὁ λόγος τοῦ τί ἦν εἶναι = "the statement of the essence," or *forma substantialis*), but in terms of the reduced notion of form characterized as external (geometrical) shape (μορφή). Because this definition of formal cause is responsible merely for the way entities appear to us in their external, physical qualities, and does not explain why they are what they are, it seems to be highly inadequate when applied to Deacon's idea of causation of absences. For his theory seems to assign to absences a much more important role than just "shaping" the external features of entities or dynamical systems. He finds absences responsible for the rising complexity of operation of emergent systems and the origin of the teleological features that these systems show at the level of reciprocal coupling between two or more morphodynamic processes. This is probably the most important metaphysical challenge for Deacon's theory of EM.

5. *The alleged dismissal of the concept of DC in the dynamical depth model of EM.* This postulate does not seem to be entirely met by Deacon. Describing characteristic features of different levels of EM, he uses causal language, resembling terminology associated with DC in the classical version of EM. Such is the character of Deacon's account of the nature of (1) the higher-order constraints preventing the disruption of the synergy between the component morphodynamic processes in teleodynamics; (2) the higher organization of teleodynamic systems influencing their constituents in a way that enables them to preserve and regenerate their own

essential constraints and to interact with the environment in a way that sustains supportive influences and compensates for unsupportive ones; (3) self-compression and self-reference of teleodynamic systems; (4) spontaneous regularizing tendencies of morphodynamic systems to become more organized and globally constrained over time; and (5) spontaneous inclinations of homeodynamic phenomena to eliminate constraints, increasing global entropy and reaching thermodynamic equilibrium. Deacon describes all these emergent phenomena in terms of causal influences, exercised by complex dynamical systems or wholes on their constituents. The nature of this causation raises questions similar to those challenging DC (see chapter 2, section 4).

6. *Rejection of substance metaphysics in favor of the process approach to reality.* Distancing himself from the mereological (and thus false) interpretation of substance metaphysics and embracing the process view on the metaphysics of nature, Deacon classifies himself among other contemporary proponents of event ontologies. However, his position on this point is not entirely clear. He sides with the Copenhagen interpretation of quantum mechanics, arguing that it proves that at the quantum level we find not solid substances, but indeterminate processes. At the same time, Deacon claims the goal of his project is to explain EM in reference to the basic physical and chemical forces that dominate at our level of scale, without references to "quantum strangeness." He rejects the Whiteheadian metaphysics of process without offering an alternative approach that would explain why events and processes should be regarded as the most basic constituents of reality and how they build up into the entities and solid structures given in sensory experience.

All these questions, addressing metaphysical aspects of Deacon's dynamical depth model of EM, are definitely crucial for the defense of its plausibility on the grounds of philosophy of science. I want to argue that the principles of the neo-Aristotelian metaphysics of dispositions and their manifestations, together with the dispositional theory of causation and some necessary references to the classical thought of Aristotle, offer a plausible and coherent answer to all six ontological challenges for Deacon's theory of EM listed here. I will treat each of them separately in the following sections.

2.1. The Engine of Dynamical Depth

As we have seen, what lies at the bottom of each stage of complexity in Deacon's dynamical depth model of EM is the spontaneity of natural processes which he calls orthograde and which he contrasts with forced contragrade changes, resulting from juxtaposition of orthograde occurrences. The question concerning the source of the spontaneity of these processes finds an answer in dispositional metaphysics, which emphasizes the natural potencies and tendencies of entities, initiating causal changes and participating in dynamic processes. Their behavior and activity is a function of manifestations of their kind-specific dispositional properties.

The source of all dynamics in the dynamical depth model of EM is, thus, located in natural potencies characteristic of all natural beings, including particles, molecules, substances, more complex inorganic structures, living organisms, and conscious organisms. It cannot be attributed to or regarded as an outcome of the laws of thermodynamics and gradients of entropy, which themselves need an explanation (again, they are best explained in terms of dependencies between natural entities and dynamical systems).

The act of locating the source of the spontaneous character of processes occurring on different levels of dynamical depth in dispositional properties of concrete entities is not trivial. It not only helps to answer the question about the possibility of various homeodynamic, morphodynamic, and teleodynamic processes but also explains the phenomenon of their specific and unique nature. Linear thermodynamic processes such as various chemical reactions, or simple nonidealized mechanical systems (e.g., a harmonic oscillator or a pendulum with friction), classified as homeodynamic, do the work required to eliminate constraints and to achieve thermodynamic equilibrium because of the particular nature and potencies of their constituents (participants). Similar is the case of self-organizing processes of morphodynamics and of the reciprocal coupling of autocatalysis and self-assembly in teleodynamics. They are unique because they tend to remain far from equilibrium due to an external perturbation (morphodynamic), or even in its absence (teleodynamics). Each example of such interactions is case specific because of the particular dispositional qualities of entities engaged in it.

One of the great advantages of Deacon's theory of EM is his emphasis on the reciprocal and simultaneous character of orthograde processes juxtaposed in an emergent transition. It is crucial in particular for the autogenesis model of teleodynamics in which it is precisely reciprocity of the two systems—self-limiting, self-undermining, and to some extent mutually exclusive, when taken separately—that decides about the occurrence of a "negentropy ratchet effect." Here the dispositional view of causation (DVC) offers an excellent theoretical base for Deacon's model of EM. It argues against Humean theory of causation of distinct events (processes), with temporal priority of one of them. It understands causal relations as outcomes of dispositions of concrete entities, manifesting themselves in causal processes, simultaneous and reciprocal. This description can be applied to teleodynamics. Both autocatalysis and self-assembly can be depicted as two vectors in one quality space, producing an effect of a causal equilibrium, characterizing stability of an autogen. In the case of the partial destruction or self-replication of the autogen, the balance of the causal equilibrium is temporarily disrupted in favor of self-assembly or autocatalysis. It is then restored again as a result of the reciprocity between the two processes (see figure 7.1).

The emphasis of the DVC on the complex nature of real causal situations and on the polygeny and pleiotropy of causes finds a good example in a teleodynamic process of autogenesis as well. Since it is described as an outcome of the two morphodynamic processes, autogenesis shows the polygeny of causes engaged in one causal situation. At the same time, due to the nature of particular dispositions characteristic of their constituents, each morphodynamic system can enter into more than one specific process of an autocell formation, which proves their pleiotropic character.

Finally, we find one more close resemblance between Deacon's model of dynamical depth and the neo-Aristotelianism of dispositional metaphysics in their common emphasis on the dynamical aspects of reality. Deacon's claim that the universe is thermodynamic inevitably alludes to the dynamic properties of all entities that populate it, entering

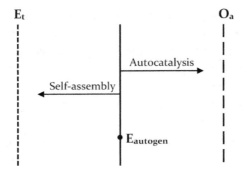

Figure 7.1. Vector model of the reciprocity and simultaneity of causes in autogenesis

E_t = thermodynamic equilibrium (maximum entropy production), which is usually reached by an unimpeded process of self-assembly; O_a = thermodynamic optimum for a transient nonequilibrium process of autocatalysis; $E_{autogen}$ = causal equilibrium of autogenesis.

causal processes described in his theory of EM. It is precisely this dynamic aspect of reality that dispositional metaphysics is interested in as well. Its contribution lies in explaining the irreducible and intrinsic character of dynamic properties, the nature of circumstances required for their manifestation, and the fact that they can, but need not, become activated. Hence, I argue that dispositional metaphysics becomes a useful tool in understanding and describing the engine of the dynamical depth model of EM.

2.2. Teleology at All Levels of Dynamical Depth

We have seen Deacon arguing in favor of a strategic meaning of teleology, which he thinks emerges at the highest level of dynamical depth and becomes the source of a primitive "self" of an autogen. Needless to say, his autogenesis theory of life's origin becomes an important development of Aristotle's philosophy of nature, which simply assumed the eternity of life. Today we postulate that life emerged out of nonliving systems, and many theorists and experimentalists are trying to explain and reproduce this process. I have mentioned that Deacon's great advantage over the alternative propositions of numerous theorists of life's origin is his emphasis on the role of teleology. At the same time, however, I disagree with his claim that teleology emerges from mere physico-dynamic processes, characterized by Aristotle's notion of efficient causation, and is specific only to the highest stage of dynamical depth.

Trying to specify the source of spontaneity of orthograde processes at the bottom level of each stage of Deacon's dynamical depth, I have suggested accepting dispositional metaphysics with its emphasis on irreducible kind-specific dispositional properties, characteristic for every nonliving and living being and decisive for the nature of all causal processes they enter. The same metaphysics acknowledges that each dispositional property is related to its proper manifestation under suppositional necessity. In other words, the property will be manifested in favorable and required circumstances, provided no impediments prevent its occurrence. I have mentioned that dispositional metaphysicians see here a natural tendency of dispositional properties to be manifested. They call this tendency "physical" or "natural intentionality" and find it a universal characteristic of all beings, living and nonliving.

I argue that this neo-Aristotelian revival of teleology, discovered and described at all levels of complexity of matter, is relevant to Deacon's model of EM, in which he frequently emphasizes natural tendencies of processes at different levels of dynamical depth to act in a specific way. It becomes clear to me that teleology is present at all stages of dynamical depth and cannot be limited only to its highest level. Naturally, I agree with Deacon that with the EM of an autocell a new type of teleology can be traced and described which is proper to autogenesis and is unlike the teleology characteristic of all lower dynamic systems. Hence, the highest level of dynamical depth can and should still be called teleodynamics, due to the novelty of its finality, providing for the primitive "self." But to say this unique type of teleology emerges from mere mechanical physico-dynamic systems is rather misleading. Contemporary metaphysics, following the teaching of Aristotle, finds teleology to be an irreducible type of feature (or type of causation), inextricably related to dispositions of concrete beings. Acceptance of this position brings us to the next important point in my argument.

2.3. The Locus of Form

In chapter 3, I discussed Deacon's tendency to locate form in the geometric properties of a probability space. I said that if form is defined as the behavior of a system, then we are left with the question about its constituents, their properties and tendencies to engage in processes described in terms of geometric properties of probability space. The question of formal principle simply goes back one level, but still remains unanswered. We could continue this reasoning going all the way down to the quantum level, always left with the same question of why the processes at the given level of complexity occur and why their participants tend to do what they do.

Acknowledging together with dispositional metaphysics that orthograde changes at any level of analysis are functions of particular and kind-specific dispositional properties of nonliving entities and organisms, rather than merely functions of the geometric properties of a probability space, we are invited to rediscover form as a metaphysical principle of unity and identity of beings. Deacon is right in pointing toward the importance of dynamical processes in his description of levels of complexity of emergent structures. His redefinition of form seems to offer a dynamic version of the structural neo-hylomorphism that sees form as structure, and matter as building materials or elements that are structured. But I argue that dispositional metaphysics, which I propose as a foundation for Deacon's model of EM, goes further, showing that the accumulation and growing complexity of quantitative aspects of dynamic systems does not by itself bring a qualitative change. Rather, dispositional metaphysics sees the qualitative aspects of dispositional properties of beings as decisive for their activity, leading to the growing complexity of nature. Some outcomes of this activity can be measurable quantitatively, but one aspect of reality does not reduce to the other. The growing complexity of essences of animate and inanimate things is mirrored in the growing complexity of measurable aspects of dynamical processes they enter, and vice versa, but we cannot conflate them, saying that qualitative change is an outcome of an accumulation of quantitative changes.

I earlier remarked (chapter 3, section 3.6) that Deacon accepts the reductionist paradigm of science. But the fact that we can isolate and measure all the internal structures, molecules, atoms, and quarks of a living organism does not entail our knowing the phenomenon of life. Deacon is aware of this and suggests bringing back teleology and formal causation. At the same time, however, he sees form as a function, and teleology as an emergent outcome of mere mechanical physico-dynamical processes. Dispositional metaphysics, by contrast, reminds us that these two features of every being are irreducible and not amenable to a mathematical description. They are rooted in the nature of animate and inanimate beings, characterized by kind-specific dispositions and their manifestations.

Knowing that in classical Aristotelianism formal causation is intrinsically related to primary matter, understood as the source of potentiality, I discussed in section 2.3 of chapter 6 and section 1.1 of the present chapter the possibility of the retrieval of Aristotelian hylomorphism in the new dispositional metaphysics. Those willing to follow this direction should find in hylomorphism another strong argument in favor of the irreducible character and interrelatedness of formal causation and teleology. But even without any reference to hylomorphism, dispositional metaphysics helps us understand that these types of causation are real and require a philosophical analysis that supplements our scientific inquiry. And yet, at the end of the day, Deacon seems to opt for metaphysical reductionism, claiming that the only type of causation different from and irreducible to physical efficient causation is the causation of absences, leading to the EM of the phenomenon of teleology. I will now try to analyze this statement from the perspective of dispositional metaphysics.

2.4. Constraints as Effects Rather Than Causes

Deacon is right in saying that constraints—that is, reduced degrees of freedom—tell us something important about the world around us. Observing dynamical and processual occurrences in nature, we can infer much about their character in reference to the way in which they are limited and constrained. Besides, even the more stable objects of our sensory experience are constrained in many ways, which becomes crucial for our perception and description of their nature.

What is peculiar about Deacon's project, however, is his assigning to constraints (absences) a real causal character. Here his theory encounters serious challenges, which I addressed in detail in chapter 3 (section 3.4). Ascribing causal activity to the more general category of absences, which Deacon sees as related to material constituents of reality, actual, and extended in space and time (features depicted in reference to his rhetorical device of the hole at the wheel's hub), raises the question about the nature of such causation. It is not clear whether it should be treated as efficient, formal, final, or maybe a new and unique type of causation.

In his attempt to offer a more specific account of the nature of the causation of absences, we saw Deacon describing their origin in terms of physical constraints, which he sees as being "contragrade, i.e., part of the efficient causality complex." According to his metaphysics, the efficient causation of constraints entails the reduction of degrees of freedom, "producing" absences. But we must not forget that constraints, as such, are always functions of limitations introduced by a mutual influence of the participants of two reciprocal processes (e.g., morphodynamic processes in autogenesis), or in a response of one and the same dynamic system to external energetic gradients. Moreover, further reflection on constraints shows they are intrinsically related to concrete entities and processes, the nature of which determines the possible degrees of freedom of a given thermodynamic system in the first place. Therefore, although I agree with Deacon that what is not present matters and that constraints are real and informative about dynamical systems and wholes, I maintain that constraints are effects of causal interactions rather than causes. Thus, I agree with one of Deacon's most recent statements on the nature of absences, in which he acknowledges that in fact they do not do any causal work (see chapter 3, section 3.4). But if this is the case, the whole dynamical depth model of EM requires a radical revision and restructuring that will provide a new answer to the question of the very source of an emergent transition.

Here dispositional metaphysics once again proves helpful. First, with its reference to intrinsic dispositions and their manifestations, it points toward the natures of beings entering causal processes as sources of constraints, which Deacon understands as reducers of various degrees of freedom. Thus, we can say it is the nature of the simultaneous and reciprocal relation of the processes of autocatalysis and self-assembly—which engage concrete molecules, characterized by kind-specific dispositions and their manifestations—that introduces the reduction of degrees of freedom typical for teleodynamics. As an effect of the processes in question, this reduction of degrees of freedom (resulting in new absences) is informative about their nature rather than causal. Second, the emphasis of dispositional metaphysics on the plural and complex character of real causal situations, expressed in vector causal modeling, helps us realize

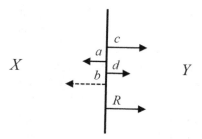

Figure 7.2. Vector model depicting "causality of absences"

The absence of the cause b changes the resultant vector R from causal equilibrium to disposing toward Y.

that we recognize the reduction of degrees of freedom only through an absence of a particular causal activity, the lack of which changes the entire causal situation. This shows once again that the notion of absence can be defined only in relation to something present (see figure 7.2).

Finally, the need for reference to the natures of objects and their dispositional causal propensities in defining the nature of constraints shows that, metaphysically speaking, Deacon's emphasis on the irreducible character of constraints must be followed by an argument about the irreducibility of natural kinds and physical intentionality. For if only constraints are irreducible, Deacon's project remains vulnerable to the criticism of Bokulich, who writes: "Recognizing that higher-level descriptions are made possible by the *constraints* that reduce the degrees of freedom of the underlying microphysics—and that the constraints themselves supervene on the microphysical facts—allows us to make sense of the physicalist claim that the microphysics fixes all higher-level facts."[7] I maintain that only by acknowledging the irreducibility of natural kinds and teleology can Deacon's theory avoid falling into metaphysical eliminativist reductionism. This argument leads to yet one more look at the issue of causation in the dynamical depth model of EM.

2.5. A Truly Pluralistic Theory of Causation

Deacon may be right in remaining skeptical about the notion of DC as defined in NP and the classical ontological theory of EM, which I thoroughly investigated in chapters 1 and 2. As long as DC is explained in terms of a new type of efficient causation, it is reducible to causes operating on the lower levels of complexity. Those who argue in favor of its irreducible, unique, and novel character may be accused of violating the rule of physical monism, introducing, on scientific grounds, a kind of causation which is mysterious and "spooky," not amenable to mathematical description. Without grounding of the classical DC-based theory of EM in a more thorough metaphysics, which I tried to do in the first half of this chapter, its project remains vulnerable to both scientific and philosophical skepticism concerning its plausibility.

To avoid the conflict faced by the proponents of the top-down version of emergentism, Deacon centers his own model of EM on dynamic interactions within systems, ascribing the crucial causal role to absences originating from constraints. I have already discussed the problem of the metaphysical character of the causality of absences. Here I want to go further, noting that Deacon's description of causation, unique and specific for each of the three levels of dynamical depth, reminds us in many ways of postulates raised by the proponents of DC in the classical theory of EM. Deacon sees the higher-order constraints in teleodynamics preventing the disruption of the synergy between autocatalysis and self-assembly and influencing the constituents of teleodynamic systems to the effect they can preserve and regenerate their own essential constraints and interact with the environment in a way sustaining supportive, and compensating for unsupportive, influences. Thus, teleodynamic systems are characterized by self-compression and self-reference. In morphodynamics Deacon points toward spontaneous regularizing tendencies of various systems to become more organized and globally constrained over time. In homeodynamics he observes a spontaneous inclination of phenomena to eliminate constraints, increasing global entropy and leading to thermodynamic equilibrium. Deacon characterizes all these emergent phenomena with reference to the language of causal influence exercised by complex dynamical systems or wholes on their constituents. But the nature of this causation remains a question.

Although it is possible to argue in defense of Deacon's description and its distinctiveness from the classical notion of DC, my reinterpretation of his theory of EM in terms of the new Aristotelianism places it relatively close to the new understanding of the main aspects of DC that I provided above. Applying dispositional metaphysics to the dynamical depth model of EM enables us to rediscover the philosophical concept of specific types of causation, derived from the fact that inanimate beings and organisms belong to natural kinds (Aristotelian formal cause), characterized by natural dispositions (tendencies) and their manifestations (Aristotelian final cause—teleology). I suggest that embracing these irreducible philosophical types of causation accommodates Deacon's description of new sorts of causes emerging on each level of dynamical depth and influencing lower-level system dynamics. Consequently, applying dispositional metaphysics to Deacon's project, we can offer a truly plural notion of irreducible types of causality that are active in the processes of growing complexity and EM. Most importantly, this list of causes goes beyond the efficient cause described in natural science and remains close to my reinterpretation of the main aspects of DC offered in this chapter.

Without doubt, my argument in favor of natural kinds, kind-specific dispositions, their manifestations, and teleology places great emphasis on the substantial aspect of reality. It has been objected in the modern critique of Aristotle that his concentration on substance undermines the importance of processual aspects of nature. Is then an attempt to defend substance metaphysics in the context of the contemporary process approach in science and philosophy possible and reliable?

2.6. Processes Not without Substances

Deacon's dynamical depth model of EM obviously shows a natural propensity toward process ontology. With dynamic systems at its center, it strongly emphasizes the constant flux of processes in nature. Without doubt, our growing awareness and knowledge of the nature of these processes has significantly influenced some recent developments in practical and theoretical science. But does this justify a metaphysical claim that processes replace substances at all levels of description, and especially at the most basic levels of micro and quantum physics? Is the radical critique and rejection of substance metaphysics, predominant in the second half of the twentieth century, coherent and plausible?

I have already discussed Deacon's criticism of substance metaphysics in chapter 3 (section 3.5). I have questioned his mereological interpretation of its premises, which contradicts the original Aristotelian definition, concentrated on substantial form as a source of unity, identity, and causal powers of living and nonliving beings. I have shown that by rejecting the mechanical explanation of whole-parts relations, Deacon opposes a reductionist account of substance metaphysics, rather than its true and original version given by Aristotle and developed in the course of the history of philosophy.

I have also mentioned the main weaknesses of event (process) ontologies (chapter 6, section 2.4). Because they overemphasize the importance of the Heraclitean flux of becoming, they are unable to give an account of and explanation for the unity and stability of the objects of our unaided and aided sensory experience. They do not provide a convincing explanation of how processes at the bottom level of reality build up into macro-scale stable objects. This observation proves that an element of the Parmenidean emphasis on the importance of the unity and stability of substance is needed for a relevant description of reality.

In the same section of chapter 6 I discussed the neo-Aristotelian character of dispositional metaphysics expressed in its recognition of the importance of process accounts in describing nature, but without rejecting the reality of substances. I suggest that this metaphysics is relevant to Deacon's theory of EM. Although his primary interest is in dynamical systems, we have seen that the spontaneity of their causal activity needs to be grounded in concrete entities, characterized by kind-specific dispositions and manifestations. Moreover, the emergent effect of teleodynamics is an autogen, which is described by Deacon and Cashman not only as "a reciprocally reinforcing system of self-propagating formative processes" but also as "a cohesive integral unit . . . a physical self that is the locus of an autonomous intrinsic end-directed dynamic."[8] This characteristic of an autocell points toward both its processual and substantial features. Its tendency of self-reparation and self-replication finds expression in the dynamic synergy of autocatalysis and self-assembly, which are kind-specific and localized in a concrete and integral unit. Both strategic tendencies of an autogen mentioned here can be classified as dispositional properties, manifested in specific conditions and grounded in its nature. This proves that both substantial and processual aspects of nature are equally real and crucial for the metaphysical base of Deacon's dynamical depth model of EM.

Conclusion

Clearly then Wisdom is knowledge about certain principles and causes. Since we are seeking this knowledge, we must inquire of what kind are the causes and the principles, the knowledge of which is Wisdom.
—Aristotle, *Meta.* 1.1.982a1–3

We began our journey noting that the interdisciplinary character of the search for causal explanation shaped the foundations of Western philosophy, empirical science, and technology. The research presented in this volume shows that the same question about the nature of causal dependencies continues to inspire this interdisciplinary conversation more than twenty-seven centuries later. Although my investigation concentrated on the philosophical aspects of causation in the systems approach to life sciences, I am aware that causal questions are being asked in the context of many other branches of science, such as health sciences (e.g., medicine, epidemiology), psychology, cognitive science, social sciences, various branches of natural sciences (other than systems biology—e.g., evolutionary biology), computer science, probability, statistics, and so on.

I outlined in the introduction the way in which the reductionist and mathematical-based approach to science gradually dismissed the robust fourfold concept of causation offered by Aristotle. As an outcome of this process, the notion of causation defined in terms of efficient (physical) pushes and pulls dominated both scientific and philosophical explanation since modernity. Today, the whole range of causal theories developed in the tradition of analytic philosophy—RVC (including INUSConVC and InfVC), CVC, ProbVC, SVC, M-basedVsC (including MVC, AVC,

ArelVC, and IntVC), and ProcVC (all discussed in chapter 4)—seem to inherit this same reductionist agenda (even if they argue, contra Hume, that the notion of physical causation has an ontological, and not merely epistemological, foundation). As a result, none of them succeeds, as they turn out to be either too inclusive, or too exclusive, or both. And yet their proponents state in their defense that these theories of causation do, in fact, explain many instances of causal situations. But even if it is true in some specific cases or refers to a more general approach of some specific disciplines of science, an attempt to provide a reductionist theory of causation that would explain the complexity of systems/organisms studied in systems biology in terms of efficient causes yields a theory that is ineffective and partial. It seems to me that the case of several other areas of science among those listed here might be similar.

One of the most important outcomes of the nonreductionist turn in the most recent science and philosophy of science is the revival of the theory of EM, which was the main object of my investigation. My critical analysis of the metaphysics of this theory led me to agree with a number of scholars (Claus Emmeche et al., Michael Silberstein, Charbel Niño El-Hani and Antonio Marcos Pereira, Alvaro Moreno and Jon Umerez), who claim that in its ontological (DC-based) version, EM can be saved only in the context of systemic causation which reaches beyond efficient causation and retrieves ideas of formal and final causes.

I found an attempt to realize this postulate in the dynamical depth model of EM proposed by Terrence Deacon. Acknowledging its novelty, originality, and intriguing character, I critically analyzed and evaluated its metaphysical foundations, its nonreductionist approach to causation, and its references to Aristotle's theory of four causes. I argued that Deacon's approach to the phenomenon of biological complexity—similar to the DC-based model of EM—needs a more thorough metaphysical foundation.

In response to the suggestion of scholars who argue in favor of reinterpreting the classical ontological view of EM in terms of Aristotle's formal and final causes, and to the ontological difficulties of Deacon's project, I have offered what I believe is the first systematic analysis and rethinking of the theory of EM and DC in the context of the old and the new Aristotelianism (i.e., the metaphysics of dispositions and their manifestations). I deliberately reached toward these two versions of Aristotle's thought to build bridges between ancient and contemporary metaphysics and to offer both of them as a reasonable and plausible philosophical proposition in the age of science. Even if the relation between these two ontologies is not intuitive and requires some adjustments, the project of bringing them together seems promising. I hope the future development of the conversation between the old and the new Aristotelianism will bring a further confirmation that Aristotle's fourfold notion of causation, his analysis of potency and act, and his understanding of change and stability in nature are by no means obsolete, static, or irrelevant to modern and contemporary science. I hope the conversation will also prove the status of the Aristotelian ontology and philosophy of nature as a valid proposition for grounding theoretical and philosophical reflection on contemporary science.

Looking once again at the main objectives of the research presented in this volume, I hope my reinterpretation and explanation of DC in terms of old and new Aristotelianism will help all the proponents of ontological EM to make their case stronger and more credible in their conversation with the many versions of ontological reductionism still present among scientists and philosophers of science. I hope it will provide them with a solid argument against Kim and other critiques of the reality and irreducibility of DC—an argument that does not refrain from interpreting DC in causal terms. I also hope my proposition of ontology underlying both substantial and processual aspects of reality, as well as the nonreductive character of formal and final causation—specific for each stage of the growing complexity of systems and structures in Deacon's model of dynamical depth—provides a necessary metaphysical ground for the defense of the philosophical consistency of his theory. Although an acceptance of the Aristotelian position would require introducing some changes in Deacon's version of EM and in his understanding of teleology, the benefits of such application prove significant. Scientific aspects of EM are enriched and grounded in a well-defined metaphysics, which makes the entire project an attractive proposition for a dialogue among the disciplines of science and philosophy.

Finally, my analysis of the phenomenon of complexity in nature, offered in this volume, concentrated mainly on its metaphysical and ontological aspects, discussed in the broader context of the philosophy of nature and the philosophy of science. Naturally, I realize it is not the only valid and important perspective for this kind of study. Neither do I find it superior to other tactics in explaining complex phenomena, including the metareflection on the methodological approach in biological sciences—offered in the new mechanical philosophy—or the actual experimental and computational practices used in both the nonreductionist systems-approach and the more reductionist attitude of molecular biology, biochemistry, chemistry, and physics (I find them all contributing in various ways to the study of complex phenomena). The inquiry into the nature of biological complexity requires a multiple-perspective and interdisciplinary approach that includes and presupposes both reductionist and nonreductionist agendas. One type of approach should not disqualify the other. I hope my project offers an important contribution to such diverse explanatory strategy.

APPENDIX 1

Potency and Act

One of the main points of interest in my project was the relation between potency and act. This is the first and principal metaphysical distinction, which—as I emphasized in section 1 of chapter 6—grounds all fundamental principles of Aristotle's metaphysics, including his notion of material and formal, as well as efficient and final causation. It also becomes a point of departure for his theory of stability and change in nature. I mentioned in section 2.1 of chapter 6 that Scholastic thinkers, who rediscovered, adapted, and further developed Aristotle's thought, introduced a number of additional metaphysical distinctions, which derive logically from the principal distinction between act and potency. Following the list of these distinctions offered by Feser in *Scholastic Metaphysics*, 38–41, I present them in figures A.1 and A.2.

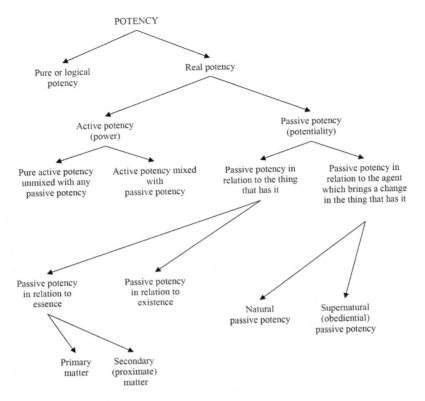

Figure A.1. Various kinds of potency

The category of potency divides, first, into the categories of pure or logical potency and real potency. Pure potency is also called objective potency, because it refers to objects of thought, without the need of them being actualized in reality—for example, a satyr (note that in this context "objective" does not mean "mind independent"). Real potency is also called a subjective potency, because it is grounded in a real subject (note that in this context "subjective" does not refer to what exists only as an object of consciousness). Next, within the category of real potency we can distinguish active potency (power; = the capacity to bring about an effect) and passive potency (potentiality in the strict sense; = the capacity to be affected). Active potency further divides into pure active potency, unmixed with any passive potency (attributed to the first mover, or God), and active potency mixed with passive potency (attributed to all other beings). Passive potency divides into passive potency considered in relation to the thing that has it and passive potency considered in relation to the agent which brings about an effect in the thing that has it. A further distinction can be made in the case of passive potency considered in relation to the thing that has it. Here, we can analyze passive potency considered in relation to the essence of a thing and passive potency considered in relation to existence of a thing. In case of the essence of material things, we can add one more distinction between primary matter (= pure potentiality to be in-formed by any form) and secondary (proximate) matter (= primary matter which has taken on a particular substantial form but is in potency to the reception of an accidental form or being in-formed by a new form in the process of substantial change). Finally, in the case of passive potency considered in relation to the agent that brings about an effect in the thing that has it, we can distinguish between natural passive potency = potentiality that can be realized due to a thing's natural capacities and an activity of an agent that is itself a mixture of active and passive potency, and supernatural (obediential) passive potency = potentiality that cannot be realized due to a thing's natural capacities and requires as an agent a purely actual divine cause (e.g., attaining the beatific vision).

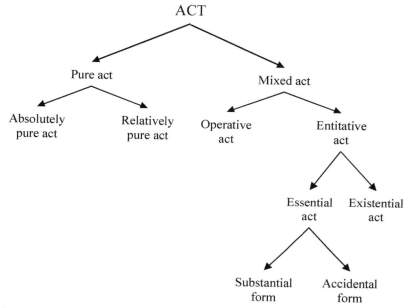

Figure A.2. Various kinds of act

The category of act divides, first, into the categories of pure act and mixed act. Within the category of pure act, we can next distinguish between absolutely pure act (= first mover or God) and relatively pure act (= an angel). The category of mixed act (all contingent things) is further divided into operative act (= a thing's operations and activities) and entitative act (= the way a given thing exists—what this thing is, statistically speaking). The latter is further divisible into essential act (= a thing's essence or nature) and existential act (= a thing's existence). Finally, the category of essential act grounds another distinction between substantial form (= a thing's "first act") and accidental form (= a thing's "second act").

APPENDIX 2

Process Metaphysics of Emergence

The debate on the metaphysical issues of EM theory is open. It has brought over the last few decades a variety of views and opinions concerning different aspects of this nonreductionist approach to reality. I referred to many of them during my analysis and argumentation. Readers familiar with the literature on EM may realize that the neo-Aristotelian constructive proposal developed in this book becomes an alternative to an interesting project introduced recently by Richard Campbell in his monograph *The Metaphysics of Emergence*.[1] I think his position requires a few words of comment here. I decided to share them in the appendix.

METAPHYSICAL IMPLICATIONS OF QUANTUM PHYSICS

What is distinctive about Campbell's view is his strong emphasis on the role of the dynamic and process aspects of reality. He claims that the entire Western tradition of metaphysics used to give priority to stable entities and their properties, treating them as primary ways of being, while the truth is that the reality is built of processes. Even after the Aristotelian view "disintegrated" in modernity, says Campbell, the new atomistic and corpuscular metaphysics of Descartes and his followers would still commit the same mistake of reifying being and prioritizing countable particular entities. Moreover, although A. N. Whitehead made an important step toward new metaphysics, his view of reality made up of momentary events ("actual occasions" building up into groupings or "societies of occasions") was still trapped in the atomistic mind-set. Answering this problem—in reference to the most contemporary physics and a number of publications authored by Johanna Seibt and Mark Bickhard—Campbell argues in favor of a new metaphysics which has the sea of quantum fields—best conceptualized as a foam of spontaneous, quantized excitations—as the most elementary aspect of reality. Trying to specify the principles of his view, Campbell says:

> *Fundamental particles* simply do not exist in any metaphysically serious sense.... Phenomena with *some* particle-like properties emerge spontaneously from quantum fields as result of quantization, but their number is not fixed; they can be created and annihilated. And certain properties emerging from relatively stable interactions amongst quantum fields constitute entities, which are *derived*, not fundamental, phenomena. That conclusion has ramifications for how we understand phenomena on the macroscopic level. For it follows that *everything* is ultimately composed of quantum fields, of various scales and complexity.
>
> If there are no fundamental particles in any metaphysically serious sense, QFT [quantum field theory] shifts the basic composition of the universe from micro-particles to quantum fields-in-process, for these fields are simply organizations of energy in process. If nothing is composed fundamentally of substance-like particles, that suggests that everything in the world has to be conceived fundamentally as *processes* of various scales and complexity, having causal efficacy in themselves.
>
> It follows that the world is composed of *organized fields in process*—all the way down, and all the way up. *Everything* there is has now to be understood as emerging from organizations of energy, excitations in the basic vacuum.
>
> ... What is standardly called a "particle" is a *quantized excitation of a field*.... Photons arise from the *quantization* of the electromagnetic field. And massive, charged "particles" such as electrons, protons, etc. arise from the quantization of their respective fields.[2]

Translating this new metaphysics into the language of EM theory, Campbell—in reference to Wimsatt—describes "EM bases" and phenomena as follows:

Emergent phenomena emerge over time in complex organizations of processes. ... The conventional notion of 'emergence bases,' when applied across the board, is deeply incoherent. The typical talk of 'constituents' belongs with the particle metaphysics. ... Using these locutions is to ignore the crucial role of organization in the emergence of those novel properties and powers which establish one or more kinds of unity and legitimate our calling the resultant an entity. It takes no account of the fact that what constitutes an entity is the emergence of the temporal properties of stability, coherence, and cohesion, which require a process framework. And it ignores the essential role of physically external relations in the self-maintenance of stable far-from-equilibrium process systems, from flames to human beings.

A property is emergent if and only if it is a system property which is necessarily dependent upon the mode of that system's organization. ... *An entity or process system is ontologically emergent if and only if it exists in the same time and place as its parts and has distinctive properties and modes of interaction which are necessarily dependent upon the mode of organization of its parts.*

What is metaphysically significant is that, in [some] cases of non-linear integration, the properties of the whole are somehow "more" than the aggregation of the properties of its parts—such system properties, and the causal powers of such a system, are emergent. Emergence should no longer be viewed as a dubious metaphysical mystery, but as explicable in terms of non-linear functions.[3]

DIFFICULTIES OF THE NEW PROCESS METAPHYSICS

Although Campbell's metaphysics and his view of EM are both interesting and intriguing, they raise some important questions. First of all, it is certainly true that metaphysicians must pay attention to the scientific description of the universe. It is also true that contemporary science tells us much more about the dynamic and processual aspects of reality than was known in the times of Aristotle. At the same time, however, we should not forget about the controversy concerning the interpretation of the mathematical formalism of quantum physics. Both scientists and philosophers of science are divided on the question whether QFT conveys a meaningful ontological truth about the reality of the world, sufficient for establishing a comprehensive and plausible metaphysics.

Adopting a full-blown realist position in this debate, Campbell is aware of the antirealist view of those who claim that—being honest about the unique character of the methodology and language of quantum physics—QFT joins other scientific theories that "should be regarded just as 'convenient fictions,' or as sets of quasi-descriptions with heuristic significance only, or instrumentally, as having only predictive value." Even a more modest position of structural realism holds that "we should commit ourselves only to the mathematical or structural content of our theories, but not to the unobservable nature of the phenomena described."[4] Campbell does not seem to fully justify his choice of scientific realism about QFT over the positions of radical and moderate antirealism (structural realism).

Nevertheless, even if we put aside the epistemological question and follow Campbell in his realism about QFT, we find it necessary—as metaphysicians—to ask some fundamental questions concerning quantum fields and their nature. We hear from Campbell on numerous occasions that quantum fields are simply organizations of energy in processes, which cannot be described with the categories of classical substance metaphysics. However, a closer analysis of Campbell's position shows that quantum fields have several important features that do fall under at least some of the Aristotelian categories. We learn, for instance, that they are extended in space and time (category of quantity) and come in different types (having different qualities), as we can distinguish "electron fields, . . . quark fields, neutrino fields, gluon fields, W and Z-boson fields, Higgs fields and a whole slew of others."[5] In other words, there is no one generic type of quantum field but numerous types of quantum fields, which translates into various results of the mathematical computational algorithm of quantization (allowing us to deduce ontological properties of what we traditionally classify as elementary particles). What is even more important, quantum fields, among other "*generic processes*, of various scales and complexity," have "causal efficacy in themselves."[6] We read in Campbell that they have the power to "organize themselves." The truth is that even "the processes constituting quantum fields always manifest some organization, however minimal."[7] These powers of organization (falling under categories of action and reaction) become the vehicle of EM, going from the most basic to the most complex phenomena, which can be described as "deriving" from the nexuses of quantum fields.

This list of basic features of quantum fields, when analyzed from the metaphysical point of view, leads to a series of fundamental questions. First of all, do we really know what they are? (A category of "process organization of energy" sounds quite enigmatic.) What makes quantum fields be extended in space and time? What makes each one of them be of a particular kind (electron, quark, neutrino quantum field, etc.)? What makes them organize into a hierarchical dynamics of quantum fields at various levels of organization of matter? In other words, what makes them enter into particular interactions with other quantum fields? Do they change in such interactions? Is it plausible to hold that under QFT *"fundamental particles* simply do not exist in any metaphysically serious sense," while emphasizing that the same theory does not lead one "to deny the existence of particular entities"?[8] How does the category of being "derived" differ from being epiphenomenal?

BACK TO THE CLASSICAL SOLUTIONS

I hold that these questions bring us back to the most fundamental query concerning the source of stability and change in nature, a question that arises in Campbell's metaphysical system just as it does in any other philosophical/scientific ontology. I think that his category of quantum fields points toward the principle of potentiality as underlying the very fabric of the cosmos. It is thus reminiscent of Aristotle's category of prime matter, except that prime matter is a strictly metaphysical category, whereas potentiality of quantum fields is still grounded/defined in terms of a physical reality. Therefore, Campbell's view of potentiality raises important questions which I have listed above. At the same time, I find his dismissal of the reality of primary matter in Aristotle too hasty and lacking a sound explanation.[9]

The fact that quantum fields are extended in space and time, fall within categories of different types, and show causal powers to interact with other quantum fields—which makes them organize into dynamic nexuses—points toward a metaphysical principle of actuality, without which these features seem to be groundless. Moreover, if Campbell wants to bring his radical process manifesto to its logical conclusion, it will be difficult for him to salvage the reality of particular entities. If it is true that even *"fundamental particles* simply do not exist in any metaphysically serious sense," all of them, as well as all other entities at the micro and macro scale, can be at most epiphenomenal. Assigning to them an identity "through change" (a "functional essence") requires a principle of actuality, providing for the changes at the global level (be it the level of an elementary particle or of any other, more complex entity/nexus of quantum fields).

I hold that Aristotle's metaphysics does have a principle of this kind—namely, the metaphysical category of substantial form. I believe his theory of hylomorphism can be reinterpreted in a way fitting with the contemporary view of reality. To put it in another way, although many of Aristotle's scientific claims are not valid today, his metaphysics proves still accurate and capable of providing a necessary and balanced theoretical background for contemporary science, which can ground the ontological reality of both stability and change. It may help us avoid the extremes of both Parmenides, who denied the reality of change, and Heraclitus, who denied the reality of stability. To show this in reference to the theory of EM was the main goal of this book.

NOTES

INTRODUCTION

1. Apart from references to primary sources, my research offered in the introduction is based on several important works on the topic of causation, including Helen Beebee, Christopher Hitchcock, and Peter Menzies, eds., *The Oxford Handbook of Causation* (New York: Oxford University Press, 2009), especially parts 1–3; Edwin Arthur Burtt, *The Metaphysical Foundations of Modern Science* (Mineola, NY: Dover, 2003; first published in 1924); Kenneth Clatterbaugh, *The Causation Debate in Modern Philosophy, 1637–1739* (New York and London: Routledge, 1999); Menno Hulswit, *From Cause to Causation: A Peircean Perspective* (Dordrecht: Kluwer Academic, 2002), chapters 1–2; Ernest Sosa and Michael Tooley, eds., *Causation* (Oxford and New York: Oxford University Press, 1993); Phyllis Illari and Federica Russo, *Causality: Philosophical Theory Meets Scientific Practice* (Oxford: Oxford University Press, 2014); John Losee, *Theories of Causality: From Antiquity to the Present* (New Brunswick, NJ, and London: Transaction Publishers, 2011); Mario Bunge, *Causality and Modern Science* (New York: Dover, 1979; first published in 1959); and a brilliant study by William A. Wallace, *Causality and Scientific Explanation*, 2 vols. (Ann Arbor: University of Michigan Press, 1972–74), to which I am especially indebted.

2. Alexander of Aphrodisias notices the difference in meaning between the two terms, saying that ἀρχή (*archē*) is that "from which," while αἰτία (*aitia*) signifies that "through which." Simplicius adds that ἀρχή has a much broader meaning than αἰτία. Latin distinguishes between *principium* and *causa*, the second term signifying a lawsuit, matter of dispute, or circumstances of an action and deriving from *cavere* (to defend) and from *cudo* (a manslaughter). See Leo J. Elders, *The Metaphysics of Being of St. Thomas Aquinas* (Leiden: E. J. Brill, 1993), 270–72.

3. This classification follows the one presented by Michael Dodds in *The Philosophy of Nature* (Oakland, CA: Western Dominican Province, 2010), 1–4.

4. It is important to distinguish between a "principle"—defined as that from which something proceeds in any way whatever—and a "cause," which is understood as that from which something proceeds with dependence in being. Hence, a

given type of cause—regarded as a special kind of principle—can be classified as a "causal principle."

5. "Thales, the founder of this type of philosophy, says the principle is water (for which reason he declared that the earth rests on water), getting the notion perhaps from seeing that the nutriment of all things is moist, and that heat itself is generated from the moist and kept alive by it (and that from which they come to be is a principle of all things). He got his notion from this fact, and from the fact that the seeds of all things have a moist nature, and that water is the origin of the nature of moist things" (Aristotle, *Meta.* 1.3.983b19–27). See also Aristotle, *De cae.* 2.13.294a28. Unless otherwise noted, quotations from Aristotle's *Physics, On the Heavens, On Generation and Corruption,* and *Metaphysics* come from McKeon's edition of *The Basic Works of Aristotle*, and quotations from his *On the Generation of Animals, Meteorology,* and *On the Parts of Animals* come from Barnes's edition of *The Complete Works of Aristotle*.

6. "'As our soul which is air,' he says, 'holds together, so wind (πνεῦμα [*pneuma*]) and air encompass the whole world'" (Aetius Doxographus, *Aetii de Placitis Reliquiae* 1.3, *Doxographi graeci* 278; I am using the translation found in Milton C. Nahm, *Selections from Early Greek Philosophy* [Englewood Cliffs, NJ: Prentice-Hall, 1964], 43). "Anaximenes of Miletus, son of Eurystratos, a companion of Anaximander, agrees with him that the essential nature of things is one and infinite, but he regards it as not indeterminate but rather determinate, and calls it air (Simplicius, *Simplici in Aristotelis Physicorum Libros Quattuor Priores* 6 r. *Doxographi graeci* 476, in Nahm, *Selections*, 43). "The air differs in rarity and in density as the nature of things is different" (ibid.). "And the form of air is as follows: When it is of a very even consistency, it is imperceptible to vision, but it becomes evident as the result of cold or heat or moisture, or when it is moved. It is always in motion; for things would not change as they do unless it were in motion. It has a different appearance when it is made more dense or thinner; when it is expanded into a thinner state it becomes fire, and again winds are condensed air, and air becomes cloud by compression, and water when it is compressed farther, and earth and finally stones as it is more condensed" (Hippolytus, *Hippolyti Philosophumena* 7, *Dox.* 560, in Nahm, *Selections*, 44).

7. "He [Anaximander] said that the first principle and element of all things is infinite, and he was the first to apply this word to the first principle; and he says that it is neither water nor any other one of the things called elements, but the infinite is something of a different nature, from which came all the heavens and the worlds in them" (Simplicius, *Simplici* 6 r., *Dox.* 476, in Nahm, *Selections*, 39). Dodds notices that if we substitute "energy" for "air," Anaximenes becomes Einstein. He adds that Heisenberg saw Anaximander's "infinite" as equivalent to "energy" in modern physics. Dodds argues, however, that "[if] energy is viewed as the universal substance underlying all things . . . Heisenberg begins to look like Thales. Instead of 'water' as the fundamental substance, we have 'matter' or 'energy.'" See Dodds, *Philosophy*

of Nature, 2–3; Albert Einstein and Leopold Infeld, *The Evolution of Physics* (New York: Simon and Schuster, 1954), 256–57; Werner Heisenberg, *Physics and Philosophy* (New York: Harper and Brothers, 1958), 61–62.

8. See the excerpts of Heraclitus's writings translated by Richmond Lattimore in Matthew Thompson McClure, *The Early Philosophers of Greece* (New York: Appleton-Century-Crofts, 1935), 119–28, especially nos. 21–22, in which he claims that "the changed states of Fire are, first, sea; half of sea is earth, and half is stormcloud. All things are exchanged for Fire and Fire for all things, as goods for gold and gold for goods." In nos. 41 and 81 he famously states that "you could not step twice in the same rivers; for other and yet other waters are ever flowing on. . . . In the same rivers we step and we do not step. We are and we are not." See also Nahm, *Selections*, 76. Dodds notices that Heraclitus can be seen as a predecessor of Whitehead's philosophy of process (Dodds, *Philosophy of Nature*, 4).

9. See McClure, *Early Philosophers*, 145–49 (especially nos. 2, 4, 8). In no. 8 we read Parmenides saying, "What *is* is without beginning, indestructible, entire, single, unshakable, endless; neither *has* it been nor *shall* it be, since now it *is*; all alike, single, solid. For what birth could you seek for it? Whence and how could it have grown? I will not let you say or think that it was from what is not; for it cannot be said or thought that anything is not. What need made it arise at one time rather than another, if it arose out of nothing and grew thence? So it must either be entirely, or not at all." For interpretations and evaluations of Parmenides' philosophy see Aristotle, *Phys.* 1.3.186a4–32; *Meta.* 1.5.986b18–987a1. Hippolytus says: "Parmenides supposes that the all is one and eternal, and without beginning and spheroidal in form; but even he does not escape the opinion of the many, for he speaks of fire and earth as the first principles of the all, of earth as matter, and of fire as agent and cause, and he says that the earth will come to an end, but in what way he does not say" (*Hippolyti* 11, *Dox.* 564, in Nahm, *Selections*, 96).

10. "In all things there is a portion of everything except [*Nous*] mind; and there are things in which there is mind also. Other things include a portion of everything, but mind is infinite and self-powerful and mixed with nothing, but it exists alone itself by itself" (excerpts in Nahm, *Selections*, 141). See also Plato, *Phaedo*, in *Plato*, vol. 1, trans. Harold North Fowler (Cambridge, MA: Harvard University Press, 1914), 97b; Aristotle, *Meta.* 1.3.984b8, 1.4.985a18.

11. Plato, *Phaedo* 96a–100b. See also Sarah Broadie, "The Ancient Greeks," in Beebee, Hitchcock, and Menzies, *Oxford Handbook of Causation*, 23–24.

12. "And these [elements] never cease changing place continually, now being all united by Love into one, now each borne apart by the hatred engendered of Strife, until they are brought together in the unity of the all, and become subject to it" (excerpts in Nahm, *Selections*, 118). See also Aristotle, *Meta.* 1.4.915a21.

13. "Those who spoke of atoms, i.e., certain infinitely numerous, indestructible, and extremely small bodies, and who assumed a void space of unbounded size, say that these atoms move at random in the void and collide with each other by

chance because of an irregular momentum, and they catch hold of each other and are entangled because of the variety of shapes. Thus they produce the world and its contents, or rather, an infinite number of worlds" (Dionysius, *On Nature*, from Eusebios, *Preparatio Evangelicae* 14.23.2.3, in Nahm, *Selections*, 158–59).

14. "Pythagoras: The universe is made from five solid figures, which are called also mathematical; of these he says that earth has arisen from the cube, fire from the pyramid, air from the octahedron, and water from the icosahedron, and the sphere of the all from the dodecahedron" (Aetius, *Aetii* 2.6, *Dox.* 334, in Nahm, *Selections*, 59). "And the Pythagoreans, also, believe in one kind of number—the mathematical; only they say it is not separate but sensible substances are formed out of it. For they construct the whole universe out of numbers—only not numbers consisting of abstract units; they suppose the units to have spatial magnitude. But how the first 1 was constructed so as to have magnitude, they seem unable to say" (Aristotle, *Meta.* 13.6.1080b16–22). See also Wallace, *Causality*, 1:18–19.

15. See Aristotle, *Meta.* 1.5.986a15–21.

16. Plato, *Timaeus*, in *Plato*, vol. 7, trans. R. G. Bury (Cambridge, MA: Harvard University Press, 1929), 28a.

17. See ibid., 50b–56d.

18. When we use terms such as "science" or "scientific" in reference to the realms of theoretical and practical knowledge in ancient Greece, we must be aware that their meaning is at most analogical when compared with the contemporary (introduced in the nineteenth century and pragmatically oriented) definitions of the same terms. Aristotle was certainly a "student of nature" or a "knower of nature," but not a "scientist" *per se*.

19. See Wallace, *Causality*, 1:20–21.

20. Aristotle, *Phys.* 2.3.194b24–28.

21. Aristotle, *Meta.* 5.2.1013a24–29.

22. Such is the mistake of some of the contemporary philosophers of science who argue for the retrieval of Aristotle's notion of causation. See, for instance, Claus Emmeche, Simo Køppe, and Frederic Stjernfeld, "Levels, Emergence, and Three Versions of Downward Causation," in *Downward Causation: Mind, Bodies and Matter*, edited by Peter Bøgh Andersen, Claus Emmeche, Niels O. Finnemann, and Peder Voetmann Christiansen (Aarhus and Oxford: Aarhus University Press, 2000), 17; Alwyn C. Scott, *The Nonlinear Universe* (Berlin: Springer, 2007), 6.

23. On numerous occasions in the entire Aristotelian corpus we find descriptions of matter which tend to suggest Aristotle sees it as being tangible. For instance, in *Meta.* 7.5.1044b34–1045a6 we read:

> It is also hard to say why wine is not said to be the matter of vinegar nor potentially vinegar (though vinegar is produced from it), and why a living man is not said to be potentially dead. In fact they are not, but the corruptions in question are accidental, and it is the *matter* of the animal that is itself in virtue of its

corruption the potency and matter of a corpse, and it is water that is the matter of vinegar. For the corpse comes from the animal, and vinegar from wine, as night from day. And *all* the things which change thus into one another must go back to their matter; e.g. if from a corpse is produced an animal, the corpse first goes back to its matter, and only then becomes an animal; and vinegar first goes back to water, and only then becomes wine.

It might be possible that Aristotle's concentration on proximate rather than primary matter is an outcome of his being faithful to the rule he himself formulates earlier on in the same seventh book of the *Metaphysics*, where he says that while, "when one inquires into the cause of something, one should, since 'causes' are spoken of in several senses, state all the possible causes, ... it is the proximate causes we must state." Hence, when asking, "What is the material cause [of man]?, we must name not fire or earth, but the matter peculiar to the thing ... [i.e.,] 'the menstrual fluid'" (*Meta.* 7.4.1044a34–1044b2). Following this example, our analysis of concrete cases of substantial change—states Aristotle—should always provide a proximate material cause, without tracing its ultimate underlying principle down to basic elements, not to mention primary matter. Such an attitude, however, does not exclude the reality of the latter.

24. Aristotle, *Phys.* 1.7.191a8–12.
25. Ibid., 1.9.192a25–33.
26. Aristotle, *Meta.* 7.3.1029a20–21, 24–25.
27. Ibid., 9.7.1049a24.
28. Aristotle, *De gen. et corr.* 2.5.332a18–19. Once again, since "there is nothing—at least, nothing perceptible—prior to these [four elements], they must be all" (ibid., 332a27–28). Consequently, for the elements to change into one another, there must be something prior and common to all of them. Primary matter, metaphysically prior—at least in some respect (in other respects substantial form is always prior to material cause)—and imperceptible, fulfills this task.
29. Ibid., 2.5.332a34–35. Aristotle continues this reflection and concludes that given there are two basic pairs of contrarieties (hot-cold and wet-dry), there must be four basic elements, each characterized by two qualities, one from each pair of contrarieties: fire (hot and dry), air (wet and hot), water (cold and wet), earth (dry and cold); two of the six possible combinations—hot and cold, and wet and dry—lead to a contradiction and cannot occur in nature (see ibid., 332b1–5). We will refer shortly to Aristotle's introduction of his list of elements and their natures in chapters 3 and 6.
30. Commenting on the tendency to treat matter as substance, Aristotle notes that "both separability and 'thisness' are thought to belong chiefly to substance. And so form and the compound of form and matter would be thought to be substance, rather than matter" (*Meta.* 7.3.1029a29–30).
31. The traditional view stating that Aristotle believed in primary matter as a single, eternal, and completely indeterminate substratum of all change in nature has

become an object of controversy among some contemporary Aristotelian scholars. To grasp the conversation—the analysis of which goes beyond our interest here—I refer the reader to the following articles: (1) challenging the traditional view, Hugh R. King, "Aristotle without *Prima Materia*," *Journal of the History of Ideas* 17 (1956): 370–89, and William Charlton, "Did Aristotle Believe in Prime Matter?," in Aristotle, *Physics: Books I and II*, trans. with introduction and notes by W. Charlton (Oxford: Clarendon Press, 1983), 129–45; (2) answering King and Charlton (successfully, in my opinion), Friedrich Solmsen, "Aristotle and Prime Matter: A Reply to Hugh R. King," *Journal of the History of Ideas* 19 (1958): 243–52, and H. M. Robinson, "Prime Matter in Aristotle," *Phronesis* 19 (1974): 168–88.

32. See Aristotle, *Phys.* 2.3.194b26; *Meta.* 5.2.1013a27. Other translators offer various interpretations of this difficult phrase. Wicksteed and Cornford read ὁ λόγος τοῦ τί ἦν εἶναι as "[something] conformity to which brings it [a given thing] within the definition of the thing we say it is," Tredennick as "the essential formula," Charlton as "the account of what the being would be," and Kirwan as "the formula of what it is to be." See their translations of the same passages from the *Physics* and the *Metaphysics*. All these suggestions seem to have a similar metaphysical import and meaning.

33. Dodds, *Philosophy of Nature*, 25. Michael Storck notes that "not only do we not sense substantial forms, but we do not measure them with scientific instruments either. We sense the size, shape, color, and so forth, of things, and we measure their frequency, mass, temperature, electrical charge, and so on. It is only through our intellect that we are able to grasp something, often not very clearly, of the substantial forms of natural things" (Michael Hector Storck, "Parts, Wholes, and Presence by Power: A Response to Gordon P. Barnes," *The Review of Metaphysics* 62 [2008]: 55).

34. Terence Irwin, *Aristotle's First Principles* (Oxford: Clarendon Press, 1988), 212. Irwin claims Aristotle links the concept of essence to the ideas of substance and subject: "What has the essence is the primary thing that persists; this is the basic subject, so it is the substance" (ibid., 216). He also shows the reference between the predication of concrete forms and the predication of universal natural kinds (the genus), which is foundational for Aristotelian moderate realism of universals.

35. See, for instance, Aristotle, *De gen. et corr.* 2.9.335b7. Joachim translates εἶδος (*eidos*) as "figure." We should note that in the logical works this word is usually understood as "species." Species are also described in this part of the Aristotelian corpus as "types" or "genera" (γένη, *genē*), saying what a given thing is.

36. See, for instance, Scott, *Nonlinear Universe*, 6. In the second book of the *Physics*, Aristotle explains briefly that the formal cause answers the question "why" something is the kind of thing it is, but without involving any motion (*Phys.* 2.7.198a17–18). In *Meta.* 1.3, he refers indirectly to the formal cause, defining it as the essence, the ultimate "why," and the ultimate principle: "Causes are spoken of in four senses. In one of these we mean the substance, i.e. the essence (for the 'why' is

reducible finally to the definition, and the ultimate 'why' is a cause and principle)" (*Meta.* 1.3.983a26–29). See also *Meta.* 7.17.1041b11–32. These passages prove once again that the meaning of "form" in Aristotle's corpus cannot be reduced or limited to the outward shape or appearance.

37. Irwin, *Aristotle's First Principles*, 100.

38. "For the action is the end, and the actuality is the action. And so even the word 'actuality' [ἐνέργεια, *energeia*] is derived from 'action' [ἔργον, *ergon*], and points to the complete reality [ἐντελέχεια, *entelecheia*]" (*Meta.* 9.8.1050a21–23). As Sachs notes, "Aristotle invents the word by combining *enteles* [ἐντελής] (complete, full-grown) with *echein* (= *hexis*, to be a certain way by the continuing effort of holding on in that condition), while at the same time punning on *endelecheia* [ἐνδελέχεια] (persistence) by inserting *telos* [τέλος] (completion). This is a three-ring circus of a word, at the heart of everything in Aristotle's thinking, including the definition of motion" (Joe Sachs, *Aristotle's Physics: A Guided Study* [New Brunswick, NJ: Rutgers University Press, 1995], 245).

39. Fran O'Rourke, "Aristotle and the Metaphysics of Evolution," *The Review of Metaphysics* 58 (2004): 17. We should be careful with O'Rourke's use of εἶδος and μορφή as terms for form (see above). He is right in saying that ἐντελέχεια is the τέλος (end) of γένεσις (*genesis*). However, I find it more appropriate to say that ἐντελέχεια is the final stage of the development of form understood as ὁ λόγος τοῦ τί ἦν εἶναι (the statement of the essence), rather than that ἐντελέχεια is εἶδος (appearance). We also need to note that since form in itself is a metaphysical principle, rather than a physical thing, it cannot develop or evolve. Thus, even if we can speak about the final stage of the realization of form in an organism, the form has been fully present in that organism since the beginning of its existence, actualizing primary matter that underlies it. See also David Bostock, "Aristotle's Theory of Form," in his *Space, Time, Matter, and Form: Essays on Aristotle's Physics* (Oxford: Clarendon Press, 2006), 79–102.

40. Aristotle, *De gen. et corr.* 1.4.319b10–18.

41. Aristotle, *Phys.* 2.3.194b29–195a2. A similar explanation can be found in Aristotle's *Metaphysics*: "'Cause' means . . . (3) That from which the change or the resting from change first begins; e.g. the adviser is a cause of the action, and the father a cause of the child, and in general the maker a cause of the thing made and the change-producing of the changing. (4) The end, i.e. that for the sake of which a thing is; e.g. health is the cause of walking. For 'Why does one walk?' we say; 'that one may be healthy'; and in speaking thus we think we have given the cause. The same is true of all the means that intervene before the end, when something else has put the process in motion, as e.g. thinning or purging or drugs or instruments intervene before health is reached; for all these are for the sake of the end, though they differ from one another in that some are instruments and others are actions" (*Meta.* 5.2.1013a29–1013b2). See also *Phys.* 2.7.198a18–20; *Meta.* 1.2.983a30–32.

42. Aristotle, *Phys.* 3.3.202a14.

43. See, for instance, Aristotle, *De part. an.* 3.2.663b12–14; 4.5.679a25–30; Aristotle, *De gen. an.* 2.4.739b27–31; 3.4.755a17–30. Bostock lists a number of scholars claiming that "Aristotle would concede (at least for the sake of argument) that a complete materialist explanation might perhaps be available, and yet still insist that a teleological account was *also* needed" (Bostock, *Space*, 58). He suggests this "seems to be roughly the position that we ourselves are in nowadays" (ibid., 60).

44. Commenting on this normative aspect of teleology, Bostock states that we can speak about a "law of goodness" in Aristotle which assumes that "there is something that counts as good, namely what is good for the animal or plant concerned." Consequently, "whatever parts a living thing needs, in order to live a life that is good for it, will for that very reason tend to be present in it (and therefore will grow as it grows). The law is limited in its application, of course, by the fact that the 'laws of matter' will only permit some kinds of parts to develop, and not others. . . . It is for him [Aristotle] a law that is basic and irreducible" (*Space*, 77–78). For more information on teleology in Aristotle's biology see Allan Gotthelf, *Teleology, First Principles, and Scientific Method in Aristotle's Biology* (Oxford: Oxford University Press, 2012).

45. Aristotle, *De part. an.* 2.1.646a31–34. Aristotle's profound conviction that "nature does nothing in vain" makes him think that there has to be a purpose for everything in nature. At the same time, he is careful to note that the final cause is not acting *sensu stricto*: "The active power is a 'cause' in the sense of that from which the process originates: but the end, for the sake of which it takes place, is not 'active.' (That is why *health* is not 'active,' except metaphorically.) For when the agent is there, the patient *becomes* something: but when 'states' [ἕξεων, *hexeōn*, dispositions] are there, the patient no longer *becomes* but already is—and 'forms' [εἴδη, *eidē*] (i.e. 'ends') [καὶ τὰ τέλη, *kai ta telē*] are a kind of 'state' [ἕξεις, *hexeis*]" (*De gen. et corr.* 1.7.342b14–18).

46. Bostock, *Space*, 71.

47. See also Bostock, *Space*, 48–78; Allan Gotthelf, "Aristotle's Concept of Final Causality," *The Review of Metaphysics* 30 (1976): 226–54; W. K. C. Guthrie, *A History of Greek Philosophy*, vol. 6, *Aristotle: An Encounter* (New York and Cambridge: Cambridge University Press, 1981), 114–18.

48. See Mark Bedau, "Where's the Good in Teleology?," *Philosophy and Phenomenological Research* 52 (1992): 783–87. See also Bedau's response to the criticism of the value analysis of teleology as viciously circular and spurious in reference to the cases of supposedly value-free teleology in ibid., 793–94.

49. Ibid., 791.

50. See William A. Wallace, *The Modeling of Nature: Philosophy of Science and Philosophy of Nature in Synthesis* (Washington, DC: Catholic University of America Press, 1996), 15–18.

51. Alwyn Scott sees final causality mainly as a desire of an intentional organism. He does not pay enough attention to the final causality inherent to all nature.

That is why he struggles with the problem of "purposive answers" in biological sciences (see Scott, *Nonlinear Universe*, 6). Emmeche, Køppe, and Stjernfelt, who argue for a retrieval of Aristotle's notion of four causes in philosophy of science, seem to think of the final cause only in terms of a natural—that is, unconscious—inclination toward maintaining the stability of "an integrated processual whole." That is why, instead of final cause, they introduce the term "functional cause" (see Emmeche, Køppe, and Stjernfelt, "Levels, Emergence," 17). Within the domain of natural science, Aristotle's notion of teleology is widely debated in the context of evolutionary biology.

52. Aristotle, *Phys.* 2.1.192b22–23.

53. The need of reference to all four causes in causal explanation is crucial, even though in many cases it might be difficult to specify them. Aristotle seems to be aware of this problem when he gives two examples of common phenomena escaping basic rules of causal explanation: "E.g. what is the cause of eclipse? What is its matter? There is none; the *moon* is that which suffers eclipse. What is the moving cause which extinguished the light? The earth. The final cause perhaps does not exist. The formal principle is the definitory formula, but this is obscure if it does not include the cause. E.g. what is eclipse? Deprivation of light. But if we add 'by the earth's coming in between,' this is the formula which includes the cause. In the case of sleep it is not clear what it is that proximately has this affection. Shall we say that it is the animal? Yes, but the animal in virtue of what, i.e. what is the proximate subject? The heart or some other part. Next, by what is it produced? Next, what is the affection—that of the proximate subject, not of the whole animal? Shall we say that it is immobility of such and such a kind? Yes, but to what process in the proximate subject is this due?" (*Meta.* 8.4.1044b9–20).

54. Irwin, *Aristotle's First Principles*, 270.

55. Ibid., 276.

56. See Aristotle, *Phys.* 2.7.198a23–26.

57. See ibid., 2.3.195a26–34; 2.5.196b24–29.

58. See ibid., 2.3.195a34–195b6. Aristotle also uses a similar example of a house builder: ibid., 2.3.196b25–29. See also 2.3.195b24; 2.5.196b27–29.

59. Ibid., 2.5.197a12–14.

60. "No incidental cause can be prior to a cause *per se*. Spontaneity and chance, therefore, are posterior to intelligence and nature. Hence, however true it may be that the heavens are due to spontaneity, it will still be true that intelligence and nature will be prior causes of this all and of many things in it besides" (ibid., 2.6.198a8–13).

61. Note that Aristotle actually has two terms for chance: (1) τύχη = luck or fortune (chance occurrences in causal situations related to human deliberating) and (2) a more general term, ταὐτομάτον (*tautomaton*) = chance (chance occurrences in causal situations related to an entity acting for its natural end [teleology]).

62. "For Aristotle it is not legitimate to view the present condition of the world as the outcome of the interaction of chains of necessary causes, as many

contemporary scientists and philosophers would hold. For Aristotle the human intellect can only trace back one chain of causes at a time, and will always have to stop the process when it reaches a free choice or a [sc. unusual] accidental cause, both of which introduce contingency into chains of causes, since the effect of free choices and [sc. unusual] accidents on the course of events is inherently unpredictable" (John Dudley, *Aristotle's Concept of Chance: Accidents, Cause, Necessity, and Determinism* [New York: State University of New York Press, 2012], 323).

63. A broader account of causation in the Middle Ages requires not only a more detailed description of the Oxford school but also an analysis of the thought of Duns Scotus and Bonaventure, also in relation to the topics of self-motion, freedom of human will, and causal accounts of perception. Such analysis goes beyond our present concern.

64. Thomas Aquinas, *De prin. nat.* 13, 14, 16. In *In II Sent.* 12.1.4 co. we find Aquinas saying: "Primary matter is that in which is the ultimate resolution of corporeal nature, which is of necessity without any form" (materia prima est illud in quo ultimo stat resolutio corporum naturalium, quod oportet esse absque omni forma) (English translation of this passage is mine). He notes that "the old natural philosophers, who held that primary matter was some actual being—for instance, fire or air, or something of that sort—maintained that nothing is generated simply, or corrupted simply; and stated that 'every becoming is nothing but an alteration,' as we read, *Phys.* 1 [4.187a30]." He obviously finds such conclusion "clearly false" (*ST* I.76.4 co.). (Unless otherwise noted, all quotations from the works of Aquinas are taken from the translations listed in the bibliography.)

65. Aquinas, *In Phys.* 2.5.183. "Quiddity" is another term used in scholastic philosophy for the essence (form) of an entity. Latin *quidditas* stands as a literal translation of Greek τὸ τί ἦν εἶναι.

66. See Aquinas, *In Meta.* 5.2.774; *ST* I.76.4 co.; *In De an.* 2.1.224.

67. See Aquinas, *In Meta.* 5.3.779. I will say more about this distinction in section 2.3 of chapter 6.

68. "Subsisting being must be one. . . . Therefore all beings apart from God are not their own being, but are beings by participation" (Aquinas, *ST* I.4.2 co.).

69. See Aquinas, *In Meta.* 5.2.776.

70. See, respectively, the following passages from Aquinas: (1) *Q. de ver.* 27.4 co. and ad 8; (2) *SCG* III.67.5; (3) *ST* I.13.5 ad 1; *In Phys.* 2.11.242. All three distinctions prove to be crucial for Aquinas's theory of divine action and his distinction between the causality of God and that of the creatures. The example of the conductor and members of the orchestra in (1) is a contemporary version of Aquinas's example of a man being begotten by man and by the sun. The latter exemplifies the general theory of lower bodies which "act through the power of the celestial bodies" (*SCG* III.67.5). This originally Aristotelian theory (see the same example of a man being begotten by man and by the sun in Aristotle, mentioned in section 2.2 above) is easily dismissed today as being based on an outdated cosmology. But

speaking philosophically, it is not entirely implausible to see the energy emitted by the sun, forces of gravitation, and other general causal principles as contributing to educing particular forms from primary matter in processes of substantial change occurring in nature.

71. See Aquinas, *In Meta.* 5.2.775; 5.3.781–82; *In Phys.* 5.11.246; *De prin. nat.* 19, 34–36.

72. Aquinas, *De mixt. elem.* 17–18.

73. See Christopher Decaen, "Elemental Virtual Presence in St. Thomas," *The Thomist* 64 (2000): 275, 287. Decaen notes that "virtual presence must at its root be a kind of potential existence. Indeed, the word *virtus* itself suggests this inasmuch as it is the translation of *dunamis* (δύναμισ) in the *De Generatione et Corruptione*.... *Dunamis* itself may be translated as 'potentiality,' 'possibility,' 'capability,' and of course 'power'" (ibid., 284). He also adds, "It is clear that he [Aquinas] does not mean that the elements are present in a manner of *pure* potentiality—the way we say prime matter is potential, and indifferent, with respect to every material form—when he says the elements remain *virtute*" (ibid., 286).

74. Other important topics related to philosophy of causation, and discussed by Aristotle and Aquinas, include (1) the reciprocal character of causation, (2) the possibility of one and the same thing being a cause of contrarieties, (3) simple and composite causes, (4) active and potential causes, (5) the principle of causation saying that what goes from potency to act requires a cause, (6) the principle of proportionate causation stating that effects must be proportionate to their causes and principles. Their analysis goes beyond the scope of this project.

75. At the rise of modern era Galileo famously states that the book of nature is "written in the language of mathematics" (Galileo Galilei, *The Assayer*, in *Discoveries and Opinions of Galileo*, trans. Stillman Drake [Garden City, NY: Doubleday, 1957], 338).

76. Francis Bacon, *The Works of Francis Bacon*, vol. 4, *Translations of the Philosophical Works 1*, ed. James Spedding, Robert L. Ellis, and Douglas D. Heath (New York: Cambridge University Press, 2011; first published in 1858), 146, 365.

77. See René Descartes, *The Philosophical Works of Descartes*, 2 vols., trans. E. S. Haldane and G. R. T. Ross (Cambridge: Cambridge University Press, 1970), 1:265; Clatterbaugh, *Causation Debate*, 18–32, 45; Hulswit, *From Cause to Causation*, 17–19. Despite contrary examples of magnetic attraction and the motions of the planets, Descartes's conviction that the cause of motion is always an effect of an impact or pressure exerted by a contiguous body (or bodies) was widely shared by his contemporaries. To explain the motions of the planets, Descartes hypothesized about a swirling vortex or invisible aether particles carrying them in their orbits around the sun. He thus understood cause as including the notion of "productive efficacy."

78. Thomas Hobbes, *English Works* (London, 1839), 3:121–22.

79. Ibid., 131.

80. Nicolas Malebranche, *Malebranche: "The Search after Truth" / "Elucidation of the Search after Truth,"* trans. T. M. Lennon and P. J. Olscamp (Columbus: Ohio State University Press, 1980), 450.

81. George Berkeley, *The Works of George Berkeley, Bishop of Cloyne*, vol. 4, ed. A. A. Luce and T. E. Jessop (Edinburgh: Thomas Nelson & Sons, 1948–57), 50.

82. David Hume, *A Treatise of Human Nature*, ed. L. A. Selby-Bigge (Oxford: Oxford University Press, 1978), 77, 266. Thus, the subjective version of Hume's theory sees a causal relation as a three-term relation. Apart from the experience of cause-and-effect events, it includes our anticipation (expectation) associated with the former leading to the latter.

83. Ibid., 172. Hume adds that this definition can be reformulated as saying that "a cause is an object precedent and contiguous to another, and so united with it, that the idea of the one determines the mind to form the idea of the other, and the impression of the one to form a more lively idea of the other" (ibid.).

84. David Hume, *An Enquiry concerning Human Understanding*, Great Books of the Western World 35 (Chicago: Encyclopedia Britannica, 1952), 477 (italics in this and all quotations are original).

85. Immanuel Kant, *Immanuel Kant's Critique of Pure Reason*, trans. N. K. Smith (London: Macmillan & Co., 1929), 218.

86. Ibid., 299.

87. Moritz Schlick, *Philosophy of Nature*, trans. A. von Zeppelin (New York: Philosophical Library, 1949), 2–3. See also Michael Stöltzner, "The Logical Empiricists," in Beebee, Hitchcock, and Menzies, *Oxford Handbook of Causation*, 108–27.

88. Niels Bohr, "Discussion with Einstein on Epistemological Problems in Atomic Physics," in *Albert Einstein: Philosopher-Scientist*, ed. P. A. Schlipp (New York: Tudor, 1949), 210.

89. Max Born, *Natural Philosophy of Cause and Chance* (Oxford: Clarendon Press, 1949), 92.

90. Friedrich Waismann, "The Decline and Fall of Causality," in *Turning Points in Physics*, ed. A. A. Crombie (Amsterdam: North-Holland Publishing, 1959), 84–154.

91. Anticipated by a number of philosophers of science in the late twentieth century (Fodor, Simon, Kaufman, Cummins, Salmon, and Cartwright), the new mechanical approach becomes a more solid theory in William Bechtel and Robert C. Richardson, *Discovering Complexity: Decomposition and Localization as Strategies in Scientific Research* (Cambridge, MA: MIT Press, 2010; first published in 1993); William Bechtel and Adele Abrahamsen, "Explanation: A Mechanist Alternative," *Studies in History and Philosophy of Science Part C: Studies in History and Philosophy of Biological and Biomedical Sciences* 36 (2005): 421–41; Stuart S. Glennan, "Mechanisms and the Nature of Causation," *Erkenntnis* 44 (1996): 49–71; Glennan, "Rethinking Mechanistic Explanation," *Philosophy of Science* 69 (2002): S342–53; Glennan, "Modeling Mechanisms," *Studies in History and Philosophy of*

Science Part C: Studies in History and Philosophy of Biological and Biomedical Sciences 36 (2005): 443–64; Glennan, "Mechanisms," in Beebee, Hitchcock, and Menzies, *Oxford Handbook of Causation*, 315–25; Peter Machamer, Lindley Darden, and Carl F. Craver (known as MDC) and their famous "Thinking about Mechanisms," *Philosophy of Science* 67 (2000): 1–25; followed by Carl F. Craver and Lindley Darden, *In Search of Mechanisms: Discoveries across the Life Sciences* (Chicago: University of Chicago Press, 2013). For a very good introduction to the new mechanical philosophy see Carl Craver and James Tabery, "Mechanisms in Science," *The Stanford Encyclopedia of Philosophy*, Spring 2017 ed., ed. Edward N. Zalta, https://plato.stanford.edu/archives/spr2017/entries/science-mechanisms/.

92. Craver and Darden, *In Search of Mechanisms*, 16.

93. Wimsatt offers four necessary and jointly sufficient conditions of aggregativity: "1. *IS* (*InterSubstitution*): Invariance under operations rearranging the parts in a system or interchanging any number of parts with a corresponding number of parts from a relevant equivalence class of parts (cf. commutativity of composition functions). 2. *QS* (*Size Scaling*): Qualitative similarity of the system property (identity, or if a quantitative property, differing only in value) under addition or subtraction of parts (cf. recursive generability of a class of composition functions). 3. *RA* (*Decomposition and ReAggregation*): Invariance of the system property under operations involving decomposition and reaggregation of parts (cf. associativity of composition functions). 4. *CI* (*Linearity*): There are no Cooperative or Inhibitory interactions among the parts of the system that affect this property" (William C. Wimsatt, "Emergence as Non-Aggregativity and the Biases of Reductionisms," in *Re-Engineering Philosophy for Limited Beings: Piecewise Approximations to Reality* [Cambridge, MA: Harvard University Press, 2007], 280–81).

94. Glennan, "Rethinking Mechanistic Explanation," S344.

95. Bechtel and Abrahamsen, "Explanation," 423.

96. Machamer, Darden, and Craver, "Thinking about Mechanisms," 3; Craver and Darden, *In Search of Mechanisms*, 15.

97. See Machamer, Darden, and Craver, "Thinking about Mechanisms," 4–6. They do not seem to develop much their theory of capacities and dispositions or the understanding of process ontology they propose.

98. Mereology is a relatively new philosophical theory of parthood relations—i.e., relations of part to a whole and of part to part in a whole (developed independently by Whitehead and Leśniewski around 1914–16 and defined more systematically by the latter in a series of articles from 1927 to 1931 entitled "On the Foundations of Mathematics"; see in Stanisław Leśniewski, *Collected Works*, vol. 1 [Dordrecht: Kluwer, 1992]). Because both parts and wholes may be defined in many different ways, the theory is applicable to a great variety of phenomena, as well as to the foundational metaphysical theories of those phenomena. Trying to apply mereology to Aristotle—whose thought is important for my project—we should note that although he is interested in the study of parts (he did write *On the Parts of Animals*),

Aristotle's fundamental explanation of substance is not so much based on the theory of "parts" and "wholes" but on "principles"—that is, formal and material causes. In other words, Aristotle's essentialism is not based on the view that objects have their parts essentially, and that a change in the number or composition of their parts leads to an essential (substantial) change (which is one way of understanding part-whole relation). Varzi notes, in reference to Aristotle's example of bronze being a "part" of the statue (which Aristotle says in reference to the material cause rather than to a "part" in the contemporary sense), that "many contemporary authors would rather construe it as a *sui generis*, non-mereological relation." See Achille Varzi, "Mereology," *The Stanford Encyclopedia of Philosophy*, Winter 2016 ed., ed. Edward N. Zalta, https://plato.stanford.edu/archives/win2016/entries/mereology/.

99. See Machamer, Darden, and Craver, "Thinking about Mechanisms," 13–14.

100. This kind of attitude allows us to classify new mechanists as followers of nonreductionist physicalism—even if they do not support it openly. I will say more about nonreductionist physicalism in the course of my investigation.

101. See Craver and Darden, *In Search of Mechanisms*, 19.

102. See Craver and Tabery, "Mechanisms in Science," section 2.3.

103. See Wesley C. Salmon, *Scientific Explanation and the Causal Structure of the World* (Princeton: Princeton University Press, 1984); Phil Dowe, "Causal Process Theories," in Beebee, Hitchcock, and Menzies, *Oxford Handbook of Causation*, 213–33. I will say more about this view of causation in chapter 4.

104. See chapter 4, section 2, for a detailed analysis of the challenges to the counterfactual theory of causation. In reference to the other—regulatory—view of causation proposed by Hume, new mechanists struggle in answering the question of how regular the relationship between the mechanism and the phenomenon must be (see Craver and Tabery, "Mechanisms in Science," section 2.1.2). I will say more about these two neo-Humean views of causation in chapter 4.

105. Carl F. Craver and William Bechtel, "Top-Down Causation without Top-Down Causes," *Biology & Philosophy* 22 (2006): 551. Machamer, Darden, and Craver ("Thinking about Mechanisms," 4) suggest that Glennan goes even further and claims that "all causal laws are explicated by providing a lower level mechanism until one bottoms out in the fundamental, non-causal laws of physics." They find this approach problematic, as they accept the reality of "direct causal laws." Their criticism is directed toward Glennan's "Mechanisms and the Nature of Causation," 52, where he states: "A mechanism underlying a behavior is a complex system which produces that behavior by . . . the interaction of a number of parts according to direct causal laws."

106. Craver and Bechtel, "Top-Down Causation," 551.

107. Naturally, I am speaking here about simultaneity of causal events relative to some frame of reference. I acknowledge that—according to Einstein's special theory of relativity—there is no objective relation of absolute simultaneity.

108. Craver and Bechtel, "Top-Down Causation," 557, 562.

109. See William C. Wimsatt, "Aggregativity: Reductive Heuristics for Finding Emergence," *Philosophy of Science* 64 (1997): S372–84.

110. Craver and Tabery, "Mechanisms in Science," section 4.2.

111. Craver and Bechtel, "Top-Down Causation," 551.

112. The most recent philosophical reflection on systems biology leads to the conclusion that reaching intelligibility in its endeavor requires a departure from the strictly mechanical approach and integrating or patching together various explanatory schemes. Besides the argument from the emergent character of biological systems, it has been emphasized that mechanistic explanations deal primarily with qualitative components of causal reactions, while the function of many molecular constituents of biological systems depends on the relative cellular concentration of different molecules. Hence, explanation in systems biology entails quantitative analysis that engages mathematical and computational tools. Moreover, the nature of many biological systems requires a description in terms of stochastic and nonlinear dynamics. Such systems may also need to be explained as involving no change-related phenomena—e.g., as characterized by stability or by an equilibrium point. Finally, the nature of these systems is often described in terms of evolutionary and developmental biology (the evo-devo field). In this context, the account of mechanistic explanation proves to be fairly limited and in need of support from other explanatory strategies. See contributions from various authors in Pierre-Alain Braillard and Christophe Malaterre, eds., *Explanation in Biology: An Enquiry into the Diversity of Explanatory Patterns in the Life Sciences* (Dordrecht: Springer, 2015).

113. Terrence Deacon, *Incomplete Nature: How Mind Emerged from Matter* (New York: W. W. Norton, 2012), 195.

114. Deacon uses the terms "homunculus" and "homuncular" in the context of preformationism—i.e., a theory, popular in antiquity and the Middle Ages, saying that animals develop from miniature versions of themselves. Sperm were believed to contain complete preformed individuals called "animalcules." In early days of modern biology, homuncular explanation was seen as an alternative to the then dominant view that some nonmaterial essence or spiritual agency was responsible for shaping a new organism from the shapeless matter. Deacon's use of the term is much broader and abstract. He refers it to Aristotle's notion of entelechy (*entelecheia*)—i.e., substantial form in a final stage of perfection of a given being—as well as to Hans Driesch's twentieth-century version of vitalism (I will say more about it in chapter 3). Moreover, applying the same concept in the context of contemporary biology, philosophy of biology, and philosophy of mind, Deacon defines it as "a form of explanation that pretends to be offering a mechanistic account of some living or mental phenomenon, but instead only appeals to another cryptically equivalent process at some lower level" (ibid., 47). It is worth noting that, understood in this way, the charge of being a homuncular explanation does not apply to Aristotle's hylomorphism or to

the modern and contemporary versions of vitalism, as they do not aspire to explain the phenomenon of life mechanistically.

INTRODUCTION TO PART 1

1. Murphy lists and defines five types of reductionism: methodological, epistemological, logical or definitional, causal, and ontological. See Nancey Murphy, "Reductionism: How Did We Fall into It and Can We Emerge from It?," in *Evolution and Emergence: Systems, Organisms, Persons*, edited by Nancey Murphy and William R. Stoeger (Oxford and New York: Oxford University Press, 2007), 23–25.

2. See Miles MacLeod and Nancy J. Nersessian, "Modeling Systems-Level Dynamics: Understanding without Mechanistic Explanation in Integrative Systems Biology," *Studies in History and Philosophy of Science Part C: Studies in History and Philosophy of Biological and Biomedical Sciences* 49 (2015): 2.

3. Timothy O'Connor and Hong Yu Wong, "Emergent Properties," *Stanford Encyclopedia of Philosophy*, Summer 2015 ed., ed. Edward N. Zalta, section 3.1, https://plato.stanford.edu/archives/sum2015/entries/properties-emergent/.

CHAPTER 1

1. See, for instance, Philip Clayton, "Conceptual Foundations of Emergence Theory," in *The Re-Emergence of Emergence: The Emergentist Hypothesis from Science to Religion*, ed. Philip Clayton and Paul Davies (New York: Oxford University Press, 2006), 4–7. I think that the Aristotelian concept of formal cause and his analysis of stability and changeability in nature do not justify making an easy connection between his philosophy and the EM theory. Aristotle's approach is a holistic one, as he begins with substances owing their unity and causal powers to substantial form in-forming primary matter present in them. To use the emergent categories of description, DC is present in beings from the start, instead of emerging only on the higher levels of complexity. Thus, as will become clear in part 2, I find it more appropriate to say that it is emergentism—introduced in the nineteenth century and developed in the twentieth—that seems to have opened the way back to a more robust essentialist ontology of Aristotle, and not vice versa. Clayton's reference to Neoplatonism is more accurate, as he acknowledges the difference in the direction of the complexity growth in Plotinus (downward movement) and in emergentism (upward movement). Still, I find the link between the two concepts hardly relevant, as in Neoplatonism DC goes from the simplicity of the One, producing the complexity of the many, whereas in emergentism it has an opposite direction, originating from the complexity of higher-level structures and exercising its influence on the more simple,

lower levels of organization. The temporalization of ontology offered by Hegel does not tie his philosophy easily with emergentism either. The idea of the unfolding of the Spirit seems to suggest that the fullness of complexity and potentiality is already present in Being, which needs to go through the process of negation to reach the synthesis. There is no easy linkage between Hegelian dialectics and the emergent metaphysics of growing complexity, which, even if defined in nonlinear terms, does not entail the process of dialectical negation.

2. For a more systematic historical analysis of British emergentism see Brian P. McLaughlin, "The Rise and Fall of British Emergentism," in *Emergence or Reduction? Essays on the Prospects of Nonreductive Physicalism*, ed. Ansgar Beckermann, Hans Flohr, and Jaegwon Kim (Berlin and New York: Walter de Gruyter, 1992), 49–93. My analysis follows the strategy and combines the accounts offered by Achim Stephan in "Emergence—a Systematic View on Its Historical Facets," in Beckerman, Flohr, and Kim, *Emergence or Reduction?*, 25–48, and by Jaegwon Kim in "Making Sense of Emergence," *Philosophical Studies: An International Journal for Philosophy in the Analytic Tradition* 95 (1999): 20–23. See also Paul Humphreys, *Emergence: A Philosophical Account* (New York: Oxford University Press, 2016), 94–120 (he pays special attention to the role played in the early part of the twentieth century by the debate on internal and external relations within the nascent logical atomism movement).

3. A list of the important works concerning EM written by the philosophers mentioned in the main text, organized alphabetically, contains Samuel Alexander, *Space, Time and Deity*, vol. 2 (London: Macmillan, 1920); Alexander Bain, *Logic*, books 2 and 3 (London: Longmans, Green, Reader, and Dyer, 1920); Charlie Dunbar Broad, *The Mind and Its Place in Nature* (London: Routledge and Kegan Paul, 1925); Mario Bunge, *The Mind-Body Problem: A Psychobiological Approach* (Oxford and New York: Pergamon Press, 1980); Bunge, *Matter and Mind: A Philosophical Inquiry* (Dordrecht: Springer, 2010); George Henry Lewes, *Problems of Life and Mind*, vol. 2 (London, 1875); John Stuart Mill, *A System of Logic* (New York, 1846); Conwy Lloyd Morgan, *Emergent Evolution* (London: Williams & Norgate, 1923); Karl Raimund Popper and John Carew Eccles, *The Self and Its Brain* (Berlin: Springer, 1977); John Jamieson Carswell Smart, "Physicalism and Emergence," *Neuroscience* 6 (1981): 109–13; Roger Wolcott Sperry, "Discussion: Macro- versus Micro-Determinism," *Philosophy of Science* 53 (1986): 265–70.

4. As an example, one can compare the following complex philosophical analysis and definition of EM presented in this chapter with the one formulated by Terrence Deacon, who simply defines EM as "unprecedented global regularity generated within a composite system by virtue of the higher-order consequences of the interactions of composite parts" (Terrence Deacon, "Emergence: The Hole at the Wheel's Hub," in Clayton and Davies, *Re-Emergence*, 122).

5. There have been several attempts at defining basic characteristics of EM in literature. My list is related to some of them. Among the most important

summaries, we find the one offered by Kim, who lists five central doctrines of EM: (1) EM of complex higher-level entities, (2) EM of higher-level properties, (3) the unpredictability of emergent properties, (4) the unexplainability/irreducibility of emergent properties, and (5) the causal efficacy of the emergents. See Kim, "Making Sense of Emergence," 20–23; Jaegwon Kim, "Emergence: Core Ideas and Issues," *Synthese* 151 (2006): 547–59. Philip Clayton finds eight core features of strong EM: (1) monism (there is one natural world made of stuff), (2) hierarchical complexity of the world, (3) temporal or emergent monism (the process of hierarchical structuring takes place in time), (4) no monolithic law of EM (EM should be viewed as a term of family resemblance), (5) patterns across levels of EM (we can name similarities shared in common by most of the various instances of EM in natural history), (6) DC, (7) emergentist pluralism (EM is not dualist in the sense that it does not privilege one particular emergent level), and (8) mind as emergent (contrary to dual-aspect monism, Clayton claims that both upward and downward influences are operative in the interaction between mental and physical properties). See Philip Clayton, *Mind and Emergence: From Quantum to Consciousness* (Oxford and New York: Oxford University Press, 2004), 60–62. Gregory R. Peterson speaks of seven interrelated characteristics of emergent entities: (1) higher-order description, (2) higher-order laws, (3) unpredictable novelty, (4) necessity of parts for the existence of the whole, (5) insufficiency of parts for the existence of the whole, (6) top-down causation, and (7) multiple realizability. See Gregory R. Peterson, "Species of Emergence," *Zygon* 41 (2006): 692–95. Stephan defines EM in terms of (1) nonadditivity of causes; (2) novelty of processes, entities, and properties; (3) nonpredictability (he distinguishes six different types of nonpredictability); (4) nondeducibility; and (5) DC. See Stephan "Emergence," 27–39, 42–45. Finally, in his monograph (*Emergence*, 26–27) Humphreys lists four criteria for EM: (1) emergent features result from something else, (2) they possess a certain kind of novelty with respect to the features from which they develop, (3) they are autonomous from the features from which they develop, and (4) they exhibit a form of holism (Humphreys discusses these criteria in ibid., 27–37). See also Rainer E. Zimmermann, *Metaphysics of Emergence, Part I: On the Foundations of Systems* (Berlin: MoMo Berlin, KonTexte, Philosophische Schriftenreihe, 2015). An abbreviated version of my own account of EM can be found in Mariusz Tabaczek, "The Metaphysics of Downward Causation: Rediscovering the Formal Cause," *Zygon* 48 (2013): 382–86.

    6. Mill, *System of Logic*, 212.

    7. Lewes, *Problems of Life and Mind*, 412.

    8. See Mark A. Bedau, "Weak Emergence," *Noûs* 31 (1997): 393, with references to other authors.

    9. See Jessica Wilson, "Metaphysical Emergence: Weak and Strong," in *Metaphysics in Contemporary Physics*, ed. Tomasz Bigaj and Christian Wüthrich (Leiden and Boston: Brill, 2015), 380–83. See also Sydney Shoemaker, "Kim on

Emergence," *Philosophical Studies* 108 (2002): 53–63. We will say a bit more about his proposition in section 3 of chapter 2.

10. Alexander, *Space, Time and Deity*, 2:45.
11. Broad, *Mind and Its Place*, 61.
12. See Jaegwon Kim, "Supervenience as a Philosophical Concept," *Metaphilosophy* 21 (1990): 5–23.
13. See Humphreys, *Emergence*, 215. Another common objection to SUP is that it tends to present the relation between the supervenient and subvenient properties as an unexplained fact about complex and dynamic systems, which leaves us with no answer to the question why these relations hold.
14. See Paul Humphreys, "Emergence, Not Supervenience," in "Proceedings of the 1996 Biennial Meetings of the Philosophy of Science Association, Part II: Symposia Papers," supplement, *Philosophy of Science* 64 (1997): S341.
15. Hong Yu Wong, "The Secret Lives of Emergents," in *Emergence in Science and Philosophy*, ed. Antonella Corradini and Timothy O'Connor (New York: Routledge, 2010), 20. McLaughlin takes a position suggesting that EM might be defined in terms of nomological SUP: "If P is a property of w, then P is emergent if and only if (1) P supervenes with nomological necessity, but not with logical necessity, on properties the parts of w have taken separately or in other combinations; and (2) some of the supervenience principles linking properties of the parts of w with w's having P are fundamental laws" (Brian P. McLaughlin, "Emergence and Supervenience," *Intellectica* 25 [1997]: 39). Contrary to his view is David Chalmers's suggestion to extend SUP to the entire universe and interpret all higher-level facts as globally (logically) supervening on the microphysical facts (i.e., to treat the entire history of the universe as one supervenient base) (see David J. Chalmers, *The Conscious Mind: In Search of a Fundamental Theory* [New York: Oxford University Press, 1996], chapter 2, 32–89). Humphreys rightly holds that "the global supervenience approach, itself predicated on an atemporal block universe ontology, converts what is an essentially diachronic type of emergence into a globally synchronic non-emergence" (Humphreys, *Emergence*, 225). Richard Campbell analyzes mereological SUP in his monograph *The Metaphysics of Emergence* (New York: Palgrave Macmillan, 2015), 241–45. He distinguishes it from Humean SUP—understood, after David Lewis, as the claim that everything supervenes on the spatiotemporal arrangement of local qualities—and global SUP, which he criticizes as "a sweeping thesis about all things and their properties everywhere everywhen" (ibid., 250–58). He finds (1) mereological SUP as plausible with respect to closed systems but not applicable in case of open systems, (2) Humean SUP as generally implausible (applied in quantum mechanics, it cannot accommodate any relation of quantum entanglement which endows the system with properties that are neither reducible to nor supervenient upon properties possibly attributable to any of its parts), and (3) global SUP as being too weak and permissive to express the physicalist thesis that somehow physical conditions determine all conditions in the world (see ibid., 257–58).

16. Note that we are using the terms "complexity" and "organization" interchangeably. Richard Campbell, following John Collier and Cliff Hooker, suggests distinguishing them:

> *Complexity* refers to the number of independent pieces of information needed to specify a system. . . . The Latin root of the word "complex" means "to mutually entwine or pleat or weave together". . . . The most fundamental sense [of complexity] is *informational complexity,* fundamental in the sense that anything which is complex in any other way must also be informationally complex. A complex object requires more information to specify it than a simple one. . . . *Organization,* on the other hand, characterizes the extent of the interrelations among the components of the system in terms of their number, scope, and dynamics (that is, the degree of non-linearity it exhibits). To explain how those components come to be organized, however, requires dynamic concepts. None of these notions—system, complexity, organization—are necessarily dynamical; they are at best abstractions from the underlying dynamics. So, rather than a new logico-mathematical theory, what is needed is a dynamical interpretation of these and related concepts which enable us to measure, or determine from measurable quantities, the amount of complexity, organization, etc. (Campbell, *Metaphysics of Emergence,* 136).

17. Alexander, *Space, Time and Deity,* 2:46.
18. Claus Emmeche, Simo Køppe, and Frederic Stjernfelt, "Explaining Emergence: Towards an Ontology of Levels," *Journal for General Philosophy of Science* 28 (1997): 96. See also ibid., 105–13.
19. See Emmeche, Køppe, and Stjernfelt, "Explaining Emergence," 15.
20. Kim, "Making Sense of Emergence," 20.
21. See Humphreys, *Emergence,* 120–26.
22. Concerning the status of laws of nature, I find interesting Campbell's remarks in his *Metaphysics of Emergence,* 230:

> Laws, then, are not simple descriptions of exceptionless regularities amongst occurrent events or facts; they describe *tendencies,* not Humean exceptionless regularities amongst actual observable phenomena. Only a few conservation laws truly state what *always* happens. As Alexander Bird has argued, laws are general relations among properties that supervene on potencies, and which have explanatory power, even if that power is inherited from the potencies. . . . Most laws are what Creary . . . calls "laws of influence." They describe the influence the instantiation of one property will have on another, but describe what actually happens only when nothing else is also impinging upon the situation.

23. See Humphreys, *Emergence,* 52.
24. Broad, *Mind and Its Place,* 77–78.

25. Ibid., 55. See also Alexander: "The existence of emergent qualities thus described is something to be noted, as some would say, under the compulsion of brute empirical fact, or, as I should prefer to say in less harsh terms, to be accepted with the 'natural piety' of the investigator. It admits no explanation" (Alexander, *Space, Time and Deity*, 2:46–47).

26. Popper, in Popper and Eccles, *Self and Its Brain*, 35.

27. Ibid., 16.

28. See Kim, "Making Sense of Emergence," 8–9.

29. Ernest Nagel, *The Structure of Science: Problems in the Logic of Scientific Explanation* (New York: Harcourt, Brace & World, 1961), 352.

30. Humphreys notes that "an alternative to Nagel reduction that is compatible with ontological emergence, although not with inferential and conceptual emergence, and which makes more of a concession to applications is Kemeny-Oppenheim reduction. A theory T_1 is Kemeny-Oppenheim reducible to theory T_2 if and only if all observational data explainable or predictable by T_1 is explainable or predictable by T_2. Any theory that is Nagel reducible is Kemeny-Oppenheim reducible, but not vice versa, because Kemeny-Oppenheim reducibility does not require the existence of bridge laws. A principle of ontological priority can then be based on Kemeny-Oppenheim irreducibility: If the existence or properties of an entity A can be explained in terms of another entity or collection of entities B, but not vice versa, then B has ontological priority over A. This is a sufficient condition for ontological priority, but it is not necessary because supervenience relations often do not permit explanations. However, it is generally taken that if A supervenes on B, but not vice versa, then B has ontological priority" (Humphreys, *Emergence*, 189).

31. Jaegwon Kim, *Mind in a Physical World: An Essay on the Mind-Body Problem and Mental Causation* (Cambridge, MA: MIT Press, 1998), 25. For the description of all three steps of Kim's functional model of reduction see his "Making Sense of Emergence," 9–18.

32. Humphreys, *Emergence*, 197.

33. Alexander, *Space, Time and Deity*, 2:8.

34. Kim, "Making Sense of Emergence," 22.

35. From the Aristotelian point of view, a symbiosis of two organisms would not imply the organizational principle of substantial form. It would be classified as an accidental arrangement—even if it is persistent. Thus, each one of the symbiotic organisms taken separately becomes an even more robust example of DC providing for the unity of all its constituent cells and their parts.

36. Donald Thomas Campbell, "'Downward Causation' in Hierarchically Organized Biological Systems," in *Studies in the Philosophy of Biology*, ed. Francisco J. Ayala and Theodosius Dobzhansky (Berkeley and Los Angeles: University of California Press, 1974), 180.

37. Ibid.

38. Ibid., 182.

39. Sperry, "Discussion," 267–68. In another article Sperry mentions another classical example of DC, i.e., a "local eddy in a stream," in which "drops of water are carried along" by the eddy "regardless of whether the individual molecules and atoms happen to like it or not." See Roger W. Sperry, "A Modified Concept of Consciousness," *Psychological Review* 76 (1969): 534.

40. Alexander Powell and John Dupré, "From Molecules to Systems: The Importance of Looking Both Ways," *Studies in History and Philosophy of Biological and Biomedical Sciences* 40 (2009): 60.

41. Michele Paolini Paoletti and Francesco Orilia, "Downward Causation: An Opinionated Introduction," in Paoletti and Orilia, *Philosophical and Scientific Perspectives*, 9. I have omitted in the quotation all the references Paoletti and Orilia provided with their list of phenomena.

Concerning DC in physics, apart from the phenomena mentioned by Paoletti and Orilia, Patrick McGivern and Alexander Rueger mention entangled quantum systems, phase transitions, and certain effects in solid-state physics. As an example of diachronic EM, they themselves discuss a much less complex phenomenon of the difference between a damped oscillating system (whose trajectory in phase space spirals down with increasing time into the origin—i.e., a "focal point attractor" in which the motion comes to rest) and an oscillating system whose damping is decreased and eventually eliminated (its trajectory turns into a closed curve—an ellipsis, or a series of concentric ellipses, also called a "center"). They claim the latter meets the requirements of strong EM—i.e., novelty of behavior and nonreducibility. To illustrate synchronic EM they analyze the case of steady heat conduction in a one-dimensional rod. See Patrick McGivern and Alexander Rueger, "Emergence in Physics," in Corradini and O'Connor, *Emergence in Science and Philosophy*, 213–32. Another account of EM and DC in physics was offered by Stewart J. Clark and Tom Lancaster. In reference to condensed matter physics, they state that "in physics, the coarse-grained variables are often viewed as causally interacting downwards on individual microscopic constituents of the matter. This may be described in terms of providing boundary conditions or constraints (such as the walls of a container containing the atoms of a gas) or more directly (such as the effective field theories described below)." They conclude by saying that "this is the physics of downward causation." See Stewart J. Clark and Tom Lancaster, "The Use of Downward Causation in Condensed Matter Physics," in Paoletti and Orilia, *Philosophical and Scientific Perspectives*, 129–30.

Concerning DC in chemistry, it has been discussed in reference to both chemical substances and the structure of separate chemical molecules. See Robin F. Hendry, "Prospects for Strong Emergence in Chemistry," in Paoletti and Orilia, *Philosophical and Scientific Perspectives*, 146–63 (Hendry provides in references a list of his numerous articles addressing this topic); Alexandru Manafu, "A Novel Approach to Emergence in Chemistry," in *Philosophy of Chemistry: Growth of a*

New Discipline, ed. Eric Scerri and Lee McIntyre, Boston Studies in the Philosophy and History of Science (Dordrecht: Springer Netherlands, 2015), 46–52 (he suggests defining chemical EM in terms of functional, multiply realized properties, such as being an acid, a base, an oxidant or a reductant, etc.; see also other essays in the same volume).

42. Alexander, *Space, Time and Deity*, 2:8–9.

43. Morgan, *Emergent Evolution*, 16.

44. See Carl Gillet, "Scientific Emergentism and Its Move beyond (Direct) Downward Causation," in Paoletti and Orilia, *Philosophical and Scientific Perspectives*, 246–55.

45. See, for instance, Robert van Gulick, "Reduction, Emergence, and the Mind/Body Problem," in Murphy and Stoeger, *Evolution and Emergence*, 59–63.

46. "Transformational emergence occurs when an individual a that is considered to be a fundamental element of a domain D transforms into a different kind of individual a^*, often but not always as a result of interactions with other elements of D, and thereby becomes a member of a different domain D^*. Members of D^* are of a different type from members of D. They possess at least one novel property and are subject to different laws that apply to members of D^* but not to members of D. The fact that a fundamental entity can be transformed makes transformational emergence essentially different from generative atomism, in which the fundamental entities are immutable. . . . An essential feature of fusion is the loss of such basal objects or properties when they fuse to produce a unified whole. Having undergone fusion, the original entities no longer exist to provide a base for the dependent properties, and causal pathways are routed through the fused emergent property, which in the cases considered will either be a fundamental$_s$ [fundamental synchronic] entity of the original domain or a member of some other domain" (Humphreys, *Emergence*, 60, 72; he discusses both types of ontological EM in detail in ibid., 60–91).

47. Ibid., 166–67. See also 40 for references to other authors discussing the inferential approach.

48. Ibid., 40; see also 180.

49. Ibid., 147.

50. Bedau is another thinker who offers a more nuanced account of weak and strong EM. He distinguishes two species (types) of weak EM, both pertaining to prediction and explanation (and thus epistemological in their nature). The first refers to situations in which description and explanation of given A's cannot be deduced from, or at least predicted in reference to, what is known about B's. The second—stronger—type of weak EM can be characterized in terms of "macrostates" of complex systems being "derivable" from "microstates" "only by simulation." So understood, weak EM is contrasted with strong EM—an explicitly ontological thesis. It points to the situations in which A's are something "in addition" to B's and contribute "causal powers" different from those contributed by B's. See Bedau, "Weak Emergence."

51. Clayton, "Conceptual Foundations," 27.

CHAPTER 2

1. Not all emergent phenomena occur on a global level, and not all of them are necessarily confined to interdisciplinary contexts (e.g., biology in relation to chemistry and physics). It is being argued that EM can happen within a single domain. Humphreys suggests, for instance, that entangled states in quantum mechanics are ontologically emergent since "the individual states before interaction do not determine the joint state after the interaction, whereas the joint state does determine the individual states. This feature makes compositional accounts of the joint system implausible" (Paul Humphreys, "Computational and Conceptual Emergence," *Philosophy of Science* 75 [2008]: 586). See also Don Howard, "Emergence in the Physical Sciences: Lessons from the Particle Physics and Condensed Matter Debate," in Murphy and Stoeger, *Evolution and Emergence*, 143–51.

2. Stephan, "Emergence," 27.

3. Robert C. Koons and Timothy Pickavance, *Metaphysics: The Fundamentals* (Oxford: Wiley-Blackwell, 2015), 121.

4. The ontological debate in analytic metaphysics is much more nuanced and includes a number of alternative versions of the main positions defined here. The reader can find a more detailed account of the entire conversation in ibid., chapter 5, 102–25; Michael J. Loux, *Metaphysics: A Contemporary Introduction* (New York and London: Routledge, 2006), chapter 3, 84–120.

5. The question whether the conglomeration of H_2O molecules is characterized by the new substantial—or rather a new accidental—form is debatable. Water molecules, taken separately, seem to possess an intrinsic unity proper for substances. It is demonstrated by modified behavior of electrons from atoms building up a given molecule, characteristic vibrational and rotational patterns proper to it (a balance of the electrostatic and quantum mechanical kinetic effects), and a characteristic V-shape with approximately 106° angle between the two O-H chemical bonds (which departs from the quantum equation for a system built of 18 protons and electrons and [typically] 8 neutrons). However, a reliance of this theory on abstractions of energy stabilization and corresponding mathematical models poses a question to what extent such description reflects an ontology of real water molecules. Taken collectively, a system of chemical equilibrium of water molecules seems to suggest that particular H_2O molecules are subsumed into a higher substantial form (i.e., the primary matter underlying them is in-formed by the new substantial form of water). But the fact that so-called pure water, chemically speaking, is a collection of H_2O, HDO (D = deuterium = hydrogen with a neutron in nucleus), D_2O, H_3O^+, OH^-, and $[H_2O]_n$ (polywater)—and all additional "ingredients" are detectable (even if only in extremely limited quantities under closely controlled laboratory conditions)—poses a question of its substantial unity. On the other hand, the fact that water is a dynamic equilibrium of all these molecules may further suggest that it does have a proper substantial form, and not merely an

accidental form. Also, the fact that various species of molecules are physically and chemically traceable in water does not exclude the possibility that they are present in it by their powers (*virtute*) while—metaphysically speaking—they are one and the same substance, that is, water (the idea of virtual presence, introduced in section 3 of the introduction, will be further discussed in section 2.3.5 of chapter 6). (I am grateful to David Chiavetta, O.P., for sharing these insights.) See also Paul Needham, "Water and the Development of the Concept of Chemical Substance," in *A History of Water*, series 2, vol. 1, *Ideas of Water from Antiquity to Modern Times*, ed. Terje Tvedt and Terje Oestigaard (London: Storbritannien, 2010), 86–123; Needham, "What Is Water?," *Analysis* 60 (2000): 13–21; Needham, "The Discovery That Water Is H_2O," *International Studies in the Philosophy of Science* 16 (2002): 205–26.

    6. Alexander, *Space, Time and Deity*, 2:46. See chapter 1, section 2.4.

    7. Interestingly, speaking of complexity of emergents, Humphreys holds, "Although it is often claimed that emergence is a result of increasing complexity in a system, the reverse happens in this example [the Greenberg-Hastings model, simulating basic Belousov-Zhabotinsky reactions, as well as neural excitation and relaxation models]. As can be seen from the outputs of the model, the initial random state of the cellular automaton is more complex than the final, patterned state. This claim does depend on the definition of complexity used, but for the measure of complexity I have adopted, the Kolmogorov complexity is low for the generated pattern and high for the random initial state. This feature is consistent with emergentist traditions in which it is complexity that generates an emergent feature but the emergent feature can itself be simple" (Humphreys, *Emergence*, 160).

    8. There is no clear and commonly accepted definition of physicalism. In its most common formulation it states that there is nothing over and above the physical and that everything supervenes on the physical. Defined thus, physicalism remains open to SUP but not to EM. However, Humphreys strives to define it in a way that does leave it open to EM. He distinguishes fundamental and nonfundamental physicalism. He says the former,

> as it is usually construed, always takes a synchronic form. . . . It identifies certain physical entities and properties as basic, these being whatever particles, forces, spaces, fields, or other entities are currently considered by physics, or would be so considered at the limit of foreseeable inquiry, to constitute the ultimate physical ontology. It then has to account for the remaining properties and objects of physics, as well as those of chemistry, biology, and other ontologically acceptable sciences. When explicit reduction or construction appears infeasible, fundamental physicalism frequently appeals to a determination relation or relations. The determination relation can be one of supervenience, . . . of realization, of determinates and determinables, of abstraction, of second-order property formation, of grounding, or one of any number of similar

determination relations. The essential feature of all these relations is that once the entities in the physical domain are present, the properties in all other domains are thereby determined. (ibid., 198–99)

Defined thus, fundamental physicalism seems to be compatible with weak (inferential and conceptual) EM, as well as diachronic strong (ontological) EM, while it stands in opposition to synchronic strong (ontological) EM. Nonfundamental physicalism, classified by many as nonreductionist physicalism, "takes as acceptable a wider domain of entities, which is usually identified with the combined domains of physics, chemistry, and perhaps some parts of biology. The boundaries of the acceptable domain vary widely; for some, the subject matters of geology, meteorology, metallurgy, astrophysics, cosmology, and so on are included even if they are irreducible to the domains of the three areas just mentioned. But always, the intentional entities of the psychological and social sciences are excluded from the domain of the physical" (ibid., 200). Described thus, nonfundamental physicalism accepts both weak (inferential and conceptual) and strong (synchronic and diachronic) EM.

Humphreys also notes that physicalism is related to naturalism (which similarly lacks a standard definition). He says it usually "contrasts the natural with the supernatural, and uses that contrast as the basis for preferring the former over the latter on the grounds that the former is amenable to scientific methods whereas the latter is not. Anything supernatural is to be excluded, and that means, in addition to theological and spiritual entities, specifically metaphysical entities." He concludes, "Naturalism is compatible with nonfundamental physicalism since biological entities are perfectly respectable natural entities and by common assent they are not fundamental physical entities" (ibid., 201).

9. See Jaegwon Kim, "'Downward Causation' in Emergentism and Nonreductive Physicalism," in *Emergence or Reduction? Essays on the Prospects of Nonreductive Physicalism*, edited by Ansgar Beckermann, Hans Flohr, and Jaegwon Kim (Berlin and New York: Walter de Gruyter, 1992), 128–29.

10. In what follows I refer to and extend Jaegwon Kim's "The Myth of Nonreductive Materialism," *Proceedings and Addresses of the American Philosophical Association* 63 (1989): 31–42.

11. See Donald Davidson, "Mental Events," in *Essays on Actions and Events* (Oxford: Oxford University Press, 1980); Kim, "Myth of Nonreductive Physicalism," 33–35. Davidson strives to solve the apparent contradiction between the two tenets of anomalous monism (i.e., physicalist ontological monism and the anomalism of the mental) by accepting the so-called token-identity thesis. It states that while mental event types are not identical with physical event types, each specific mental event token is identical with an underlying physical event token. It is rather doubtful, however, that his strategy is entirely plausible. See Kim's criticism of Davidson's position in Jaegwon Kim, *Supervenience and Mind* (Cambridge: Cambridge University Press, 1993), 22.

12. Putnam's idea of "physical realization" is followed by Fodor, Boyd, LePore, and Loewer. See Hilary Putnam, "The Nature of Mental States," in *Mind, Language, and Reality: Philosophical Papers* (Cambridge: Cambridge University Press, 1975), 2:433–35; Kim, "Myth of Nonreductive Physicalism," 36–39. Humphreys distinguishes heterogenous realizability ("when the type of entity that realizes the second-order property varies widely") and homogenous realizability ("when there is more than one type of realizer, but they can all be unified under a single nontrivial type at the same level as the more specific realizers"). See Humphreys, *Emergence*, 228.

13. Kim, "Myth of Nonreductive Physicalism," 40. See also Kim, "Supervenience," 23–27.

14. See Sydney Shoemaker, "Kim on Emergence," 54.

15. Menno Hulswit, "How Causal Is Downward Causation?," *Journal for General Philosophy of Science* 36 (2006): 264.

16. Parts of this section come from Tabaczek, "Metaphysics of Downward Causation," 386–94.

17. Campbell, "Downward Causation," 179–80.

18. See Michael Polanyi, *Knowing and Being* (Chicago: University of Chicago Press, 1969); Clayton, "Conceptual Foundations," 15–19.

19. Arthur Peacocke, "God's Interaction with the World: The Implications of Deterministic 'Chaos' and of Interconnected and Independent Complexity," in *Chaos and Complexity: Scientific Perspectives on Divine Action*, ed. Robert J. Russell, Nancey Murphy, and Arthur Peacocke (Vatican City State: Vatican Observatory; Berkeley, CA: Center for Theology and the Natural Sciences, 1995), 273.

20. See Nancey Murphy, "Emergence and Mental Causation," in Clayton and Davies, *Re-Emergence of Emergence*, 229; George F. R. Ellis, "Science, Complexity, and the Nature of Existence," in Murphy and Stoeger, *Evolution and Emergence*, 115; Paul Davies, "The Physics of Downward Causation," in Clayton and Davies, *Re-Emergence of Emergence*, 49.

21. See Alicia Juarrero, *Dynamics in Action: Intentional Behavior as a Complex System* (Cambridge, MA: MIT Press, 1999), 132–33. As we will see, Terrence Deacon introduces the same idea of constraints, but in addition makes a link between constraints and absences, assigning to them a causal role. The originality of his project consists also in the fact that he redefines the whole theory of causation in nature with reference to Aristotle, concentrating not only on human intentionality (the main concern of Juarrero) but on the general "teleogenesis" in nature, which he defines in terms of "ententionality." He also rejects the language of DC, replacing it with the formative causation of absences, specific to each level of the dynamical depth. Nevertheless, I claim that his description of the formative and teleological aspects of causation at the higher levels of organization of matter is embedded in terminology close to DC. That is why I have listed Deacon's proposition on my list of theories concerning the nature of the causal factor in DC. I will analyze his view in detail in the next chapter.

22. Deacon, "Emergence," 130.
23. Robert van Gulick, "Who's in Charge Here? And Who's Doing All the Work?," in Murphy and Stoeger, *Evolution and Emergence*, 83.
24. Austin Farrer, *The Freedom of the Will*, The Gifford Lectures (London: Adam and Charles Black, 1958), 57.
25. Deacon, "Emergence," 137.
26. Murphy, "Reductionism," 27.
27. See Emmeche et al., "Levels, Emergence," 18, 24. We will discuss their distinction of the three kinds of EM in section 7 of this chapter.
28. Van Gulick, "Who's in Charge Here?," 64–65.
29. See Murphy, "Emergence and Mental Causation," 228; Murphy, "Reductionism," 26–27.
30. Hulswit, "How Causal," 284. His radical claim finds support in Bedau, who asserts, "All the evidence today suggests that strong emergence is scientifically irrelevant.... There is no evidence that strong emergence plays any role in contemporary science.... Strong Emergence starts where scientific explanation ends" (Mark Bedau, "Downward Causation and the Autonomy of Weak Emergence," *Principia: An International Journal of Epistemology* 6 [2002]: 11).
31. Kim, "Making Sense of Emergence," 28.
32. Ibid., 29. Interestingly, Kim embraces here Aristotle's rule stating that an action always proceeds from an act, which is a source of properties and causal/determinative powers.
33. See section 2.2. of chapter 1 for the view of Broad, who, contrary to Kim, emphasizes the synchronic more strongly than the diachronic aspect of EM.
34. "Synchronic approaches have misrepresented what lies at the core of emergence.... They should be deemphasized in favor of diachronic approaches that give priority to the temporal evolution of the system. The traditional contrast between emergence and reduction, with its emphasis on synchronic reductions, fails to capture an important class of diachronically emergent phenomena, and we should not insist that examples of emergence are always a result of failures of synchronic reduction" (Humphreys, *Emergence*, 8). "Those who take emergence to be synchronic are assuming that the emergent feature is simultaneously present with that from which it is said to emerge.... By contrast, the scientific literature on emergence often concerns diachronic emergence, where the emergent phenomena develop over time from their precursors. Evolutionary emergence must be distinguished from ontological emergence. The evolution of one species into another does not satisfy that criterion: a creature of a species which has evolved from another does not occupy the same place at the same time as a creature of the species from which it evolved.... Consequently, any putative theory of any phenomenon—and any metaphysical schema—must be able, at least in principle, to account for both the diachronic (historical) and synchronic (ontological) emergence of that phenomenon since the Big Bang" (Campbell, *Metaphysics*, 194, 197).

35. The literal symbols used by Kim in the original version of his argument were changed to apply it to the concrete example of pain and escape reaction. See Kim, "Making Sense of Emergence," 31–33; Kim, "Emergence: Core Ideas," 557–58. The argument presupposes that causal overdetermination is metaphysically incoherent. While commonly accepted, this conviction finds its critics. See, for instance, Ted Sider's "What's So Bad about Overdetermination?," *Philosophy and Phenomenological Research* 67 (2003): 719–26. Kim's position reflects three main objections concerning DC when referred to the most popular theoretical framework for causation: (1) causation is a temporally extended relation, and hence its relata are not synchronous, (2) entities related by causation are usually in a different (even if adjacent) spatial location, and (3) causal relata are distinct entities. Carl Gillett notes that all three rules are being challenged in EM and DC. See Gillett, "Scientific Emergentism," 250.

36. Kim, "Making Sense of Emergence," 33.

37. Erasmus Mayr, "Powers and Downward Causation," in Paoletti and Orilia, *Philosophical and Scientific Perspectives*, 81.

38. See Scott, *Nonlinear Universe*, 287–88.

39. Timothy O'Connor and Hong Yu Wong, "The Metaphysics of Emergence," *Noûs* 39 (2005): 665–68.

40. Ibid., 670.

41. Ibid. Commenting on O'Connor and Wong's response to Kim, Humphreys (*Emergence*, 240–44) states that its key idea is that lower-level constituents—when causally interacting within the system that reaches a certain level of complexity—possess latent dispositions that result in manifest and indecomposable higher-level emergent properties. He thinks that through its requirement that an emergent property must be possessed by a composite system, the position in question retains a connection to the generative atomism (a persistent view that the world is nothing but spatiotemporal arrangements of fundamental physical objects and properties). He thinks it might be treated as a special case of fusion EM (see chapter 1, section 2.8).

42. Kim, *Supervenience and Mind*, 356. It is interesting what Robin F. Hendry (who asks whether DC in chemistry is real) says about causal closure of physics:

> Closure is a thesis that concerns the relationship of physics to everything else, so to find evidence for it we must look beyond the internal structure of physical theories and see how they are applied to the special sciences. Of all the special sciences, chemistry has the closest relationship to physics, which as we have seen is embodied in two great scientific achievements. Firstly, there is the twentieth-century discovery that chemical substances can be individuated, and their behaviour understood, in terms of their structures at the atomic scale. Secondly there is the fact that non-relativistic quantum mechanics provides a "theory of everything" for molecules, an all-encompassing framework within which to understand their dynamical behaviour. Yet neither of these facts

entails closure. In short, chemistry is where one might expect to find the imperial ambitions of physics fully played out, if they are played out anywhere. It is where we might expect to see some evidence for closure. Yet, as I have argued above, strong emergence is a plausible interpretation of the evidence offered by the explanatory relationships between physics and chemistry, which must surely weaken the case for closure" (Hendry, "Prospects for Strong Emergence in Chemistry," 160–61).

43. See Wong, "Secret Lives of Emergents," 10–14. Interestingly, Kim himself states—at some point—that "the usual notion of overdetermination involves two or more separate and independent causal chains intersecting at a common effect" (e.g., "two bullets hitting the victim's heart at the same time, the short circuit and the overturned lantern causing a house fire, and so on"). In reference to the specific and unique relation between neuronal (physical) and mental levels of causation he adds: "That is not the kind of situation we have here. In this sense, this is not a case of genuine causal overdetermination." Nevertheless, rather than question the plausibility of his own argument, he concludes by saying with an even stronger conviction that in these situations "exclusion applies in a straightforward way." See Jaegwon Kim, *Physicalism, or Something Near Enough* (Princeton: Princeton University Press, 2005), 48.

44. Cynthia Macdonald and Graham Macdonald, "Emergence and Downward Causation," in *Emergence in Mind*, ed. Cynthia Macdonald and Graham Macdonald (New York: Oxford University Press, 2010), 156.

45. Ibid., 157. It seems to me that, in a similar vein, Kim's position can also be challenged on its assumption that neural states N_1 and N_2 are just basal configurations of cells indispensable for the emergence of properties P and ER, and describable without any reference to an organism as a whole. These neural configurations or states are only capable of being configurations or states in a meaningful way when they are integrated into the "operation" of the system that they belong to. It does not seem plausible to claim that we can have basal neural configurations underlying the properties of pain and escape reaction apart from a living organism. Hence, the configurations themselves seem to have an emergent character.

46. Gillett, "Scientific Emergentism," 257. Paoletti and Orilia list four accounts of DC that might classify as noncausal: (1) higher-level entities select the powers to be activated at the lower levels of complexity of a given entity; (2) higher-level entities constrain what happens at the lower levels by imposing certain limits on the lower-level outcomes, by reducing the degrees of freedom of lower-level parameters, etc. (Terrence Deacon, whose notion of EM will be discussed in chapter 3, makes this account causal, as he speaks about causation of absences); (3) higher-level entities structure lower-level interactions in a specific way; (4) higher-level entities provide lower-level entities with novel powers. See Paoletti and Orilia, "Downward Causation," 7.

47. Gillett, "Scientific Emergentism," 258.

48. Ibid., 257.

49. Francesco Orilia and Michele Paolini Paoletti, "Three Grades of Downward Causation," in Paoletti and Orilia, *Philosophical and Scientific Perspectives*, 26. They refer to Jaegwon Kim, "Causation, Nomic Subsumption, and the Concept of Event," *Journal of Philosophy* 70 (1973): 217–36; Jaegwon Kim, "Events as Property Exemplifications," in *Action Theory: Proceedings of the Winnipeg Conference on Human Action, Held at Winnipeg, Manitoba, Canada, 9–11 May 1975*, ed. M. Brand and Douglas Walton (Reidel, 1976), 310–26; and his *Supervenience and Mind*.

50. See Orilia and Paoletti, "Three Grades of Downward Causation," 30–37. I find the second proposition, which defines causal relata as generic events, a less interesting and less plausible solution to the problem in question. Hence, I will not discuss it in more detail.

51. See John Heil, *The Nature of True Minds* (Cambridge and New York: Cambridge University Press, 1992); Douglas Ehring, "Mental Causation, Determinables, and Property Instances," *Noûs* 30 (1996): 461–80; David Robb, "The Properties of Mental Causation," *The Philosophical Quarterly* 47 (1997): 178–94.

52. Orilia and Paoletti, "Three Grades of Downward Causation," 35.

53. Ibid.

54. Erasmus Mayr, "Powers and Downward Causation," 87.

55. See Anna Marmodoro, "Power Mereology: Structural Powers *versus* Substantial Powers," in Paoletti and Orilia, *Philosophical and Scientific Perspectives*, 110–27, especially 116–18, 120–22. Humphreys's concept of fusion EM (see chapter 1, section 2.8) offers a similar answer to Kim's argument. He claims that having undergone fusion, the original lower-level entities no longer exist to provide a base for the higher-level properties, and hence the new causal pathways are routed through the fused emergent property. However, he does not refer to substantial language which poses a question concerning the ontological identity of emergents in his account. See Humphreys, *Emergence*, 70–72.

56. Rani Lill Anjum and Stephen Mumford, "Emergence and Demergence," in Paoletti and Orilia, *Philosophical and Scientific Perspectives*, 102. They do not seem to be clear on whether demergence is a causal or rather a noncausal notion.

57. I find it puzzling that although several contributors to the volume edited by Paoletti and Orilia refer to Aristotelian powers, none of them strives to ground them in the classical or the new version of Aristotelian metaphysics of substance, especially in reference to Aristotle's formal and final types of causation. My project is meant to fill this lacuna.

58. On the retrieval of teleology in biology see Theodosius Dobzhansky, *Genetics of the Evolutionary Process* (New York: Columbia University Press, 1970); Francisco J. Ayala, "Teleological Explanations in Evolutionary Biology," in *Nature's Purposes: Analyses of Function and Design in Biology*, edited by Colin Allen, Marc Bekoff, and George Lauder (Cambridge, MA: MIT Press, 1998), 29–49; Ernst Mayr,

"Teleological and Teleonomic: A New Analysis," in *Evolution and the Diversity of Life: Selected Essays* (Cambridge, MA, and London: Belknap Press of Harvard University Press, 1976), 383–404; Denis Walsh, "Teleology," in *The Oxford Handbook of Philosophy of Biology*, edited by Michael Ruse (Oxford and New York: Oxford University Press, 2008), 113–37; Mark Perlman, "The Modern Philosophical Resurrection of Teleology," in *Philosophy of Biology: An Anthology*, ed. Alex Rosenberg and Robert Arp (Oxford: Blackwell, 2010), 149–63.

59. See Emmeche, Køppe, and Stjernfelt, "Levels, Emergence," 18–23. As an example of an erroneous understanding of DC as SDC, they refer to a living cell defined as an emergent entity. In literature a cell is described as causing changes in the molecules constitutive to it, making them specifically "biological," substantially different from other molecules of the nonliving matter. "But if we imagine a microscopic view of this alleged causal process"—say Emmeche, Køppe, and Stjernfelt—"we will be unable to find any effective causality in the scenario. First, the process does not take place in time; second, the two events in question do not even possess the ability of causing each other. Of course, it is evident that the biological cell 'governs' or 'influences' the biochemical processes taking place in it—but at the same time the cell remains in itself a biochemical construct.... The cell consists of biochemical processes, we could say, but this is a non-temporal (mereological) relation and therefore non-causal in the efficient-causality use of the word. So even the idea of an upward efficient cause (or 'strong' upward causation) from biochemistry to cell is wrong because of this; what we could say instead is that the molecules and the biochemical reactions in question *constitute* the cell, that is, they are the material and formal causes of the cell" (ibid., 20).

60. Ibid., 25.
61. Ibid.
62. Ibid., 17.
63. Ibid., 26–31.
64. See Charbel Niño El-Hani and Antonio Marcos Pereira, "Higher-Level Descriptions: Why Should We Preserve Them?," in Andersen et al., eds., *Downward Causation*, 134–35. As one can see, Emmeche et al. offer a more nuanced distinction of different kinds of DC. They associate SDC with a strictly efficient type of causation. Thus, what they classify as MDC becomes a description of the new top-down causation, which is regarded by other thinkers as an indispensable characteristic of strong (ontological) EM.
65. Michael Silberstein, "In Defence of Ontological Emergence and Mental Causation," in Clayton and Davies, *Re-Emergence of Emergence*, 218.
66. Alvaro Moreno and Jon Umerez, "Downward Causation at the Core of Living Organization," in Andersen et al., *Downward Causation*, 107.
67. Ibid., n. 2.
68. See Scott, *Nonlinear Universe*, 5–7.

69. See Ellis, "Science, Complexity," 118–22.

70. Hilary Putnam seems to support the nonreductionist stance in metaphysics with some references to Aristotle, even if he is less engaged in the debate on EM and DC. Concerning the principle of the causal closure of the physical, he claims that accepting some form of a pluralistic and nonreductionist ontology enables us to support the idea that a given physical event can be caused by irreducible nonphysical events, and vice versa. Debating the charge of causal overdetermination in case of causality of the mental, he contends that it is not necessarily problematic, once we acknowledge that causal histories analyzed in various explanations of the same event may involve different contexts of explanation, with no unifying ontological background, where the conflict could really arise. See Hilary Putnam, *The Threefold Cord: Mind, Body, and World* (New York: Columbia University Press, 1999), 215. Moreover, even if he was a stern realist about the scientific worldview, Putnam was convinced that our ontology cannot be limited to entities and properties described by natural science: "I do indeed deny that the world can be completely described in the language game of theoretical physics; not because there are regions in which physics is false, but because, to use Aristotelian language, the world has many levels of form, and there is no realistic possibility of reducing them all to the level of fundamental physics" (Putnam, "From Quantum Mechanics to Ethics and Back Again," in *Philosophy in an Age of Science: Physics, Mathematics, and Skepticism* [Cambridge, MA: Harvard University Press, 2012], 65). He thought that the view that any single vocabulary could suffice to give a complete and robust description of the world is "a metaphysical fantasy," "a utopian program," nothing more than science fiction.

71. Hulswit, "How Causal," 278–79.

72. Ibid., 283.

73. Mark H. Bickhard and Donald T. Campbell, "Emergence," Lehigh University, October 15, 2016, http://www.lehigh.edu/~mhb0/emergence.html, section on supervenience.

74. See Hulswit, "How Causal," 276–77.

75. Emmeche et al., "Levels, Emergence," 31–32.

76. Hulswit, "How Causal," 278.

CHAPTER 3

1. "I propose that we use the term *ententional* as a generic adjective to describe all phenomena that are intrinsically incomplete in the sense of being in relation to, constituted by, or organized to achieve something non-intrinsic" (Deacon, *Incomplete Nature*, 27).

2. See Terrence W. Deacon and Spyridon Koutroufinis, "Complexity and Dynamical Depth," *Information* 5 (2014): 404–23.

3. Paul Cassell, "Incomplete Deacon: Why New Research Programs in the Sciences and Humanities Should Emerge from Terrence Deacon's *Incomplete Nature*," *Religion, Brain & Behavior* 3 (2013): 13.

4. Brian Green, "Catholic Thomistic Natural Law and Terrence Deacon's *Incomplete Nature*: A Match Made in Heaven," *Theology and Science* 13 (2015): 89–95.

5. Robert Cummings Neville, "Teleodynamic Remarks about Two Cultures," *Religion, Brain & Behavior* 3 (2013): 20.

6. Terrence Deacon and Tyrone Cashman, "Teleology versus Mechanism in Biology: Beyond Self-Organization," in *Beyond Mechanism: Putting Life Back into Biology*, edited by Brian G. Henning and Adam C. Scarfe (Lanham, MD: Lexington Books, 2013), 287.

7. Deacon and Cashman, "Teleology versus Mechanism," 290. See also Deacon and Koutroufinis, "Complexity and Dynamical Depth," 407–8; Deacon, *Incomplete Nature*, chapter 12. Deacon and Cashman notice that the teleological character of the physical work required in construction of an organism "is ignored in theories of evolution that are limited to natural selection logic alone" (Deacon and Cashman, "Teleology versus Mechanism," 291). They observe that "most modern introductory textbooks of biology and evolutionary theory define 'evolution' as change in allele or gene frequency. Notice that, besides reducing genetics to transcribed DNA sequences, this definition completely ignores any reference to function or adaptation, except in accounting for the differential copying of genes. Abstracted from this analysis are the processes responsible for the organism's growth and maintenance, the work required to copy (reproduce) its genetic information, and the means by which it responds to it[s] environment" (ibid., 289).

8. Ibid., 291.

9. Ibid., 292.

10. Terrence Deacon, Alok Srivastava, and Augustus Bacigalupi, "The Transition from Constraint to Regulation at the Origin of Life," *Frontiers in Bioscience* 19 (2014): 947.

11. Deacon and Cashman, "Teleology versus Mechanism," 295.

12. Ibid., 296.

13. Deacon, "Emergence," 124.

14. Deacon, *Incomplete Nature*, 145.

15. See Terrence W. Deacon, "The Hierarchic Logic of Emergence: Untangling the Interdependence of Evolution and Self-Organization," in *Evolution and Learning: The Baldwin Effect Reconsidered*, edited by Bruce H. Weber and David J. Depew (Cambridge, MA: MIT Press, 2003), 273–308.

16. Deacon, *Incomplete Nature*, 168–80. It should be noticed that in his article from 2006, "Emergence: The Hole at the Wheel's Hub," Deacon uses all three terms (see, for instance, "Emergence," 126–27), which he rejects later on in *Incomplete Nature*. However, even there he refers to supervenience, which he now defines

in dynamic terms (see, for instance, *Incomplete Nature*, 226, and 255). He seems to finally dispense with the supervenient terminology in his most recent publications.

17. Deacon, *Incomplete Nature*, 193.
18. Deacon, "Emergence," 143.
19. Deacon, *Incomplete Nature*, 194–95.
20. Ibid., 203.
21. Deacon and Koutroufinis, "Complexity and Dynamical Depth," 413.
22. Ibid.
23. Deacon, "Hierarchic Logic of Emergence," 288. See also Deacon, "Emergence," 126–30; In "Complexity and Dynamical Depth" we find other examples of homeodynamic systems such as chemical reactions, simple nonidealized mechanical systems (e.g., a harmonic oscillator or a pendulum with friction), compound mechanical systems (e.g., clockwork), and computational devices. See Deacon and Koutroufinis, "Complexity and Dynamical Depth," 413.
24. Deacon, *Incomplete Nature*, 223, 226.
25. See Deacon, "Emergence," 127.
26. See Deacon, *Incomplete Nature*, 237–38; "Emergence," 130; Deacon and Koutroufinis, "Complexity and Dynamical Depth," 414.
27. See Deacon and Koutroufinis, "Complexity and Dynamical Depth," 412, 414.
28. See Deacon, *Incomplete Nature*, 254–55, 261.
29. See Deacon, "Emergence," 130–37; *Incomplete Nature*, 239–61.
30. Deacon, "Emergence," 136. See also Deacon "Hierarchic Logic of Emergence," 293–97.
31. See Deacon and Koutroufinis, "Complexity and Dynamical Depth," 416–17.
32. Deacon, *Incomplete Nature*, 270.
33. Deacon and Koutroufinis, "Complexity and Dynamical Depth," 417–18.
34. Deacon, *Incomplete Nature*, 275.
35. Ibid., 266.
36. Ibid., 276.
37. See Deacon, "Emergence," 139.
38. The term "autogenesis" has been used in biology to refer to spontaneous generation, abiogenesis, and the EM of eukaryotic cells. Deacon uses it "for a very precise type of reciprocally interdependent interaction between different kinds of self-organizing process" (Deacon and Cashman, "Teleology versus Mechanism," 305n24).
39. See ibid., 298.
40. See Terrence W. Deacon, "Reciprocal Linkage between Self-Organizing Processes Is Sufficient for Self-Reproduction and Evolvability," *Biological Theory* 1 (2006): 140.
41. See Deacon and Cashman, "Teleology versus Mechanism," 302; Deacon, *Incomplete Nature*, 317; "Emergence," 141.

42. Deacon and Cashman, "Teleology versus Mechanism," 300. Deacon and Cashman emphasize that "in autogenesis, it is not just constituents that are joined in a reciprocally productive loop, but the constraints that each process generates, because each of these processes generates the boundary constraints that make the other process possible" (ibid., 299). See also Deacon, *Incomplete Nature*, 305–8.

43. See ibid., 271; Deacon and Cashman, "Teleology versus Mechanism," 300; Deacon, Srivastava, and Bacigalupi, "Transition from Constraint," 952. At one point, Deacon adds that "what constitutes an autogenic 'self' cannot then be identified with any particular substrate, bounded structure, or energetic process. Indeed, in an important sense, the self that is created by the teleodynamics of autogens is only a virtual self, and yet is at the same time the locus of real physical and chemical influences" (*Incomplete Nature*, 311).

44. Deacon, "Reciprocal Linkage," 143; see also ibid., 137–38. In *Incomplete Nature* Deacon adds that "[The] retained foundation of reproduced constraints is effectively the precursor to genetic information (or rather, the general property that genetic information also exhibits. . . . Whether it is embodied in specific information-bearing molecules (as in DNA) or merely in the molecular interaction constraints of a simple autogenetic process, information is ultimately constituted by preserved constraints" (*Incomplete Nature*, 317–18).

45. See Deacon, "Reciprocal Linkage," 142. In *Incomplete Nature* we find Deacon saying that "natural selection is ultimately an operation that differentially preserves certain alternative forms of morphodynamic processes compared to others, with respect to their synergy with one another, and with respect to the boundary conditions that enable them" (*Incomplete Nature*, 315).

46. See ibid., 315–19.

47. Deacon and Cashman, "Teleology versus Mechanism," 302. The phenomenon of autogenesis becomes for Deacon a point of departure for a further analysis in which he explains the EM of information, significance, self-differentiation, sentience, and consciousness. See Deacon, *Incomplete Nature*, chapters 11–17. But before he develops his argument, Deacon emphasizes once again the importance of autogens saying that "even these simple molecular systems have crossed a threshold in which we can say that a very basic form of value has emerged, because we can describe each of the component autogenic processes as there for the sake of the autogen integrity, or for the maintenance of that particular form of autogenicity" (ibid., 322).

48. Deacon and Cashman, "Teleology versus Mechanism," 303. "An intrinsic tendency to realize a general type of end state that is beneficial to the source of that tendency is the essence of final causality (i.e., teleology). It is what warrants calling this emergent disposition of life and mind 'teleodynamics.' And it is the defining attribute of biological selfhood" (ibid.).

49. See Deacon, "Emergence," 146–48.

50. In his paper written with Koutroufinis, Deacon suggests that their concept of dynamical depth applied to three levels of nested dynamics can be expanded to "additional higher order dynamics layered on top of what is characteristic of simple organisms. Complex organisms may occupy different levels of dynamical depth during different adaptive phases or life stages and may even subdivide into components of different dynamical depth" (Deacon and Koutroufinis, "Complexity and Dynamical Depth," 419). Teleodynamic unit systems (e.g., organisms) are themselves involved in homeodynamic, morphodynamic, and teleodynamic patterns of interaction (see ibid., 419–21).

51. Deacon, *Incomplete Nature*, 35–36; Deacon, "Emergence," 114–17.

52. See Richard Dawkins, *The Blind Watchmaker* (New York: W. W. Norton & Company, 1986).

53. Deacon, *Incomplete Nature*, 36–37. See also ibid., 132–35. Deacon is also critical about another kind of causal reductionism in evolutionary biology, which was proposed by Pittendrigh and supported by Ernst Mayr. The latter found it necessary to refer to end-directedness and information-based terminology in the modern evolutionary synthesis, but wanted to avoid the metaphysical baggage of teleology by replacing it with teleonomy—i.e., a process owing its goal-directedness to the operation of a program (see Mayr, "Teleological and Teleonomic," 387–90, 403). Deacon shows the inevitable character of the question about the source of this "program," which brings the conversation back to the issue of teleology.

54. See Deacon, *Incomplete Nature*, 23.

55. See ibid., 34, 59, 210; Deacon, "Emergence," 113–14.

56. "Of course there is so much else to distinguish this analysis from that of Aristotle (including ignoring his material causes)" (ibid., 148).

57. Deacon, *Incomplete Nature*, 34. Deacon's position is most likely grounded in the standard interpretation of Aristotle's *Phys.* 2.3.194b24–25 and *Meta.* 5.2.1013a24–26, which was discussed in the introduction.

58. Scott, *Nonlinear Universe*, 6.

59. See Emmeche, Køppe, and Stjernfelt, "Levels, Emergence," 17.

60. Deacon, *Incomplete Nature*, 167–68. "At the lowest level of scale there are only quantum fields, not indivisible point particles, or distinguishable stable extended configurations. . . . Quantum interactions become 'classical' in the Newtonian sense in interactions involving macroscopic 'measurement,' and only then do they exhibit mereologically identifiable properties" (ibid., 167).

61. Ibid., 168. Explaining his methodology and approach to EM in an article written together with Jeremy Sherman, Deacon says, "And we avoid passing the explanatory buck to strange aspects of physics, such as quantum effects. These are not considered in our approach to the emergence of *telos* for two reasons: first, because we believe we can show that the basic physical and chemical forces that dominate at our level of scale are sufficient to explain the emergence of life and the forms of *telos*

that life exhibits, and more important, because this indulges in an implicitly mysterian maneuver to claim to have explained one mystery by invoking another" (Jeremy Sherman and Terrence W. Deacon, "Teleology for the Perplexed: How Matter Began to Matter," *Zygon* 42 [2007]: 879).

62. Wallace, "A Place for Form in Science: The Modeling of Nature," *Proceedings of the American Catholic Philosophical Association* 69 (1995): 40. In our private conversation Robert Russell noticed that Wallace's remark about modern science treating matter as "the passive and inert component" is not entirely true. Wallace claims that matter—or more properly mass as its property—was regarded as being both passive (which explains why it continues in motion—cf. Newton's first law of motion) and active (it resists change—Newton's third law of motion).

63. It might be possible to refer mass, energy, and quantum field processes to the Aristotelian-Thomistic idea of signate matter (*materia signata*)—i.e., matter that is numerically distinct in different individuals but is the same in quality or character for cognition. As such, signate matter is regarded as a principle of individuation for form, in reference to Aquinas stating: "Nature or quiddity [in substances composed of matter and form] is received in designated matter. . . . And because of the division of designated matter, the multiplication of individuals in one species is here possible" (*De ente* 4.98). I will refer shortly to the concept of *materia signata* one more time in section 2.3.6 of chapter 6, discussing Michael Rea's version of neo-hylomorphism.

64. See Deacon, *Incomplete Nature*, 162–63. Speaking about fusion, Humphreys says, "I want to emphasize here that it is the fusion operation on the property instances that has the real importance for emergence. . . . *By a fusion operation, I mean a real physical operation, and not a mathematical or logical operation on predictive representations of properties.* . . . The key feature of [fusion] is that it is a unified whole in the sense that its causal effects cannot be correctly represented in terms of the separate causal effects of [its constituents.] Moreover, within the fusion . . . the original property instances . . . no longer exist as separate entities and they do not have all of their i-level causal powers available for use at the (i +1)st level" (Paul Humphreys, "How Properties Emerge," in *Emergence: Contemporary Readings in Philosophy and Science*, edited by Mark A. Bedau and Paul Humphreys [Cambridge, MA: MIT Press], 117). Note that Humphreys's view is reminiscent of the scholastic notion of virtual presence of simple beings and substances in those in-formed by higher forms.

65. See Deacon, *Incomplete Nature*, 210–13.

66. Ibid., 230–31. We find a similar idea of form in Deacon's approval of the position of Emmeche et al., who argue for interpreting DC as a case of formal causation. Deacon sees it as "the systemic 'geometric' position" within the dynamical network of the biomolecules building a living cell and generating specific higher constraints (see ibid., 231–32). See also ibid., 212–13, 338.

67. See Deacon, Srivastava, and Bacigalupi, "Transition from Constraint," 951–52, 954–55.

68. "In our rephrasing of formal and efficient cause, the incessant change of state is efficiently caused, but the asymmetry of those changes is formally caused" (Deacon, *Incomplete Nature*, 232). In his response to the common tendency to reduce form to structure, Jude Dougherty defends hylomorphism thus: "Have we reached the point where we are willing to say that the structures postulated or described by the sciences are really nothing other than Aristotle's form? Form is the principle of actuality in the essence, but it is not act without limit. Matter limits form and in doing so determines the resulting structure. In Aristotelian terms, structure cannot be understood simply as form. Structure results from the union of form and matter, and as such is in the order of accident, although it may be said to flow necessarily from the principle of actuality in the essence" (Jude P. Dougherty, *The Nature of Scientific Explanation* [Washington, DC: Catholic University of America Press, 2013], 43).

69. See Deacon, *Incomplete Nature*, 231, 234.

70. In my private correspondence with Koutroufinis and Deacon, Koutroufinis noticed that the analysis of contragrade changes on the level of teleodynamics allows for reinterpreting efficient causality in terms that are closer to Aristotle's understanding of this type of causation—i.e., as referring to autonomous agency rather than reducing efficient causality to blind material/physicochemical actions. He points out that Deacon's project emphasizes a fundamental difference between Aristotle and modern philosophy in defining efficient causality. In contrast to homeo- and morphodynamics, teleodynamics allows for thinking about material entities not as interacting particles but as systems generating significant parts of their constraints—that is, entities with a partial autonomy. Although I generally agree with Koutroufinis's remark, nevertheless, I find Deacon's notion of efficient causality still at risk of falling into reductionism. After all, any contragrade change in his project is defined in terms of a juxtaposition of two or more lower orthograde changes. This is an explanation which—although developed in the context of the study of dynamic systems—remains mechanical in its nature. I think Deacon's theory of the origin and persistence of a teleodynamic system in reference to a contragrade change has the same mechanical flavor, unless he is willing to ground it in the ontological nature of entities which are responsible for orthograde changes, juxtaposed in an emergent transition to a new kind of entity (dynamical system), characterized by the new and unique nature.

71. See Deacon, *Incomplete Nature*, 54.

72. See introduction, section 2.1.

73. See Deacon, *Incomplete Nature*, 54, 64–65.

74. It is worth noticing that for Aristotle both formal and final types of causation are responsible for qualitative changes. Deacon's understanding of form in terms of the geometry of the vector probability space and the laws of thermodynamics might suggest that he sees formal change as explainable quantitatively.

75. See John Farrell, "Book of the Year: The Return of Teleology," Forbes, December 26, 2012, http://www.forbes.com/sites/johnfarrell/2012/12/26/book-of-the-year-the-return-of-teleology/.

76. "Being organized for the sake of achieving a specific end is implicit in Aristotle's phrase 'final cause.' Of course, there cannot be a literal ends-causing-the-means process involved, nor did Aristotle imply that there was" (Deacon, *Incomplete Nature*, 109). "This physical disposition to develop toward some target state of order merely by persisting and replicating better than neighbouring alternatives is what justifies calling this class of physical processes *teleodynamic*, even if it is not directly and literally a 'pull' from the future. . . . The 'constitutive absences' characteristic of both life and mind are the sources of this apparent 'pull of yet unrealized possibility' that constitutes function in biology and purposive action in psychology. The point is that absent form can indeed be efficacious, in the very real sense that it can serve as an organizer of thermodynamic processes" (Deacon, "Emergence," 143–44).

77. See Deacon, *Incomplete Nature*, 275.

78. See ibid., 22–24.

79. See ibid., 43.

80. Ibid., 140–41.

81. Deacon, Srivastava, and Bacigalupi, "Transition from Constraint," 952.

82. See Aristotle, *Phys.* 2.3.195a11–14; Aristotle, *Meta.* 5.2.1013b11–16. Aristotle does not further develop his thought on this topic.

83. Aristotle, *Phys.* 1.7.190b10–13.

84. Deacon, *Incomplete Nature*, 12, 14, 45.

85. See Terrence Deacon, "Making Sense of Incompleteness: A Response to Commentaries," *Religion, Brain & Behavior* 3 (2013): 42–43, where he is responding to Peter Bokulich, "Missing or Modal? Where Emergence Finds the Physical Facts," *Religion, Brain & Behavior* 3 (2013): 6–12.

86. The problem of reifying "absence" in Deacon's project is more evident in relation to Aristotle's "privation." Privation is simply a mode of nonbeing, a lack of something (e.g., nonround). Deacon's constitutive absence is really a structure of some kind that eliminates a possibility. The system/structure then produces "what was not cancelled." His "absence" seems to involve actuality/presence of some kind, while Aristotle's privation is simply a way of not being (nonround), and thus it cannot be causal.

87. Terrence Deacon and Tyrone Cashman, "Steps to a Metaphysics of Incompleteness," *Theology and Science* 14 (2016): 419.

88. Ibid., 420.

89. Terrence W. Deacon and Tyrone Cashman, "Deacon and Cashman Respond to Green, Pryor, Tabaczek, and Moritz," *Theology and Science* 14 (2016): 472.

90. Ibid., 473.

91. Deacon, *Incomplete Nature*, 203.

92. This term appears for the first time in Deacon and Cashman's "Steps to a Metaphysics of Incompleteness."

93. Deacon, *Incomplete Nature*, 165.

94. Ibid., 168. Bickhard seems to make a classical mistake of thinking of Aristotle's primary matter as particles (stuff). The truth is that while primary matter

participates in the organization of a substance, it does not itself have any organization (any actuality). Thus, the particles of the substance are not primary matter. We could call them "elements." Having their own structure, they participate in the organization of the whole, similar to elements virtually present in an organism—e.g., calcium in a human being.

95. In his *De gen. et corr.* 2.3.330b31–35, following the list of elements offered by Empedocles, Aristotle says, "The 'simple' bodies, since they are four, fall into two pairs which belong to the two regions, each to each: for Fire and Air are forms of the body moving towards the 'limit,' while Earth and Water are forms of the body which moves towards the 'centre.'" In *De cae.* 1.2 he adds the fifth element—i.e., aether. What is crucial for him is that "'element' means the primary component immanent in a thing and indivisible in kind into other kinds" (*Meta.* 5.2.1014a26–27). In other words, each element is organized in larger and smaller chunks of it and thus divisible. Yet dividing it does not effect in obtaining parts that belong to different kinds—i.e., having different substantial forms. Moreover, metaphysically speaking, each element has an indispensable intrinsic "organization" in terms of primary matter and substantial form, which enable it to act, react, go through changes, and form complex entities. Naturally, the question remains whether we should treat elementary particles described by quantum physics as ultimate and most basic elements, replacing not only Aristotle's four elements but also chemical elements classified in Mendeleev's periodic table. We will come back to this issue in chapter 6.

96. Deacon, *Incomplete Nature*, 180.

97. Ibid., 79. See Cummings Neville, "Teleodynamic Remarks," 21–24, and Deacon's response in "Making Sense of Incompleteness," 45–49.

98. Deacon, "Emergence," 148.

99. Deacon, *Incomplete Nature*, 203.

100. Thus, while he distances himself from the position of eliminative materialism, what Deacon might be offering—at the end of the day—is yet another version of reductionist materialism. Taking on account the fact that all levels of emergents in his understanding bottom down in basic thermodynamics and that teleology is for Deacon an entirely emergent phenomenon, the idea of a Laplacian demon predicting everything (including teleological phenomena) from an atelic starting point is highly possible in his worldview.

101. See Sherman and Deacon, "Teleology for the Perplexed," 882–83.

102. Interestingly, speaking of the necessary increase of entropy, contemporary physics does not seem to take it to be a fundamental law of nature anymore. It sees it rather as the by-product of the fact that our universe happened to start out in a state of extremely low entropy. See Koons and Pickavance, *Metaphysics*, 61.

103. It would be interesting to see Deacon's response to the project of the new mechanical philosophy. It seems to me that he would remain rather skeptical about its claim that all causation in complex systems has an intralevel character. Deacon's

theory of EM—with its emphasis on teleology and formal causes—seems to have an ontological, rather than merely mechanistic or organizational, character.

INTRODUCTION TO PART 2

1. Willard V. O. Quine, "Two Dogmas of Empiricism," *The Philosophical Review* 60 (1951): 34.

2. See Nancy Cartwright and John Pemberton, "Aristotelian Powers: Without Them, What Would Modern Science Do?," in *Powers and Capacities in Philosophy: The New Aristotelianism*, edited by Ruth Groff and John Greco (New York: Routledge, 2013), 104.

CHAPTER 4

1. While my analysis of causation in analytic philosophy is based on a systematic investigation of particular causal theories, another strategy is also possible which concentrates on specific metaphysical queries concerning various aspects of discussed theories. It includes (1) questions about the ontological category and number of causal relata and (2) questions about the nature and metaphysical basis for causal connections and direction. This strategy was followed by Jonathan Schaffer in "The Metaphysics of Causation," *Stanford Encyclopedia of Philosophy*, Fall 2016 ed., ed. Edward N. Zalta, https://plato.stanford.edu/archives/fall2016/entries/causation-metaphysics/.

2. Stated slightly differently, RVC tells us that C is a cause of E iff C belongs to a minimal set of conditions that are jointly sufficient for E, given the laws.

3. The version of RVC referring to the laws of nature (nomic regularity theory) states that an event of type C causes an event of type E just in case C and E actually occur and the laws of nature imply that C-type events are regularly followed by E-type events.

4. They particularly oppose the idea of grounding RVC in powers, which is for them too metaphysically heavyweight a means for supporting the existence and operation of regularities. See Stathis Psillos, "Regularity Theories," in Beebee, Hitchcock, and Menzies, *Oxford Handbook of Causation*, 136–39.

5. Here I follow the classification suggested by Hulswit in *From Cause to Causation*, 48–53. He mentions that (1) is supported by Collingwood, Hart and Honoré, Nagel, and Hartshorne; that (2) is supported by Mill, Braithwaite, Hart and Honoré, Hempel, Popper, and Hondreich; and that (4) was proposed by Mackie. See ibid. for references. See also Illari and Russo, *Causality*, 27–34; Douglas Kutach, *Causation* (Cambridge, UK, and Malden, MA: Polity Press, 2014), 24–27.

6. Losee, *Theories of Causality*, 100.

7. Philipp Frank, *Philosophy of Science: The Link between Science and Philosophy* (Englewood Cliffs, NJ: Prentice-Hall, 1957).

8. See ibid., 266.

9. Losee, *Theories of Causality*, 86.

10. See ibid., 85–86.

11. Frank, *Philosophy of Science*, 274.

12. Psillos, "Regularity Theories," 141.

13. Ibid., 143.

14. See ibid., 148–50.

15. See Norman Swartz, "A Neo-Humean Perspective: Laws as Regularities," in *Laws of Nature: Essays on the Philosophical, Scientific and Historical Dimensions*, ed. Friedel Weinert (Berlin and New York: Walter de Gruyter, 1995), 86–88.

16. Hume, *Enquiry concerning Human Understanding*, 477. I have already alluded to this concept of causation in the introduction (section 4.2).

17. David Lewis, a main proponent of this theory, published extensively on the topic. See, for instance, *Counterfactuals* (Oxford: Blackwell, 1973); "Causation," in *Journal of Philosophy* 70 (1973): 556–67, reprinted in Sosa and Tooley, *Causation*, 193–204. See also John Leslie Mackie, "Causes and Conditionals," in *American Philosophical Quarterly* 2/4 (1965): 245–64; Aidan Lyon, "Causality," in *British Journal for the Philosophy of Science* 18 (1967): 1–20.

18. Lewis, "Causation," 194.

19. Laurie Ann Paul, "Counterfactual Theories," in Beebee, Hitchcock, and Menzies, *Oxford Handbook of Causation*, 160.

20. Edward Feser, *Scholastic Metaphysics: A Contemporary Introduction* (Heusenstamm: Editiones Scholasticae, 2014), 239. Referring to Aristotelian essentialism, Feser offers another argument: "The appeal to possible worlds also gets things the wrong way around in a further respect than the one mentioned above. For it is the essence of water, or of a tree, a dog, or a human being, that determines what will be true of these things in various possible worlds. It is not what is true of them in various possible worlds that determines their essences" (ibid., 240). For further criticism of possible worlds modality see Ellis, *Scientific Essentialism*, 270–74; David S. Oderberg, *Real Essentialism* (New York: Routledge, 2007), 1–6.

21. For more information on the modality debate in analytic philosophy see, for instance, Koons and Pickavance, *Metaphysics*, chapter 7, 154–81; Loux, *Metaphysics*, chapter 5, 153–86. Sophie Allen analyzes an attempt to explicate modality in terms of categorical or dispositional properties (a distinction which I will introduce and investigate in chapter 5). See Sophie Allen, *A Critical Introduction to Properties* (London and New York: Bloomsbury, 2016), 181–87.

22. Losee, *Theories of Causality*, 156.

23. In response to the challenge of preemption Lewis, in his later version of CVC, suggests complicating the theory and replacing the simple counterfactual

dependence between C and E with a chain of counterfactual dependence between C and E (i.e., a chain of a type $c_1 \rightarrow c_2 \rightarrow c_3 \rightarrow \ldots \rightarrow c_n \rightarrow E$), which does not seem to solve the problem. See Lewis, "Causation."

24. Jaegwon Kim, "Causes and Counterfactuals," in Sosa and Tooley, *Causation*, 206. See chapter 5, section 5, for graphic models of overdetermination and early and late preemption. Paul offers possible solutions to the problems of CVC. See Paul, "Counterfactual Theories," 173–82. See also Illari and Russo, *Causality*, 86–98; Kutach, *Causation*, 64–70.

25. See Ned Hall, "Two Concepts of Causation," in *Causation and Counterfactuals*, ed. John David Collins and Edward Jonathan Hall (Cambridge, MA: MIT Press, 2004), 225–76.

26. David Lewis, "Causation as Influence," *The Journal of Philosophy* 97 (2000): 195–96.

27. See John Haldane, "Privative Causality," *Analysis* 67 (2007): 183–85.

28. See Patrick Suppes, *A Probabilistic Theory of Causality* (Amsterdam: North Holland Publishing, 1970). See Jon Williamson, "Probabilistic Theories," in Beebee, Hitchcock, and Menzies, *Oxford Handbook of Causation*, 191–93. Among other supporters of ProbVC we find Michael Tooley (see his *Causation: A Realist Approach* [Oxford: Oxford University Press, 1987]) and Peter Menzies (see his "Probabilistic Causation and Causal Processes: A Critique of Lewis," *Philosophy of Science* 56 [1989]: 642–63).

29. See Losee, *Theories of Causality*, 107.

30. See Menzies, "Probabilistic Causation and Causal Processes," 645–47. Salmon offers another fictitious example of the decay of an excited atom, while Elliott Sober imagines a squirrel intercepting a golfer's putt by kicking the golf ball, which ricochets off a tree and reaches the cup. See Wesley C. Salmon, *Causality and Explanation* (Oxford: Oxford University Press, 1998), 222–23; Elliott Sober, "Two Concepts of Cause," *PSA: Proceedings of the Biennial Meeting of the Philosophy of Science Association* (1984): 406–7.

31. Sosa and Tooley, *Causation*, 23.

32. See Jonathan Schaffer, "Overlappings: Probability-Raising without Causation," *Australasian Journal of Philosophy* 78 (2000): 45–46.

33. To find more on ProbVC see Hulswit, *From Cause to Causation*, 58–59; Illari and Russo, *Causality*, 75–85; Kutach, *Causation*, 98–109; and Losee, *Theories of Causality*, 105–12.

34. See Elizabeth Anscombe, "Causality and Determination," in Sosa and Tooley, *Causation*, 88–104 (originally published in 1971); Hulswit, *From Cause to Causation*, 60–61.

35. Curt J. Ducasse, "On the Nature and the Observability of the Causal Relation," *Journal of Philosophy* 23 (1926): 65.

36. Ibid., 67.

37. Psillos, "Regularity Theories," 146.

38. See John Leslie Mackie, *The Cement of the Universe: A Study of Causation* (Oxford: Clarendon Press, 1974), 137.

39. Georg Henrik von Wright, "On the Logic and Epistemology of the Causal Relation," in Sosa and Tooley, *Causation*, 118.

40. See Losee, *Theories of Causality*, 144.

41. Peter Menzies and Huw Price, "Causation as a Secondary Quality," *The British Journal for the Philosophy of Science* 44 (1993): 189.

42. Robin G. Collingwood, "On the So-Called Idea of Causation," *Proceedings of the Aristotelian Society* 38 (1937–38): 89.

43. Douglas Gasking, "Causation and Recipes," *Mind*, n.s., 64 (1955): 483.

44. Georg Henrik von Wright, *Explanation and Understanding* (London: Routledge, 1971), 70. Hulswit (*From Cause to Causation*, 56) defines the cause in AVC as "*an event or state that we can produce and prevent at will, or otherwise manipulate, in order to produce or prevent a certain other event as an effect.*" See also James F. Woodward, "Agency and Interventionist Theories," in Beebee, Hitchcock, and Menzies, *Oxford Handbook of Causation*, 238.

45. Menzies and Price, "Causation as a Secondary Quality," 190.

46. Ibid., 191.

47. Ibid., 197.

48. Ibid., 198.

49. See Donald Gillies, "An Action-Related Theory of Causality," *The British Journal for the Philosophy of Science* 56 (2005): 823–42.

50. James Woodward, "Causation and Manipulability," *Stanford Encyclopedia of Philosophy*, Winter 2016 ed., ed. Edward N. Zalta, https://plato.stanford.edu/archives/win2016/entries/causation-mani/, section 8. See also Woodward, *Making Things Happen: A Theory of Causal Explanation* (New York: Oxford University Press, 2003), chapters 2–3. An intervention that excludes common causes of X and Y, accidental correlations with Y, etc. and shows specifically causal dependency between X and Y is often called "an ideal intervention."

51. Woodward, "Causation and Manipulability," section 5.

52. See Judea Pearl, *Causality: Models, Reasoning, and Inference* (New York: Cambridge University Press, 2000), 27.

53. Ibid., 70.

54. Woodward, "Causation and Manipulability," section 5.

55. Woodward, "Causation and Manipulability," section 11.

56. Again, see my argument against quantification over possible worlds in section 2 of this chapter. Woodward, together with some other thinkers, seems to make the mistake of classifying the statements mentioned in his definition of IntVC as counterfactual and building bridges from IntVC to CVC. Even if both theories include the possible worlds modality, we must not forget that Woodward's definition of IntVC operates on subjunctive conditionals, while Humean CVC on counterfactuals (i.e., contrary to fact conditionals of the type: if it were not the case that p then q

would not have happened). The latter assumes that *q* has already happened, while the former does not. Counterfactuals are thus a special class of, and should be differentiated from, subjunctive conditionals.

57. Pearl, *Causality*, 350.

58. Woodward, "Causation and Manipulability," section 12.

59. Woodward, "Agency and Interventionist Theories," 258n14. Woodward adds that it may be the case that causal claims can still play a central role in fundamental physics, but IntVC fails to capture that role. Therefore, causal claims at the quantum level might need a different causal theory. And yet he finds such a scenario peculiar, considering that "interventionist accounts are successful at elucidating causal claims in the special sciences" (ibid.).

Paoletti and Orilia ("Downward Causation," 6) note that IntVC has been used to respond to Kim's argument from causal exclusion. They mention Woodward, *Making Things Happen*; James Woodward, "Mental Causation and Neural Mechanisms," in *Being Reduced: New Essays on Reduction, Explanation, and Causation*, ed. Jakob Hohwy and Jesper Kallestrup (Oxford: Oxford University Press, 2008), 218–62; Woodward, "Interventionism and Causal Exclusion," *Philosophy and Phenomenological Research* 91 (2015): 303–47; Peter Menzies and Christian List, "Nonreductive Physicalism and the Limits of the Exclusion Principle," *Journal of Philosophy* 106 (2009): 475–502; Menzies and List, "The Causal Autonomy of Special Sciences," in *Emergence in Mind*, ed. Graham Macdonald and Cynthia Macdonald (Oxford and New York: Oxford University Press, 2010), 108–28; Larry Shapiro, "Lessons from Causal Exclusion," *Philosophy and Phenomenological Research* 81 (2010): 594–604; and Panu Raatikainen, "Causation, Exclusion, and the Special Sciences," *Erkenntnis* 73 (2010): 349–63. At the same time, they rightly state that "if successful, the response only shows that there is irreducible mental-to-mental causation (i.e., higher-level-to-higher-level causation). It does not concern mental-to-physical causation (i.e., downward causation)."

60. See Salmon, *Scientific Explanation*; Phil Dowe, "Causal Process Theories," in Beebee, Hitchcock, and Menzies, *Oxford Handbook of Causation*, 213–33.

61. Wesley C. Salmon, "Causality: Production and Propagation," in Sosa and Tooley, *Causation*, 158. This paper was first published in *PSA 1980*, vol. 2, ed. P. Asquith and R. Giere (East Lansing, MI: Philosophy of Science Association, 1981).

62. See ibid., 156–57.

63. Ibid., 169.

64. Salmon introduces also the third category of "perfect forks"—i.e., situations in which probabilities take on limiting values, and it is impossible to tell from the statistical relationships alone whether the fork should be considered interactive or conjunctive. See ibid., 168.

65. See ibid., 155, 170.

66. Phil Dowe, *Physical Causation* (Cambridge: Cambridge University Press, 2000), 90.

67. See Willard Quine, *The Roots of Reference* (La Salle, IL: Open Court, 1974); D. Fair, "Causation and the Flow of Energy," *Erkenntnis* 14 (1979): 219–50; J. Aronson, "On the Grammar of 'Cause,'" *Synthese* 22 (1971): 414–30.

68. Wesley C. Salmon, "Causality and Explanation: A Reply to Two Critiques," *Philosophy of Science* 64 (1997): 462. "A process transmits a conserved quantity between A and B ($A \neq B$) if and only if it possesses this quantity [in fixed amount] at A and at B and at every stage of the process between A and B without any interactions in the open interval (A, B) that involve an exchange of the particular conserved quantity" (ibid.).

69. Dowe, *Physical Causation*, 118.

70. "Other noninteracting intersections include radio waves entering a building and light coming through the glass window" (ibid., 121).

71. Phil Dowe, "Good Connections: Causation and Causal Processes," in *Causation and Laws of Nature*, ed. H. Sankey (Dordrecht: Kluwer, 1999), 257–61. He lists other examples of prevention and omission. Prevention: flipping a switch to turn off a light (Ehring), or a black patch in the sky being a cause of a visual experience (Goldman). Omission: Big Ben's failure to deliver its usual midnight chime being a cause of man's reaction, or a fieldsman's lack of judgment being a cause of a ball running out of bounds—all of which do not include a conserved quantity transfer.

72. See Losee, *Theories of Causality*, 125; Dowe, *Physical Causation*, 169–71.

73. See Dowe, "Causal Process Theories," 221–24, for references to critiques of ProcVC. See also Illari and Russo, *Causality*, 111–19; Kutach, *Causation*, 41–50.

74. See Peter Godfrey-Smith, "Causal Pluralism," in Beebee, Hitchcock, and Menzies, *Oxford Handbook of Causation*, 335.

75. See Christopher Hitchcock, "Causal Modeling," in Beebee, Hitchcock, and Menzies, *Oxford Handbook of Causation*, 299–314.

CHAPTER 5

1. Stephen Dean Mumford, "Causal Powers and Capacities," in Beebee, Hitchcock, and Menzies, *Oxford Handbook of Causation*, 266.

2. See David Malet Armstrong, *A World of States of Affairs* (Cambridge and New York: Cambridge University Press, 1997); Armstrong, "Defending Categoricalism," in *Properties, Powers, and Structures: Issues in the Metaphysics of Realism*, edited by Alexander Bird, B. D. Ellis, and Howard Sankey (New York: Routledge, 2012), 27–33.

3. See Alexander Bird, "Limitations of Power," in *Powers and Capacities in Philosophy: The New Aristotelianism*, edited by Ruth Groff and John Greco (New

York: Routledge, 2013), 29–30. See also Ruth Groff commenting on Bird's position in her "Whose Powers? Which Agency?," in *Powers and Capacities in Philosophy: The New Aristotelianism*, edited by Ruth Groff and John Greco (New York: Routledge, 2013), 222; Feser, *Scholastic Metaphysics*, 59–60.

4. See George Molnar, *Powers: A Study in Metaphysics*, edited by Stephen Mumford (Oxford and New York: Oxford University Press, 2003), 57.

5. See Stephen Mumford, *Dispositions* (Oxford and New York: Oxford University Press, 1998), 10.

6. Molnar suggests treating properties as "tropes," which are genuine, mind-independent, and nonrepeatably particular ("unit properties"). He claims this distinguishes "properties metaphysics" from both classical realism and classical nominalism (see Molnar, *Powers*, 22–23).

7. See Bird, "Limitations of Power," 27. He also suggests replacing the term "powers" with "potency" and explains, "It is in the nature (essence) of a potency/power P that there is a specific dispositional character $D_{S,M}$ such that, necessarily, for any particular, x, that possesses P, $D_{S,M}x$ holds [x is disposed to manifestation M in response to stimulus S]" (ibid.).

8. Mumford, "Causal Powers and Capacities," 268.

9. See Stephen Mumford and Rani Lill Anjum, *Getting Causes from Powers* (Oxford and New York: Oxford University Press, 2011), 175–94.

10. See Mumford, *Dispositions*, 74–75.

11. See ibid., chapter 3, and Mumford and Anjum, *Getting Causes from Powers*, 190–93, where they distinguish simple, revised, ceteris paribus, and modally strengthened condition analyses of dispositions and test them in terms of their ability to accommodate the possibility of prevention. Molnar (see *Powers*, chapter 4) offers a lengthy reflection on the conditional analysis of powers as well. He defines and distinguishes naïve, causal, and reformed conditional analyses. He challenges each one of them in terms of (1) the loss of intrinsicality (the conditional does not seem to be saying anything about what it is in the object that makes it react upon the stimulus); (2) their inapplicability to powers whose manifestations are spontaneous (not in response to stimuli); (3) finkishness (the case of the nonpermanent powers that objects can acquire or lose in response to a subtractive interferer that is active at the moment they are tested); (4) the assumption of the universal quantification which is vulnerable to exceptions whose range is not specified (the problem which is not resolved by either statistical or defeasible quantification that uses a ceteris paribus clause). Following Mumford and Anjum, Molnar summarizes his investigation by saying that powers cannot be analyzed as conditionals. Koons and Pickavance add to the argument from finkishness cases of mimicking (e.g., using a powerful sonic beam to break a vase which otherwise is not fragile) and masking (e.g., filling a fragile vase with an adhesive plastic which prevents its breaking if struck). Again, both cases offer additional arguments against grounding dispositionalism on subjunctive conditionals. See Koons and Pickavance, *Metaphysics*, 50.

12. Marmodoro comments on these aspects of powers metaphysics in "Power Mereology," 113–14.

13. Mumford, *Dispositions*, 77.

14. Ibid. "Categorical" ascriptions are usually classified as qualitative and used interchangeably with "real," "actual," "unconditional," "episodic," or "occurrent."

15. See Kristina Engelhard, "Categories and the Ontology of Powers. A Vindication of the Identity Theory of Properties," in Marmodoro, *Metaphysics of Powers*, 42–43; Mumford, *Dispositions*, 172–76. Allen concentrates her analysis of the problem of grounding of powers around the question whether properties have their causal roles necessarily or contingently. She thus classifies the main competitive positions in this debate as dispositionalism and contingentism (see Allen, *Critical Introduction to Properties*, chapter 7, 139–65).

16. See Molnar, *Powers*, chapter 10.

17. See Brian D. Ellis, "The Categorical Dimensions of Causal Powers," in *Properties, Powers, and Structures: Issues in the Metaphysics of Realism*, edited by Alexander Bird, B. D. Ellis, and Howard Sankey (New York: Routledge, 2012), 20; Ellis, "Causal Powers and Categorical Properties," in Marmodoro, *Metaphysics of Powers*, 138; Ellis, *Scientific Essentialism* (Cambridge and New York: Cambridge University Press, 2001), 217–18. See also Groff, "Whose Powers? Which Agency?," 213. At the same time, however, in agreement with other powers theorists, Ellis acknowledges that at the bottom level of reality, as described in quantum physics, powers are groundless. Even if we discover in the future that present elementary particles are further complex and therefore reducible to yet more basic constituents, the latter will be classified as groundless dispositions (see Ellis, "Categorical Dimensions of Causal Powers," 24).

18. Edward Jonathan Lowe, *The Four-Category Ontology: A Metaphysical Foundation for Natural Science* (New York: Oxford University Press, 2006), 139.

19. See ibid. Oderberg criticizes Lowe's suggestion in *Real Essentialism*, 132–33.

20. See David Malet Armstrong, *A Materialist Theory of the Mind* (London: Routledge & K. Paul, 1968), 86; Armstrong, "Defending Categoricalism," 28–29; Feser, *Scholastic Metaphysics*, 79–80.

21. See Charles Burton Martin, *The Mind in Nature* (Oxford and New York: Clarendon Press, 2008); John Heil, "Powerful Qualities," in Marmodoro, *Metaphysics of Powers*, 58–72. Engelhard ("Categories and the Ontology of Powers," 51–54) discusses counterarguments to the identity theory and proposes Lowe's four-category ontology as a solution.

22. See Quine, *Roots of Reference*.

23. Mumford, *Dispositions*, 192.

24. See Mumford, "Causal Powers and Capacities," 268; Mumford and Anjum, *Getting Causes from Powers*, 3–4; Stephen Dean Mumford, "The Power of Power,"

in *Powers and Capacities in Philosophy: The New Aristotelianism*, edited by Ruth Groff and John Greco (New York: Routledge, 2013), 12–13.

25. See Bird, "Limitations;" Alexander Bird, "Monistic Dispositional Essentialism," in *Properties, Powers, and Structures: Issues in the Metaphysics of Realism*, edited by Alexander Bird, B. D. Ellis, and Howard Sankey (New York: Routledge, 2012), 11–26; Bird, "Causation and the Manifestation of Powers," in Marmodoro, *Metaphysics of Powers*, 35–41; Karl R. Popper, *The Logic of Scientific Discovery* (New York: Basic Books, 1959), 424; David Hugh Mellor, "Counting Corners Correctly," *Analysis* 42 (1982): 96–97; Mellor, "In Defense of Dispositions," *Philosophical Perspectives* 12 (1974): 283–312; Sydney Shoemaker, "Causal and Metaphysical Necessity," *Pacific Philosophical Quarterly* 79 (1998): 59–77; Shoemaker, "Causality and Properties," in *Time and Cause: Essays Presented to Richard Taylor*, edited by Richard Taylor and Peter van Inwagen (Dordrecht and Boston: Reidel, 1980), 109–35. In *Powers* (chapter 11) Molnar examines at length major objections to pan-dispositionalism.

26. Mumford states that dispositions are ascribed to at least three classes of things: objects, substances, and persons. In *Dispositions* he lists five examples of dispositions which show the diversity and flexibility of powers ascriptions. He describes fragility, belief, bravery, thermostats, and divisibility by 2 (see *Dispositions*, 5–11).

27. See Mumford, "Causal Powers and Capacities," 270–71. Although an analysis of dispositions in terms of subjunctive conditionals helps in specifying the necessary character of their relation to manifestations, such an analysis cannot be treated as the basic definition of powers, for reasons mentioned above.

28. See Jennifer McKitrick, "Manifestations as Effects," in Marmodoro, *Metaphysics of Powers*, 73–83; John Heil, *The Universe as We Find It* (Oxford: Clarendon Press, 2012), 121.

29. See Toby Handfield, "Dispositions, Manifestations, and Causal Structure," in Marmodoro, *Metaphysics of Powers*, 106–32.

30. See Molnar, *Powers*, 183–87.

31. Brian D. Ellis, *The Philosophy of Nature: A Guide to the New Essentialism* (Montreal and Ithaca, NY: McGill-Queen's University Press, 2002), 48.

32. Stephen Mumford and Rani Lill Anjum, "Causal Dispositionalism," in *Properties, Powers, and Structures: Issues in the Metaphysics of Realism*, edited by Alexander Bird, B. D. Ellis, and Howard Sankey (New York: Routledge, 2012), 104.

33. Although the ontological distinction and priority of powers over their manifestations is essential in powers metaphysics, it does not contradict their simultaneity in time. In the standard example of a billiard ball a impacting and causing movement of a billiard ball b Hume distinguishes two separate events influencing one another, separated in time and linked by our impression of constant conjunction. But the movement of a is not a cause of any change in b before they interact. A powers view of causation emphasizes the simultaneity of cause and effect without assuming they are instantaneous. Quite the contrary, it notes they are often extended in

time. It seems to rely upon the reality of causal processes—i.e., temporally extended "things" or "situations" that unite cause and effect into a single, undivided whole. See Mumford and Anjum, *Getting Causes from Powers*, chapter 5, 106–29; Koons and Pickavance, *Metaphysics*, 69.

34. Mumford and Anjum, *Getting Causes from Powers*, 116.

35. As I have said before, the same problem of generalization of causal claims challenges ProbVC, MVC, and ProcVC. It also becomes unavoidable in SVC, which by definition questions its plausibility. DVC seems to offer a solution in the context of these causal theories as well.

36. See Bird, "Causation and the Manifestation of Powers," 161–62.

37. Mumford and Anjum list and answer four major objections to the argument against absolute necessity of causal dependencies in nature. See Stephen Mumford and Rani Lill Anjum, "A Powerful Theory of Causation," in Marmodoro, *Metaphysics of Powers*, 147–51. They also acknowledge that the argument in favor of the suppositional nature of necessity in causation has its source in Aquinas's philosophy of nature (see ibid., 143).

38. See Nancy Cartwright, *Hunting Causes and Using Them: Approaches in Philosophy and Economics* (Cambridge and New York: Cambridge University Press, 2007), 11–23.

39. Neil Williams goes even further, arguing for what he calls "power holism." He claims that powers make up a system in which they collectively determine each other's natures and combine to produce manifestations, which are thus understood as mutual products of powers in distinct objects or parts of objects working together. See Neil E. Williams, "Puzzling Powers: The Problem of Fit," in Marmodoro, *Metaphysics of Powers*, 94–98.

40. See Molnar, *Powers*, 194–99; John Dupré, *The Disorder of Things: Metaphysical Foundations of the Disunity of Science* (Cambridge, MA: Harvard University Press, 1993), 123–24.

41. Molnar, *Powers*, 195. He also distinguishes the polygenic-pleiotropic division from the issue of "single-track" powers (highly specific or determinate—e.g., fragility of an object breaking upon some specific impact) and "multi-track" powers (highly generic or determinable—e.g., high temperature can be manifested in various ways). He finds this classification offered by Mackie questionable. See ibid., 198–99; John Leslie Mackie, *Truth, Probability and Paradox: Studies in Philosophical Logic*, Clarendon Library of Logic and Philosophy (Oxford: Clarendon Press, 1973), 122.

42. The dispositionalist approach to probability seems to follow the Aristotelian emphasis on the reference to *per se* causes in dealing with chance events. Acknowledging the ontological character of chance (probabilistic) occurrences, dispositional metaphysics analyzes them in reference to kind-specific dispositions of entities and their manifestations, which are suppositional in nature. See introduction, section 2.4.

43. As we have seen, the same problem of dependency on the possible worlds modality refers to AVC and IntVC (see sections 5.2 and 5.3 of chapter 4).

44. See Aristotle's example of the captain of the ship discussed in chapter 3, section 3.4.

45. See Mumford and Anjum, "Powerful Qualities," 153–55; *Getting Causes from Powers*, 143–48.

46. See Mumford, *Dispositions*, chapter 10; Mumford, "Power of Power," 17–19; Bird, "Limitations of Power," 35–38; Cartwright and Pemberton, "Aristotelian Powers," 105–8; Molnar, *Powers*, 199; Illari and Russo, *Causality*, 150–60.

47. Mumford and Anjum, *Getting Causes from Powers*, 19. Neuron diagrams—also called "Bayesian networks"—are not the only way of philosophical and scientific causal modeling. Among other methods we find (1) structural models—a type of regression models expressed in the form of lists of equations or a recursive decomposition characterizing a system; (2) multilevel (hierarchical) models—statistical modeling tools analyzing, in a single modeling framework expressed in equations, both individual-level and aggregate variables; (3) contingency tables (category data analysis)—a form of statistical modeling that orders variables in the data in rows and columns, analyzing relations between them; and other methods. See Illari and Russo, *Causality*, 64–67.

48. See Mumford and Anjum, "Causal Dispositionalism"; Mumford and Anjum, *Getting Causes from Powers*; Mumford and Anjum, "Powerful Theory of Causation"; Lawrence Brian Lombard, *Events: A Metaphysical Study* (London and Boston: Routledge & Kegan Paul, 1986).

49. See Mumford and Anjum, *Getting Causes from Powers*, 38–44.

50. Ibid., 43.

51. See ibid., 32.

CHAPTER 6

1. Cartwright and Pemberton, "Aristotelian Powers," 93.

2. Aristotle analyzes act and potency in book 9 of the *Metaphysics*, which follows books 7 and 8, dedicated to a description of being as indicated by substance and other categories. He thus introduces the topic of book 9: "We have treated of that which is primarily and to which all the other categories of being are referred—i.e. of substance. For it is in virtue of the concept of substance that the others also are said to be—quantity and quality and the like; for all will be found to involve the concept of substance, as we said in the first part of our work. And since 'being' is in one way divided into individual thing, quality, and quantity, and is in another way distinguished in respect of potency and complete reality, and of function, let us now add a discussion of potency and complete reality" (*Meta.* 9.1.1045b27–35).

3. Feser, *Scholastic Metaphysics*, 33.

4. Aristotle, *Meta.* 5.12.1019a19–20.
5. Aristotle, *Meta.* 9.1.1046a2.
6. Aristotle, *Meta.* 9.2.1046a35–1046b3.
7. See Aristotle, *Meta.* 9.2.1046b4–24, 9.5.1047b35–1048a24; Charlotte Witt, *Ways of Being: Potentiality and Actuality in Aristotle's Metaphysics* (Ithaca, NY: Cornell University Press, 2003), 64–66. Witt, together with the majority of analytic metaphysicians, translates Aristotle's δύναμις not as "potency" but as "power." While the term "power" is applicable to active potencies, it is rather confusing when referred to passive potencies. I will say more on this topic in one of the following sections.
8. Aristotle, *Meta.* 9.3.1046b29–31.
9. Aristotle, *Meta.* 9.3.1047a15–17. See the entire chapter 4 of book 9 for a more detailed analysis of arguments against the Megarians.
10. Aristotle classifies nature in the same genus as potency: "And I mean by potency not only that definite kind which is said to be a principle of change in another thing or in the thing itself regarded as other, but in general every principle of movement or of rest. For nature also is in the same genus as potency; for it is a principle of movement—not, however, in something else but in the thing itself qua itself" (*Meta.* 9.8.1049b5–11).
11. See Witt, *Ways of Being*, 20–21.
12. Irwin, *Aristotle's First Principles*, 228.
13. Naturally, this conclusion refers only to contingent reality. It does not apply to Aristotle's unmoved mover or God, who is pure actuality, with no admixture of potentiality.
14. For more information about the scholastic theory of distinctions go to Feser, *Scholastic Metaphysics*, 72–79.
15. Aristotle, *Meta.* 9.8.1049b18–19.
16. Aristotle, *Meta.* 9.8.1049b24–25.
17. Aristotle, *Meta.* 9.8.1050a4–6. Note that when he speaks about a man's having the form lacking in a boy, Aristotle thinks about ἐντελέχεια (entelechy), which is a final stage of development and realization of form, and not a form in general. The former is developed and achieved at a certain moment in time. The latter is the same and fully present throughout the whole process of development of an organism.
18. Aristotle, *Meta.* 9.8.1050b 6–7. Witt offers a commentary on all three arguments in *Ways of Being*, chapter 4, 75–96.
19. Referring to Scholastic analyses, Feser notes that the distinction between potency and act leads to a series of further distinctions which are strategic for the Aristotelian-Thomistic metaphysics. I present his analysis of these distinctions in appendix 1.
20. John Locke, *An Essay concerning Human Understanding*, edited by A. C. Fraser (Chicago: Encyclopedia Britannica, 1952), book 2, chapter 23, para. 15.

21. Cartwright and Pemberton, "Aristotelian Powers," 96. See also Nancy Cartwright, *The Dappled World: A Study of the Boundaries of Science* (New York: Cambridge University Press, 1999), 49.
22. Feser, *Scholastic Metaphysics*, 61–62.
23. We find an allusion to the same distinction—this time seen from the point of view of potentiality—in *Phys.* 8.4.255b18–24: "One who is learning a science potentially knows it in a different sense from one who while already possessing the knowledge is not actually exercising it."
24. Stephen Mumford, *David Armstrong* (New York: Routledge, 2007), 88.
25. Molnar, *Powers*, 101, 126.
26. Charles Burton Martin, "Final Replies to Place and Armstrong," in D. M. Armstrong et al., *Dispositions: A Debate*, ed. Tim Crane (London: Routledge, 1996), 176.
27. See Oderberg, *Real Essentialism*, 138.
28. Molnar, *Powers*, 61.
29. John Heil, *From an Ontological Point of View* (Oxford: Clarendon Press, 2005), 222.
30. Ullin T. Place, "Dispositions as Intentional States," in Armstrong et al., *Dispositions*, 24.
31. See Franz Brentano, *Psychology from an Empirical Standpoint*, trans. Antos C. Rancurello, D. B. Terrell, and Linda L. McAlister (London and New York: Routledge, 1973; first published in 1874 as *Psychologie vom empirischen Standpunkte*), 211–14.
32. See Molnar, *Powers*, 63–66.
33. Alexander Bird, *Nature's Metaphysics: Laws and Properties* (Oxford: Clarendon Press, 2007), 125.
34. See ibid., 121–23.
35. Oderberg, *Real*, 137.
36. See Molnar, *Powers*, 71. Feser shows the imprecision of the terminology distinguishing consciousness and intentionality. Recognizing the importance of the notion of physical intentionality in analytic metaphysics, he concludes his analysis by saying, "The intentionality of perceptual states and propositional attitudes is too often taken matter-of-factly as the paradigm of directedness, so that the deck is unwittingly stacked in advance in favor of the presumption that there is something inherently 'mental' about directedness. And while Molnar and other powers theorists have tried to introduce a broader diet of examples, that directedness is inherently mental is treated as the default position, which believers in 'physical intentionality' have the burden of moving away" (Feser, *Scholastic Metaphysics*, 104).
37. The retrieval of teleology in dispositionalism differentiates it from the teleological skepticism of the new mechanical philosophers. Craver classifies functional descriptions as perspectival means of situating parts within higher-level mechanisms. Thus, he sees teleology as imposed upon the world by an intentional

describer, rather than as feature of the world itself. Others characterize functional descriptions as aspects of etiological explanation of the presence of parts of an organism, which are ultimately reducible to efficient (physical) causal stories (Garson). See Craver and Tabery, "Mechanisms in Science," section 4.5.

38. Alex Rosenberg, *The Atheist's Guide to Reality: Enjoying Life without Illusions* (New York: W. W. Norton, 2011), 179.

39. The contemporary versions of hylomorphism are proposed as an alternative answer to the question of the reality and the nature of substance and composition in analytic disputes on mereology. Their proponents address the variety of opinions ranging from (1) Lewis's doctrine of unrestricted composition (mereological universalism), which assumes that any combination of things in the world constitutes a further thing (e.g., my right hand and the large hand of the clock on the tower of the Notre Dame Basilica of the Sacred Heart), to (2) Peter van Inwagen's doctrine of restricted composition, which assumes that necessary and sufficient conditions cannot be given for the cohesion that is supposed to bind atoms into inanimate complexes, and thus the only existing substances are elementary particles (mereological simples) and living organisms; and (3) early Peter Unger's doctrine of mereological nihilism, which states that composition never occurs, and consequently leads him to admit and say, "I do not exist." See David Lewis, *On the Plurality of Worlds* (Oxford: Blackwell, 1986); Peter van Inwagen, *Material Beings* (Ithaca, NY, and London: Cornell University Press, 1990); Peter Unger, *Ignorance: A Case for Scepticism* (Oxford: Oxford University Press, 1975).

40. See Aquinas, *In Meta.* 5.3.779.

41. That is why, as I mentioned in the introduction, some thinkers tend to interpret Aristotle's notion of form merely as a geometrical shape. Aquinas says that here form may be called species: "The form corresponding to such a matter can be called the species" (ibid.). It seems that the category of species has a general and unqualified meaning for him at this point.

42. Kathrin Koslicki, "Aristotle's Mereology and the Status of Form," *Journal of Philosophy* 103 (2006): 724. See also Kathrin Koslicki, *The Structure of Objects* (Oxford: Oxford University Press, 2010), chapter 6.

43. See Koslicki, *Structure of Objects*, chapter 7.

44. Mark Johnston, "Hylomorphism," *Journal of Philosophy* 103 (2006): 658. He also gives an alternative definition of genuine parts and principles of unity: "*What it is for*. . . (the item is specified here) . . . *to be is for* . . . (some parts are specified here) . . . *to have the property or stand in the relation* . . . (the principle of unity is specified here)" (ibid.).

45. Ibid., 659.

46. Ibid., 663–64.

47. Kit Fine, "Things and Their Parts," *Midwest Studies in Philosophy* 23 (1999): 62.

48. Ibid.

49. Ibid., 70.

50. William Jaworski, *Structure and the Metaphysics of Mind: How Hylomorphism Solves the Mind-Body Problem* (Oxford: Oxford University Press, 2016), 8.

51. Ibid., 5. He formulates the "embodiment thesis"—i.e., "the claim that the capacities of structured wholes are essentially embodied in their parts" (ibid., 6).

52. Ibid., 18–19. In his description of EM and DC Jaworski claims that "emergentists are . . . free to endorse panpsychism" (ibid., 273), which I find questionable as none of them actually do so.

53. Ibid., 277–78.

54. Naturally, the dilemma occurs only if one assumes physicalism and causal closure when formulating the second assumption. Suggesting that the whole is more than its parts while being composed of smaller elements is not controversial in itself. It is not entirely clear to me whether Koons accepts physicalism and physical causal closure in complex dynamic systems.

55. Robert Koons, "Staunch vs. Faint-Hearted Hylomorphism: Toward an Aristotelian Account of Composition," *Res Philosophica* 91 (2014): 171.

56. Edward Jonathan Lowe, "A Neo-Aristotelian Substance Ontology: Neither Relational nor Constituent," in *Contemporary Aristotelian Metaphysics*, ed. Tuomas E. Tahko (Cambridge: Cambridge University Press, 2011), 230–31.

57. Ibid., 230, 236. This view might be defended if it was based on an implicit reference to primary matter. Yet Lowe does not refer to this concept in his works.

58. Ibid., 230.

59. Ibid., 237.

60. Thus, metaphysically speaking, we are not dealing with an electron anymore, but a hydrogen atom in which the electron in question is present virtually (by power), and not actually. I will come back to the concept of virtual presence in section 2.3.5, below.

61. Aristotle, *Meta.* 7.10.1035b35–1036a1, 7.11.1036a29. Aristotle adheres to the view that particulars cannot be defined in 7.10.1036a2–9: "When we come to the concrete thing, e.g. *this* circle, i.e. one of the individual circles . . . of these there is no definition"; and in 7.13.1039b27–1040a7: "There is neither definition of nor demonstration about sensible individual substances, because they have matter whose nature is such that they are capable of both being and of not being. . . . And so when one of the definition-mongers defines any individual, he must recognize that his definition may always be overthrown; for it is not possible to define such things."

62. Aristotle, *Meta.* 7.4.1030a6.

63. Gordon P. Barnes, "The Paradoxes of Hylomorphism," *The Review of Metaphysics* 56 (2003), 506.

64. Ibid., 507.

65. See Francis Suárez, *On the Formal Cause of Substance: Metaphysical Disputation* 15, trans. John Kronen and Jeremiah Reedy (Milwaukee: Marquette University Press, 2000), 77–79.

66. See Barnes, "Paradoxes of Hylomorphism," 509–15.
67. Ibid., 517.
68. Ibid., 517, 520–21. Barnes claims that such an intrinsic property is emergent and possible due to substantial form. But he does not offer a more precise definition and analysis of EM (see ibid., 521).
69. Referring to his example of a human being, Storck rightly states that "the difficulty with Barnes's multiplicity of substantial forms is that it does not permit the powers of the elements which compose the human body to really belong to the human body. . . . Biologically speaking, the powers of the elements are not just in the same place as the human body; indeed, they are its powers. . . . Thus, the powers of the elements, powers which Barnes uses to argue that the elements are actually present, need to be powers not of atoms that 'coincide' with the human body, but of the human body itself" (Storck, "Parts, Wholes," 57). In this context, it becomes apparent that the theory of presence by power (*virtute*) offers a much more plausible solution to the puzzle of many elements present in a whole. We will discuss it in more detail below, in section 2.3.5.
70. Anna Marmodoro, "Aristotle's Hylomorphism without Reconditioning," *Philosophical Inquiry* 37 (2013): 15. See also Theodore Scaltsas, *Substances and Universals in Aristotle's Metaphysics* (Ithaca, NY: Cornell University Press, 1994).
71. Marmodoro, "Aristotle's Hylomorphism without Reconditioning," 16–18. See also *Meta.* 8.6.1045b8–18.
72. Marmodoro, "Aristotle's Hylomorphism without Reconditioning," 19. Strictly speaking, primary matter is not potentially the form. Primary matter *is* the principle of potentiality.
73. *Meta.* 5.2.1014a26–27. Aquinas follows Aristotle in his definition of elements, both in *In Meta.* 5.4.795–807, and in *De prin. nat.* 19.
74. See Aquinas, *De mixt. elem.*, 15–18, and my comments on this point in section 3 of the introduction. Aquinas rejects the two other options: (1) the view of Avicenna, who thought that the elements remain in compounds with their substantial forms, but that their active and passive qualities change into some sort of mean qualities, and (2) the position held by Averroës that the elements persist in compounds with their substantial forms, which nonetheless have been changed into some sort of a mean form (see ibid., 2–14; *ST* I.76.4 ad 4). In Aristotle's account we read: "Since, however, some things *are-potentially* while others *are-actually*, the constituents combined in a compound can 'be' in a sense and yet 'not-be.' The compound may *be-actually* other than the constituents from which it has resulted; nevertheless each of them may still *be-potentially* what it was before they were combined, and both of them may survive undestroyed. (For this was the difficulty that emerged in the previous argument: and it is evident that combining constituents not only coalesce, having formerly existed in separation, but also can again be separated out from the compound.) The constituents, therefore, neither (*a*) *persist actually*, as 'body' and 'white' persist: nor (*b*) are they *destroyed* (either of them or

both), for their 'power of action' is preserved" (*De gen. et corr.* 1.10.327b24–32; see also 334b17–20, 24–30).

75. See, for example, Joseph Bobik, *Aquinas on Matter and Form and the Elements: A Translation and Interpretation of the "De principiis naturae" and the "De mixtione elementorum" of St. Thomas Aquinas* (Notre Dame, IN: University of Notre Dame Press, 1998), 245–60; Nancy Cartwright's assertion quoted below in section 2.3.7.

76. Both "production" and corruption of a given compound are cases of substantial change. The case of reclaiming of elements from a complex substance in a process that does not effect in the corruption of the latter is slightly different. What seems to be a substantial change "from the point of view" of a reclaimed element looks like an accidental change "from the point of view" of the compound. Moreover, from the perspective of contemporary science, the question remains whether elements existing potentially or virtually in mixed substances are the same in number with the elements that went into the change and will be yielded out of those substances upon their corruption. Decaen (see "Elemental Virtual Presence," 296) claims Aquinas would probably answer in the negative. But it need not be damaging for his theory since even the modern atomistic view hesitates to speak about numeral identity of particles throughout their various alterations and interactions at quantum level, when they are not actually being measured. It becomes apparent in the case of quantum entanglement, which results in a unified system that in some sense contains two electrons, and yet we are unable to point toward a distinct identity associated with either one of them. Quantum entanglement requires a replacement of classical statistics (assuming four possible states with an equal probability of 0.25 for a two-electron system) with Einstein-Bose statistics (allowing for just three possible states with an equal 1/3 probability: two electrons up, two electrons down, and one electron in each state). Some find it to be an argument in favor of EM. See Koons and Pickavance, *Metaphysics*, 147–48.

77. In his *In Meta.* 7.17.1674 Aquinas states: "Flesh is not merely fire and earth, or the hot and the cold, by whose power the elements are mixed, but there is also something else by which flesh is flesh." This "something else" is obviously the substantial form of an organism.

78. Offering this example as an answer to Lowe's claim that the constitution (coming into being) of a hydrogen atom is nothing more than a combination of one proton and one electron, both of which nevertheless remain unchanged, I follow his apparent acceptance of protons and electrons as separate physical entities.

79. I do not find any direct nor indirect references to the theory of virtual presence in the versions of neo-hylomorphism offered by Koslicki, Jaworski, and Lowe.

80. At the same time, my evaluation shows that both contemporary philosophy of science and at least some of the neo-hylomorphic propositions discussed here could accommodate and find their completion and ground in the language of virtual presence, primary matter, substantial form, and substantial change.

81. See Michael C. Rea, "Hylomorphism Reconditioned," *Philosophical Perspectives* 25 (2011): 341–58.

82. Ibid., 342.

83. Ibid., 345.

84. Ibid., 351. "One power P unites some other powers just in the case that P is so connected to the other powers that its manifestation depends upon the cooperative manifestation of the united powers and, furthermore, the latter do not confer any powers on the object that has P that are both intrinsic to the object and independent of P" (Ibid., 348–49).

85. According to Aristotle, "When we have the whole, such and such a form in this flesh and in these bones, this is Callias or Socrates; and they are different in virtue of their matter (for that is different), but the same in form; for their form is indivisible" (*Meta.* 7.8.1034a5–7). Referring to Avicenna, Aquinas further develops Aristotle's teaching, saying that "the principle of individuation is not matter taken in just any way whatever, but only designated matter [*materia signata*]. And I call that matter designated which is considered under determined dimensions" (*De ente* 1.23). In other words, for Aquinas individuation comes from matter designated by quantity.

86. Rea, "Hylomorphism Reconditioned," 349. The preceding quotation from ibid., 352.

87. Ibid., 342.

88. Marmodoro, "Aristotle's Hylomorphism," 13.

89. Ibid., 14.

90. Rea, "Hylomorphism Reconditioned," 342.

91. It is worth noticing that Marmodoro has recently offered her own version of powers ontology, which she classifies as "power structuralism." I have mentioned some of its aspects in chapter 2, section 6.6. Trying to define its basic premises (with some indirect references to her version of neo-hylomorphism), she states:

> My ontology has instances of physical powers as its building blocks: power tropes (e.g., an electric charge here and now) from which everything else is composed or derived. Such building blocks are the sparse fundamental properties in nature, as defined by David Lewis (e.g., mass, spin, charge); only that on my view—but not on Lewis's—they are powers, essentially defined by the type of change (namely, interactions or noninteractive activities) that they or their possessors can bring about in the world. So, for instance, I take an electron to be composed of the power tropes mass, spin and charge; mass, spin and charge are the building blocks that make up the electron; they characterise the electron as its properties, and enable it to behave causally as it does in its environment. But the electron is not reducible to its compresent mass, spin and charge. These powers are not simply compresent in the electron; they are structured in relation to one another and, as I will argue, they furthermore compose into a single entity, the electron. The electron is the composition of the structured mass, spin and charge. (Marmodoro, "Power Mereology," 110–11)

92. He says at one point, "Given that I am assuming dispositional monism (which, again, posits dispositions 'all the way down'), it is not clear that the act-potency distinction can be meaningfully drawn" (Rea, "Hylomorphism Reconditioned," 348).

93. Quine challenges essentialism through paradoxical thought experiments such as the one concerning the essence of a person who is both a good mathematician and a cyclist, or his famous puzzle concerning the number of the planets. He also asserts that the reason for holding on to or abandoning a given proposition should always depend on its changing our entire system of beliefs (logical and mathematical truths thus seem necessary, while essentialism seems spurious). Finally, he suggests that our tendency to classify things into natural kinds is a product of natural selection. Popper claims that Aristotelian knowledge of essences depends on the faculty of our intellectual intuition, which is subjective and varies. He also sees essentialism as committed to the doctrine that science aims at an ultimate explanation. He thinks accepting essentialism hampers scientific progress. Finally, in the same vein as Quine, he states that we should take seriously questions of fact rather than speculations about words and their meanings. Wittgenstein, in his famous theory of games and "family resemblances," notes that similarities among things may lead us to assume one attribute must be shared by all of the members of a "game," but it is an illusion, which also has its consequence in attributing essences to things. His argument is based on the charge of vagueness of essentialist claims. See Willard Van Orman Quine, *Word and Object* (Cambridge, MA: MIT Press, 1960); Quine, *Ontological Relativity, and Other Essays* (New York: Columbia University Press, 1969); Quine, *Quiddities* (Cambridge, MA: Harvard University Press, 1987); Karl Raimund Popper, *The Open Society and Its Enemies*, vol. 2, *The High Tide of Prophecy* (New York: Harper & Row, 1962); Popper, *Conjectures and Refutations: The Growth of Scientific Knowledge* (London: Routledge & K. Paul, 1963); Popper, *Objective Knowledge: An Evolutionary Approach* (Oxford: Clarendon Press, 1979); Popper, *Unended Quest: An Intellectual Autobiography* (London and New York: Routledge, 2002); Ludwig Wittgenstein, *Philosophical Investigations* (New York: Macmillan, 1968). All arguments listed and referenced here are also discussed and answered by Feser in *Scholastic Metaphysics*, 216–23, and Oderberg in *Real Essentialism*, 21–43.

94. Cartwright, *Dappled World*, 102. See my comments on treating elementary particles as basic elements in chapter 3, section 3.5, note 95, and in section 2.3.5 of the present chapter.

95. Ellis, *Scientific Essentialism*, 59.

96. See Richard Boyd, "Realism, Anti-Foundationalism and the Enthusiasm for Natural Kinds," *Philosophical Studies* 61 (1991): 127–48.

97. See Craver and Tabery, "Mechanisms in Science," section 4.4.

98. Heil, *Ontological Point of View*, 221.

99. Kripke's "rigid designator" is an expression referring to the same thing in every possible world in which it exists. He sees proper names and natural-kind terms

as rigid designators. "Sodium chloride" and "NaCl" are examples of rigid designators. Since sodium chloride = NaCl in the actual world, the same will be true in every possible world. Thus, we can say sodium chloride is "essentially" NaCl. From the Aristotelian point of view, determining essence in reference to possible worlds modality is problematic. Possible worlds exist merely as objects of thought, which contradicts Aristotelian moderate realism, which describes essences as real, existent, and mind independent. To treat possible worlds as concrete entities (as Lewis) or as Platonic abstract entities means a departure from realism to nominalism.

100. To take sodium chloride as an example, what makes it a natural kind for Kripke is that every bit of sodium chloride has the same internal structure—i.e., that of a crystal in which each Na^+ ion is surrounded by six Cl^- ions, while each Cl^- ion is surrounded by six Na^+ ions. This is the essence of sodium chloride. Thus, to be a sodium chloride is to have this internal structure in any possible world. See Saul A. Kripke, *Naming and Necessity* (Cambridge, MA: Harvard University Press, 1980), 120, 126. According to Putnam, two objects have the same essence iff they have the same, or very similar, hidden structures. See Hilary Putnam, "The Meaning of Meaning," in *Mind, Language and Reality: Philosophical Papers* (Cambridge: Cambridge University Press, 1975), 2:241.

101. Oderberg, *Real Essentialism*, 131. Commenting on the distinction between the "categorical" and the "dispositional," Oderberg says, "In fact, neither term is a happy one. 'Categorical' could mean 'real,' 'actual,' 'unconditional,' 'episodic,' 'occurrent,' all of which are distinct aspects of phenomena . . . [while] 'disposition,' as well as sometimes connoting potentiality, also connotes actuality" (Ibid., 131–32). He claims hylomorphism resolves the problem of this distinction.

102. Ibid., 140. Naturally, being in a state of pure actuality is impossible to all contingent beings, not for God.

103. According to another interpretation of powers metaphysics, powers reveal substantial forms of beings, while groundless dispositions at the bottom level of the complexity of matter, described by quantum physics, can be thought as referring to primary matter (in line with Heisenberg's interpretation of quantum mechanics). But one may object that primary matter is still more basic than quantum indeterminacy, which is said to "produce" quantum effects. Primary matter cannot produce anything by itself. As a principle of pure potentiality, it needs to be in-formed by substantial form. Hence, Oderberg says "there is the worry that scientific essentialists tend to treat the intrinsic properties of fundamental particles as all dispositional [e.g., Ellis, *Scientific Essentialism*, 135], in which case what is left of the notion of intrinsic essence? Isn't the use of the term 'intrinsic' for such properties merely honorific? . . . What this shows is that scientific essentialists are wrong to suggest that the essence of anything, fundamental particles included, is purely dispositional, even if some of a thing's properties are ungrounded potentialities, and even if the essence itself involves some ungrounded potentiality. . . . There is no pure potentiality, i.e. potentiality existing in the absence of actuality to

shape and determine it. The same applies to particles such as electrons" (Oderberg, *Real Essentialism*, 146).

104. See ibid., 62–85.

105. Ibid., 143; see also 130–43. Although the term "hylomorphism" is most commonly used in English to express Aristotelian composition of ὕλη (*hylē*) and μορφή (*morphē*), it is also acceptable to use "hylemorphism" instead, as Oderberg does in the passage quoted here.

106. I refer here to Christopher Shields, "Causal Processes: An Aristotelian Alternative to an Unwarranted Humean Hegemony" (paper presented at Neo-Aristotelianism: A Conference on Neo-Aristotelian Metaphysics, Ethics, and Politics, Chicago, September 30–October 1, 2016), and to our personal conversation. To support his definition of efficient causation Shields refers to *Phys.* 3.3.202a13–20, where Aristotle says, "It is the fulfilment (ἐντελέχεια [*entelecheia*] = actualization) of this potentiality by the action of that which has the power of causing motion; and the actuality (ἐνέργεια [*energeia*]) of that which has the power of causing motion is not other than the actuality of the movable; for it must be the actualization (ἐντελέχεια) of *both*. A thing is capable of causing motion because it *can* do this, it is a mover because it actually (τῷ ἐνεργεῖν [*tō energein*]) *does* it. But it is on the movable that it is capable of acting (ἐνεργητικόν [*energētikon*]). Hence there is a single actuality of both alike, just as one to two and two to one are the same interval, and the steep ascent and the steep descent are one—for these are one and the same, although their account (λόγος) is not one. So it is with the mover and the moved."

107. Among proponents of entity realism, we find Ian Hacking, *Representing and Intervening: Introductory Topics in the Philosophy of Natural Science* (Cambridge: Cambridge University Press, 1983); Nancy Cartwright, *How the Laws of Physics Lie* (Oxford: Clarendon Press, 1983). Structural realism is supported by John Worrall, "Structural Realism: The Best of Both Worlds?," *Dialectica* 43 (1989): 99–124; James Ladyman, "Structural Realism," *Stanford Encyclopedia of Philosophy*, winter 2016 ed., ed. Edward N. Zalta, https://plato.stanford.edu/archives/win2016/entries/structural-realism/.

108. Anjan Chakravartty, "Dispositions for Scientific Realism," in *Powers and Capacities in Philosophy: The New Aristotelianism*, edited by Ruth Groff and John Greco (New York: Routledge, 2013), 118.

109. Oderberg, *Real Essentialism*, 144. See also Chakravartty, "Dispositions for Scientific Realism," 118–22.

110. Allen, *Critical Introduction to Properties*, 179, 176. For a more elaborate version of the same argument see Stephen Mumford, *Laws in Nature* (New York: Routledge, 2004).

111. Commenting on the arbitrary character of the neo-Humean theory choice, Koons and Pickavance refer to Lewis's attempt to provide a more rigid and accurate strategy, which suggests that "we rigidly build our own present, actual preferences into the definition of the 'best' theory, and then use that rigidified conception

of what's best in defining the laws of nature.... If we rigidify the reference of 'law of nature' in this way, then the fact that at other times, in other cultures, or in possible scenarios, scientists have different standards of good theories is completely irrelevant.... Nonetheless, the fact remains that there is something problematically anthropocentric (people-centered) about the neo-Humeist's account of the laws of nature," which "from the point of scientific realism ... are what they are, independently of our preferences" (Koons and Pickavance, *Metaphysics*, 57, 56).

CHAPTER 7

1. Although not every advocate of essentialism is a hylomorphist, every hylomorphist is an essentialist. The argumentation presented by many dispositionalists suggests a reference to both essentialism and hylomorphism.
2. See chapter 2, note 5, on whether the conglomeration of H_2O molecules is characterized by the new substantial, or rather a new accidental, form.
3. See *In Meta.* 5.2.775; *De prin. nat.* 28.
4. *In Phys.* 5.11.246.
5. *De prin. nat.* 34; *In Phys.* 2.11.242.
6. Aquinas acknowledges this difficulty when he speaks of immobile—that is, inanimate—beings: "Since the form is the cause of existing absolutely, the other three are causes of existence insofar as something receives existence. Hence in immobile things the other three causes are not considered, but only the formal cause is considered" (*In Phys.* 2.10.240). Moreover, he emphasizes that sometimes, for practical reasons, "the 'why' is reduced to the first moving cause. Thus, why does someone fight? Because he has stolen. For this is what brought on the fight. Sometimes it is reduced to the final cause, as if we should ask for the sake of what does someone fight, and the answer is that he might rule. Sometimes it is reduced to the material cause, as when it is asked why this body is corruptible, and the answer is because it is composed of contraries" (ibid., 239). These reflections of Aquinas echo, in a way, considerations brought by Aristotle and mentioned above in section 2.3 of the introduction (note 53).
7. Bokulich, "Missing or Modal?," 12.
8. Deacon and Cashman, "Teleology versus Mechanism," 302.

APPENDIX 2

1. I referred to Campbell's book in the first part of this project. See especially chapter 1, notes 15 and 16.
2. Ibid., 68–70, 58, 65 (italics original). Campbell introduces and criticizes the metaphysics of particular entities in the second chapter of his book. In chapter 3 he analyzes conceptual shifts in physics and develops his own metaphysical theory.

3. Ibid., 203, 208, 210. I have neither time nor space to discuss in detail other aspects of Campbell's position, including his intriguing explanation of identity through change (in reference to the concepts of homometry and autometry), his metaphysical taxonomy of emergent entities and evolutionary taxonomy of types of action and life, and his analysis of the EM of human mental activity. With regard to these issues I refer readers to Campbell's book (chapters 5, 6, 7, and 10).

4. Ibid., 68.

5. Ibid., 58 (Campbell refers here to David Tong's "Lectures on Quantum Field Theory" given at the University of Cambridge in 2006).

6. Ibid., 134. Campbell defines generic processes as those that "occur in particular spatiotemporal regions, like particulars, but can occur in different spatiotemporal regions, like universals" (ibid., 79).

7. Ibid., 138.

8. Ibid., 68, 72.

9. See ibid., 27–28.

BIBLIOGRAPHY

Alexander, Samuel. *Space, Time and Deity*. Vol. 2. London: Macmillan, 1920.
Allen, Sophie. *A Critical Introduction to Properties*. London and New York: Bloomsbury, 2016.
Anjum, Rani Lill, and Stephen Mumford. "Emergence and Demergence." In Paoletti and Orilia, *Philosophical and Scientific*, 92–109.
Anscombe, Elizabeth. "Causality and Determination." In *Causation*, edited by Ernest Sosa and Michael Tooley, 88–104. Oxford and New York: Oxford University Press, 1993.
Aquinas, Thomas. *De ente et essentia*. In *Opera omnia iussu Leonis XIII P. M. edita*, 43:131–57. Roma: Editori di San Tommaso, 1976. Translated by Joseph Bobik as *Aquinas on Being and Essence: A Translation and Interpretation* (Notre Dame, IN: University of Notre Dame Press, 1965).
———. *De mixtione elementorum ad magistrum Philippum de Castro Caeli*. In *Opera omnia iussu Leonis XIII P. M. edita*, 43:315–81. Roma: Editori di San Tommaso, 1976. Translated by Joseph Bobik as *Aquinas on Matter and Form and the Elements: A Translation and Interpretation of the De Principiis Naturae and the De Mixtione Elementorum of St. Thomas Aquinas* (Notre Dame, IN: University of Notre Dame Press, 1998).
———. *De principiis naturae*. In *Opera omnia iussu Leonis XIII P. M. edita*, 43:39–47. Rome: Typographia polyglotta, 1976. Translated by Robert P. Goodwin as *The Principles of Nature*, in *Selected Writings of St. Thomas Aquinas* (New York: Bobbs-Merrill, 1965), 7–28.
———. *In Aristotelis librum De anima commentarium*. Vol. 45/1 of *Opera omnia*. Rome: Typographia polyglotta, 1984. Translated by Kenelm Foster and Silvester Humphries as *Commentary on Aristotle's De Anima in the Version of William of Moerbeke and the Commentary of St. Thomas Aquinas* (London: Routledge and Kegan Paul, 1951).
———. *In Metaphysicam Aristotelis commentaria*. Turin and Rome: Marietti, 1926. Translated by John Rowan as *Commentary on the Metaphysics of Aristotle*. 2 Vols. (Chicago: Regnery Press, 1961).
———. *In octo libros Physicorum Aristotelis expositio*. Turin and Rome: Marietti, 1965. Translated by Richard J. Blackwell, Richard J. Spath, and W. Edmund

Thirlkel as *Commentary on Aristotle's Physics* (Notre Dame, IN: Dumb Ox Books, 1999).

———. *Quaestiones disputatae de veritate*. Vol. 22/1–3 of *Opera Omnia*. Rome: Typographia polyglotta, 1972–76. Translated by Robert W. Mulligan, S.J., et al. as *Truth*. 3 vols. (Albany, NY: Preserving Christian Publications, 1993).

———. *Scriptum super Libros Sententiarum Magistri Petri Lombardi Episcopi Parisiensis*. Edited by P. Mandonnet. Paris: P. Lethielleux, 1929–47.

———. *Summa contra gentiles*. 3 vols. Turin and Rome: Marietti, 1961. Translated by Anton C. Pegis et al. as *On the Truth of the Catholic Faith: Summa Contra Gentiles*. 4 vols. (Garden City, NY: Image Books, 1955–57).

———. *Summa theologiae*. Rome: Editiones Paulinae, 1962. Translated by the Fathers of the English Dominican Province as *Summa Theologica*. 3 vols. (New York: Benzinger Bros., 1946).

Aristotle. *De Anima (On the Soul)*. Translated by J. A. Smith. In *The Basic Works of Aristotle*, edited by Richard McKeon, 533–603. New York: The Modern Library, 2001.

———. *De caelo (On the Heavens)*. Translated by J. L. Stocks. In *The Basic Works of Aristotle*, edited by Richard McKeon, 398–466. New York: The Modern Library, 2001.

———. *Generation of Animals*. Translated by Arthur Platt. In *The Complete Works of Aristotle: The Revised Oxford Translation*, vol. 1, edited by Jonathan Barnes, 1111–1218. Princeton: Princeton University Press, 1984.

———. *De generatione et corruptione (On Generation and Corruption)*. Translated by Harold H. Joachim. In *The Basic Works of Aristotle*, edited by Richard McKeon, 465–531. New York: The Modern Library, 2001.

———. *Metaphysica (Metaphysics)*. Translated by W. D. Ross. In *The Basic Works of Aristotle*, edited by Richard McKeon, 681–926. New York: The Modern Library, 2001.

———. *The Metaphysics*. Translated by Hugh Tredennick. London and Cambridge, MA: W. Heinemann, 1936.

———. *Metaphysics, Books Γ, Δ, and E*. Translated by Christopher Kirwan. Oxford: Clarendon Press, 1971.

———. *Meteorology*. Translated by W. Webster. In *The Complete Works of Aristotle: The Revised Oxford Translation*, vol. 1, edited by Jonathan Barnes, 555–625. Princeton: Princeton University Press, 1984.

———. *Parts of Animals*. Translated by William Ogle. In *The Complete Works of Aristotle: The Revised Oxford Translation*, vol. 1, edited by Jonathan Barnes, 994–1086. Princeton: Princeton University Press, 1984.

———. *Physica (Physics)*. Translated by R. K. Gaye. In *The Basic Works of Aristotle*, edited by Richard McKeon, 213–394. New York: Modern Library, 2001.

———. *The Physics*. Translated by Philip H. Wicksteed and Francis M. Cornford. London and New York: W. Heinemann, 1929.

———. *Physics: Books I and II*. Translated by W. Charlton. Oxford: Clarendon Press, 1970.

Armstrong, David Malet. "Defending Categoricalism." In *Properties, Powers, and Structures: Issues in the Metaphysics of Realism*, edited by Alexander Bird, B. D. Ellis, and Howard Sankey, 27–33. New York: Routledge, 2012.

———. *A Materialist Theory of the Mind*. London: Routledge & K. Paul, 1968.

———. *A World of States of Affairs*. Cambridge and New York: Cambridge University Press, 1997.

Aronson, J. "On the Grammar of 'Cause.'" *Synthese* 22 (1971): 414–30.

Ayala, Francisco J. "Teleological Explanations in Evolutionary Biology." In *Nature's Purposes: Analyses of Function and Design in Biology*, edited by Colin Allen, Marc Bekoff, and George Lauder, 29–49. Cambridge, MA: MIT Press, 1998.

Bacon, Francis. *The Works of Francis Bacon*. Vol. 4, *Translations of the Philosophical Works 1*. Edited by James Spedding, Robert L. Ellis, and Douglas D. Heath. New York: Cambridge University Press, 2011. First published in 1858.

Bain, Alexander. *Logic*. Books 2 and 3. London: Longmans, Green, Reader, and Dyer, 1920.

Barnes, Gordon P. "The Paradoxes of Hylomorphism." *The Review of Metaphysics* 56 (2003): 501–23.

Bechtel, William, and Adele Abrahamsen. "Explanation: A Mechanist Alternative." *Studies in History and Philosophy of Science Part C: Studies in History and Philosophy of Biological and Biomedical Sciences* 36 (2005): 421–41.

Bechtel, William, and Robert C. Richardson. *Discovering Complexity: Decomposition and Localization as Strategies in Scientific Research*. Cambridge, MA: MIT Press, 2010. First published in 1993.

Bedau, Mark A. "Downward Causation and the Autonomy of Weak Emergence." *Principia: An International Journal of Epistemology* 6 (2002): 5–50.

———. "Weak Emergence." *Noûs* 31 (1997): 375–99.

———. "Where's the Good in Teleology?" *Philosophy and Phenomenological Research* 52 (1992): 781–806.

Beebee, Helen, Christopher Hitchcock, and Peter Charles Menzies, eds. *The Oxford Handbook of Causation*. Oxford and New York: Oxford University Press, 2009.

Berkeley, George. *The Works of George Berkeley, Bishop of Cloyne*. Edited by A. A. Luce and T. E. Jessop. 9 vols. Edinburgh: Thomas Nelson & Sons, 1948–57.

Bickhard, Mark H., and Donald T. Campbell. "Emergence." Lehigh University, October 15, 2016, http://www.lehigh.edu/~mhb0/emergence.html.

Bird, Alexander. "Causation and the Manifestation of Powers." In Marmodoro, *Metaphysics of Powers*, 35–41.

———. "Limitations of Power." In *Powers and Capacities in Philosophy: The New Aristotelianism*, edited by Ruth Groff and John Greco, 25–47. New York: Routledge, 2013.

---. "Monistic Dispositional Essentialism." In *Properties, Powers, and Structures: Issues in the Metaphysics of Realism*, edited by Alexander Bird, B. D. Ellis, and Howard Sankey, 11–26. New York: Routledge, 2012.

---. *Nature's Metaphysics: Laws and Properties*. Oxford: Clarendon Press, 2007.

Bobik, Joseph. *Aquinas on Matter and Form and the Elements: A Translation and Interpretation of the "De principiis naturae" and the "De mixtione elementorum" of St. Thomas Aquinas*. Notre Dame, IN: University of Notre Dame Press, 1998.

Bohr, Niels. "Discussion with Einstein on Epistemological Problems in Atomic Physics." In *Albert Einstein: Philosopher-Scientist*, edited by P. A. Schlipp, 201–41. New York: Tudor Publishing, 1949.

Bokulich, Peter. "Missing or Modal? Where Emergence Finds the Physical Facts." *Religion, Brain & Behavior* 3 (2013): 6–12.

Born, Max. *Natural Philosophy of Cause and Chance*. Oxford: Clarendon Press, 1949.

Bostock, David. *Space, Time, Matter, and Form: Essays on Aristotle's "Physics."* Oxford: Clarendon Press, 2006.

Boyd, Richard. "Realism, Anti-Foundationalism and the Enthusiasm for Natural Kinds." *Philosophical Studies* 61 (1991): 127–48.

Braillard, Pierre-Alain, and Christophe Malaterre, eds. *Explanation in Biology: An Enquiry into the Diversity of Explanatory Patterns in the Life Sciences*. Dordrecht: Springer, 2015.

Brentano, Franz. *Psychology from an Empirical Standpoint*. Translated by Antos C. Rancurello, D. B. Terrell, and Linda L. McAlister. London and New York: Routledge, 1973. First published in 1874 as *Psychologie vom empirischen Standpunkte*.

Broad, Charlie Dunbar. *The Mind and Its Place in Nature*. London: Routledge and Kegan Paul, 1925.

Broadie, Sarah. "The Ancient Greeks." In Beebee, Hitchcock, and Menzies, *Oxford Handbook of Causation*, 21–39.

Bunge, Mario. *Causality and Modern Science*. New York: Dover, 1979. First published in 1959.

---. *Matter and Mind: A Philosophical Inquiry*. Dordrecht: Springer, 2010.

---. *The Mind-Body Problem: A Psychobiological Approach*. Oxford and New York: Pergamon Press, 1980.

Burtt, Edwin Arthur. *The Metaphysical Foundations of Modern Science*. Mineola, NY: Dover, 2003. First published in 1924.

Campbell, Donald Thomas. "'Downward Causation' in Hierarchically Organized Biological Systems." In *Studies in the Philosophy of Biology*, edited by Francisco J. Ayala and Theodosius Dobzhansky, 179–86. Berkeley and Los Angeles: University of California Press, 1974.

Campbell, Richard. *The Metaphysics of Emergence*. New York: Palgrave Macmillan, 2015.

Cartwright, Nancy. *The Dappled World: A Study of the Boundaries of Science*. New York: Cambridge University Press, 1999.
———. *How the Laws of Physics Lie*. Oxford: Clarendon Press, 1983.
———. *Hunting Causes and Using Them: Approaches in Philosophy and Economics*. Cambridge and New York: Cambridge University Press, 2007.
Cartwright, Nancy, and John Pemberton. "Aristotelian Powers: Without Them, What Would Modern Science Do?" In *Powers and Capacities in Philosophy: The New Aristotelianism*, edited by Ruth Groff and John Greco, 93–112. New York: Routledge, 2013.
Cassell, Paul. "Incomplete Deacon: Why New Research Programs in the Sciences and Humanities Should Emerge from Terrence Deacon's *Incomplete Nature*." *Religion, Brain & Behavior* 3 (2013): 12–19.
Chakravartty, Anjan. "Dispositions for Scientific Realism." In *Powers and Capacities in Philosophy: The New Aristotelianism*, edited by Ruth Groff and John Greco, 113–27. New York: Routledge, 2013.
Chalmers, David J. *The Conscious Mind: In Search of a Fundamental Theory*. New York: Oxford University Press, 1996.
Charlton, William. "Did Aristotle Believe in Prime Matter?" In *Physics: Books I and II*, by Aristotle, translated with introduction and notes by W. Charlton, 129–45. Oxford: Clarendon Press, 1983.
Clark, Stewart J., and Tom Lancaster. "The Use of Downward Causation in Condensed Matter Physics." In Paoletti and Orilia, *Philosophical and Scientific Perspectives*, 129–45.
Clatterbaugh, Kenneth. *The Causation Debate in Modern Philosophy, 1637–1739*. New York and London: Routledge, 1999.
Clayton, Philip. "Conceptual Foundations of Emergence Theory." In Clayton and Davies, *Re-Emergence of Emergence*, 1–31.
———. *Mind and Emergence: From Quantum to Consciousness*. Oxford and New York: Oxford University Press, 2004.
Clayton, Philip, and Paul Davies, eds. *The Re-Emergence of Emergence: The Emergentist Hypothesis from Science to Religion*. Oxford and New York: Oxford University Press, 2006.
Collingwood, Robin G. "On the So-Called Idea of Causation." *Proceedings of the Aristotelian Society* 38 (1937–38): 85–112.
Craver, Carl F., and William Bechtel. "Top-Down Causation without Top-Down Causes." *Biology & Philosophy* 22 (2006): 547–63.
Craver, Carl F., and Lindley Darden. *In Search of Mechanisms: Discoveries across the Life Sciences*. Chicago: University of Chicago Press, 2013.
Craver, Carl, and James Tabery. "Mechanisms in Science." In *Stanford Encyclopedia of Philosophy*. Spring 2017 ed. Edited by Edward N. Zalta. https://plato.stanford.edu/archives/spr2017/entries/science-mechanisms/.
Cummings Neville, Robert. "Teleodynamic Remarks about Two Cultures." *Religion, Brain & Behavior* 3 (2013): 19–25.

Davidson, Donald. "Mental Events." In *Essays on Actions and Events*. Oxford: Oxford University Press, 1980.

Davies, Paul. "The Physics of Downward Causation." In Clayton and Davies, *Re-Emergence of Emergence*, 35–52.

Dawkins, Richard. *The Blind Watchmaker*. New York: W. W. Norton, 1986.

Deacon, Terrence. "Emergence: The Hole at the Wheel's Hub." In Clayton and Davies, *Re-Emergence of Emergence*, 111–50.

———. "The Hierarchic Logic of Emergence: Untangling the Interdependence of Evolution and Self-Organization." In *Evolution and Learning: The Baldwin Effect Reconsidered*, edited by Bruce H. Weber and David J. Depew, 273–308. Cambridge, MA: MIT Press, 2003.

———. *Incomplete Nature: How Mind Emerged from Matter*. New York: W. W. Norton, 2012.

———. "Making Sense of Incompleteness: A Response to Commentaries." *Religion, Brain & Behavior* 3 (2013): 37–52.

———. "Reciprocal Linkage between Self-Organizing Processes Is Sufficient for Self-Reproduction and Evolvability." *Biological Theory* 1 (2006): 136–49.

Deacon, Terrence W., and Tyrone Cashman. "Deacon and Cashman Respond to Green, Pryor, Tabaczek, and Moritz." *Theology and Science* 14 (2016): 464–81.

———. "Steps to a Metaphysics of Incompleteness." *Theology and Science* 14 (2016): 401–29.

———. "Teleology versus Mechanism in Biology: Beyond Self-Organization." In *Beyond Mechanism: Putting Life Back into Biology*, edited by Brian G. Henning and Adam C. Scarfe, 287–308. Lanham, MD: Lexington Books, 2013.

Deacon, Terrence W., and Spyridon Koutroufinis. "Complexity and Dynamical Depth." *Information* 5 (2014): 404–23.

Deacon, Terrence W., Alok Srivastava, and Augustus Bacigalupi. "The Transition from Constraint to Regulation at the Origin of Life." *Frontiers in Bioscience* 19 (2014): 945–57.

Decaen, Christopher. "Elemental Virtual Presence in St. Thomas." *The Thomist* 64 (2000): 271–300.

Descartes, René. *The Philosophical Works of Descartes*. 2 vols. Translated by E. S. Haldane and G. R. T. Ross. Cambridge: Cambridge University Press, 1970.

Dobzhansky, Theodosius. *Genetics of the Evolutionary Process*. New York: Columbia University Press, 1970.

Dodds, Michael J. *The Philosophy of Nature*. Oakland, CA: Western Dominican Province, 2010.

———. "Top Down, Bottom Up or Inside Out? Retrieving Aristotelian Causality in Contemporary Science." Paper presented at the annual Thomistic Institute sponsored by the Jacques Maritain Center, University of Notre Dame, July 25, 1997. Accessed October 15, 2016. http://www3.nd.edu/~afreddos/papers/dodds.htm.

Dougherty, Jude P. *The Nature of Scientific Explanation*. Washington, DC: Catholic University of America Press, 2013.

Dowe, Phil. "Causal Process Theories." In Beebee, Hitchcock, and Menzies, *Oxford Handbook of Causation*, 213–33.

———. "Good Connections: Causation and Causal Processes." In *Causation and Laws of Nature*, edited by H. Sankey, 247–64. Dordrecht: Kluwer, 1999.

———. *Physical Causation*. Cambridge: Cambridge University Press, 2000.

Ducasse, Curt John. "On the Nature and the Observability of the Causal Relation." *Journal of Philosophy* 23 (1926): 57–68.

Dudley, John. *Aristotle's Concept of Chance: Accidents, Cause, Necessity, and Determinism*. New York: State University of New York Press, 2012.

Dupré, John. *The Disorder of Things: Metaphysical Foundations of the Disunity of Science*. Cambridge, MA: Harvard University Press, 1993.

Ehring, Douglas. "Mental Causation, Determinables, and Property Instances." *Noûs* 30 (1996): 461–80.

Einstein, Albert, and Leopold Infeld. *The Evolution of Physics*. New York: Simon and Schuster, 1954.

Elders, Leo. *The Metaphysics of Being of St. Thomas Aquinas*. Leiden: Brill, 1993.

El-Hani, Charbel Niño, and Antonio Marcos Pereira. "Higher-Level Descriptions: Why Should We Preserve Them?" In *Downward Causation: Mind, Bodies and Matter*, edited by Peter Bøgh Andersen, Claus Emmeche, Niels O. Finnemann, and Peder Voetmann Christiansen, 118–42. Aarhus and Oxford: Aarhus University Press, 2000.

Ellis, Brian D. "The Categorical Dimensions of Causal Powers." In *Properties, Powers, and Structures: Issues in the Metaphysics of Realism*, edited by Alexander Bird, B. D. Ellis, and Howard Sankey, 11–26. New York: Routledge, 2012.

———. "Causal Powers and Categorical Properties." In Marmodoro, *Metaphysics of Powers*, 133–42.

———. *The Philosophy of Nature: A Guide to the New Essentialism*. Montreal and Ithaca, NY: McGill-Queen's University Press, 2002.

———. *Scientific Essentialism*. Cambridge and New York: Cambridge University Press, 2001.

Ellis, George F. R. "Science, Complexity, and the Nature of Existence." In *Evolution and Emergence: Systems, Organisms, Persons*, edited by Nancey Murphy and William R. Stoeger, 113–40. Oxford and New York: Oxford University Press, 2007.

Emmeche, Claus, Simo Køppe, and Frederic Stjernfelt. "Explaining Emergence: Towards an Ontology of Levels." *Journal for General Philosophy of Science* 28 (1997): 83–119.

———. "Levels, Emergence, and Three Versions of Downward Causation." In *Downward Causation: Mind, Bodies and Matter*, edited by Peter Bøgh Andersen, Claus Emmeche, Niels O. Finnemann, and Peder Voetmann Christiansen, 13–34. Aarhus and Oxford: Aarhus University Press, 2000.

Engelhard, Kristina. "Categories and the Ontology of Powers: A Vindication of the Identity Theory of Properties." In Marmodoro, *Metaphysics of Powers*, 41–57.

Fair, D. "Causation and the Flow of Energy." *Erkenntnis* 14 (1979): 219–50.

Farrell, John. "Book of the Year: The Return of Teleology." Forbes, December 26, 2012. http://www.forbes.com/sites/johnfarrell/2012/12/26/book-of-the-year-the-return-of-teleology/.

Farrer, Austin. *The Freedom of the Will*. The Gifford Lectures. London: Adam and Charles Black, 1958.

Feser, Edward. *Scholastic Metaphysics: A Contemporary Introduction*. Heusenstamm: Editiones Scholasticae, 2014.

Fine, Kit. "Things and Their Parts." *Midwest Studies in Philosophy* 23 (1999): 61–74.

Frank, Philipp. *Philosophy of Science: The Link between Science and Philosophy*. Englewood Cliffs, NJ: Prentice-Hall, 1957.

Galileo Galilei. *The Assayer*. In *Discoveries and Opinions of Galileo*, translated by Stillman Drake, 229–80. Garden City, NY: Doubleday, 1957.

Gasking, Douglas. "Causation and Recipes." *Mind*, n.s., 64 (1955): 479–87.

Gillett, Carl. "Scientific Emergentism and Its Move beyond (Direct) Downward Causation." In Paoletti and Orilia, *Philosophical and Scientific Perspectives*, 242–62.

Gillies, Donald. "An Action-Related Theory of Causality." *The British Journal for the Philosophy of Science* 56 (2005): 823–42.

Glennan, Stuart S. "Mechanisms." In Beebee, Hitchcock, and Menzies, *Oxford Handbook of Causation*, 315–25.

———. "Mechanisms and the Nature of Causation." *Erkenntnis* 44 (1996): 49–71.

———. "Modeling Mechanisms." *Studies in History and Philosophy of Science Part C: Studies in History and Philosophy of Biological and Biomedical Sciences* 36 (2005): 443–64.

———. "Rethinking Mechanistic Explanation." *Philosophy of Science* 69 (2002): S342–53.

Godfrey-Smith, Peter. "Causal Pluralism." In Beebee, Hitchcock, and Menzies, *Oxford Handbook of Causation*, 326–37.

Gotthelf, Allan. "Aristotle's Concept of Final Causality." *The Review of Metaphysics* 30 (1976): 226–54.

———. *Teleology, First Principles, and Scientific Method in Aristotle's Biology*. Oxford: Oxford University Press, 2012.

Green, Brian. "Catholic Thomistic Natural Law and Terrence Deacon's *Incomplete Nature*: A Match Made in Heaven." *Theology and Science* 13 (2015): 89–95.

Groff, Ruth. "Whose Powers? Which Agency?" In *Powers and Capacities in Philosophy: The New Aristotelianism*, edited by Ruth Groff and John Greco, 207–27. New York: Routledge, 2013.

Guthrie, W. K. C. *A History of Greek Philosophy*. Vol. 6, *Aristotle: An Encounter*. New York and Cambridge: Cambridge University Press, 1981.

Hacking, Ian. *Representing and Intervening: Introductory Topics in the Philosophy of Natural Science*. Cambridge: Cambridge University Press, 1983.

Haldane, John. "Privative Causality." *Analysis* 67 (2007): 180–86.

Hall, Ned. "Two Concepts of Causation." In *Causation and Counterfactuals*, edited by John David Collins and Edward Jonathan Hall, 225–76. Cambridge, MA: MIT Press, 2004.

Handfield, Toby. "Dispositions, Manifestations, and Causal Structure." In Marmodoro, *Metaphysics of Powers*, 106–32.

Heil, John. *From an Ontological Point of View*. Oxford: Clarendon Press, 2005.

———. *The Nature of True Minds*. Cambridge and New York: Cambridge University Press, 1992.

———. "Powerful Qualities." In Marmodoro, *Metaphysics of Powers*, 58–72.

———. *The Universe as We Find It*. Oxford: Clarendon Press, 2012.

Heisenberg, Werner. *Physics and Philosophy*. New York: Harper and Brothers, 1958.

Hendry, Robin F. "Prospects for Strong Emergence in Chemistry." In Paoletti and Orilia, *Philosophical and Scientific Perspectives*, 146–63.

Hitchcock, Christopher. "Causal Modeling." In Beebee, Hitchcock, and Menzies, *Oxford Handbook of Causation*, 299–314.

Hobbes, Thomas. *English Works*. 11 vols. London, 1839.

Howard, Don. "Emergence in the Physical Sciences: Lessons from the Particle Physics and Condensed Matter Debate." In *Evolution and Emergence: Systems, Organisms, Persons*, edited by Nancey Murphy and William R. Stoeger, 141–57. Oxford and New York: Oxford University Press, 2007.

Hulswit, Menno. *From Cause to Causation: A Peircean Perspective*. Dordrecht: Kluwer Academic, 2002.

———. "How Causal Is Downward Causation?" *Journal for General Philosophy of Science* 36 (2006): 261–87.

Hume, David. *An Enquiry concerning Human Understanding*. Great Books of the Western World 35. Chicago: Encyclopedia Britannica, 1952.

———. *A Treatise of Human Nature*. Edited by L. A. Selby-Bigge. Oxford: Oxford University Press, 1978.

Humphreys, Paul. "Computational and Conceptual Emergence." *Philosophy of Science* 75 (2008): 584–94.

———. *Emergence: A Philosophical Account*. New York: Oxford University Press, 2016.

———. "Emergence, Not Supervenience." In "Proceedings of the 1996 Biennial Meetings of the Philosophy of Science Association," part 2, "Symposia Papers." Supplement, *Philosophy of Science* 64 (1997): S337–45.

———. "How Properties Emerge." In *Emergence: Contemporary Readings in Philosophy and Science*, edited by Mark A. Bedau and Paul Humphreys, 111–26. Cambridge, MA: MIT Press, 2008.

Illari, Phyllis, and Federica Russo. *Causality: Philosophical Theory Meets Scientific Practice*. Oxford: Oxford University Press, 2014.

Inwagen, Peter van. *Material Beings*. Ithaca, NY, and London: Cornell University Press, 1990.
Irwin, Terence. *Aristotle's First Principles*. Oxford: Clarendon Press, 1988.
Jaworski, William. *Philosophy of Mind: A Comprehensive Introduction*. Malden, MA: John Wiley & Sons, 2011.
———. *Structure and the Metaphysics of Mind: How Hylomorphism Solves the Mind-Body Problem*. Oxford: Oxford University Press, 2016.
Johnston, Mark. "Hylomorphism." *Journal of Philosophy* 103 (2006): 652–98.
Juarrero, Alicia. *Dynamics in Action: Intentional Behavior as a Complex System*. Cambridge, MA: MIT Press, 1999.
Kant, Immanuel. *Immanuel Kant's Critique of Pure Reason*. Translated by N. K. Smith. London: Macmillan & Co., 1929.
Kim, Jaegwon. "Being Realistic about Emergence." In Clayton and Davies, *Re-Emergence of Emergence*, 189–202.
———. "Causation, Nomic Subsumption, and the Concept of Event." *Journal of Philosophy* 70 (1973): 217–36.
———. "Causes and Counterfactuals." In *Causation*, edited by Ernest Sosa and Michael Tooley, 205–7. Oxford and New York: Oxford University Press, 1993.
———. "'Downward Causation' in Emergentism and Nonreductive Physicalism." In *Emergence or Reduction? Essays on the Prospects of Nonreductive Physicalism*, edited by Ansgar Beckermann, Hans Flohr, and Jaegwon Kim, 119–38. Berlin and New York: Walter de Gruyter, 1992.
———. "Emergence: Core Ideas and Issues." *Synthese* 151 (2006): 547–59.
———. "Events as Property Exemplifications." In *Action Theory: Proceedings of the Winnipeg Conference on Human Action, Held at Winnipeg, Manitoba, Canada, 9–11 May 1975*, edited by M. Brand and Douglas Walton, 310–26. Dordrecht: Reidel, 1976.
———. "Making Sense of Emergence." *Philosophical Studies: An International Journal for Philosophy in the Analytic Tradition* 95 (August 1999): 3–36.
———. *Mind in a Physical World: An Essay on the Mind-Body Problem and Mental Causation*. Cambridge, MA: MIT Press, 1998.
———. "The Myth of Nonreductive Materialism." *Proceedings and Addresses of the American Philosophical Association* 63 (1989): 31–47.
———. *Physicalism, or Something Near Enough*. Princeton: Princeton University Press, 2005.
———. *Supervenience and Mind*. Cambridge: Cambridge University Press, 1993.
———. "Supervenience as a Philosophical Concept." *Metaphilosophy* 21 (1990): 1–27.
King, Hugh R. "Aristotle without *Prima Materia*." *Journal of the History of Ideas* 17 (1956): 370–89.
Koons, Robert. "Staunch vs. Faint-Hearted Hylomorphism: Toward an Aristotelian Account of Composition." *Res Philosophica* 91 (2014): 151–77.

Koons, Robert C., and Timothy Pickavance. *Metaphysics: The Fundamentals*. Oxford: Wiley-Blackwell, 2015.
Koslicki, Kathrin. "Aristotle's Mereology and the Status of Form." *Journal of Philosophy* 103 (2006): 715–36.
———. *The Structure of Objects*. Oxford: Oxford University Press, 2010.
Kripke, Saul A. *Naming and Necessity*. Cambridge, MA: Harvard University Press, 1980.
Kutach, Douglas. *Causation*. Cambridge, UK, and Malden, MA: Polity Press, 2014.
Ladyman, James. "Structural Realism." In *Stanford Encyclopedia of Philosophy*. Winter 2016 ed. Edited by Edward N. Zalta. https://plato.stanford.edu/archives/win2016/entries/structural-realism/.
Leśniewski, Stanisław. *Collected Works*. Vols. 1–2. Dordrecht: Kluwer, 1992.
Lewes, George Henry. *Problems of Life and Mind*. Vol. 2. London, 1875.
Lewis, David. "Causation." In *Journal of Philosophy* 70 (1973): 556–67. Reprinted in *Causation*, edited by Ernest Sosa and Michael Tooley (Oxford and New York: Oxford University Press, 1993), 193–204.
———. "Causation as Influence." *The Journal of Philosophy* 97 (2000): 182–97.
———. *Counterfactuals*. Oxford: Blackwell, 1973.
———. *On the Plurality of Worlds*. Oxford: Blackwell, 1986.
Locke, John. *An Essay concerning Human Understanding*. Edited by A. C. Fraser. Chicago: Encyclopedia Britannica, 1952.
Lombard, Lawrence Brian. *Events: A Metaphysical Study*. London and Boston: Routledge & Kegan Paul, 1986.
Losee, John. *Theories of Causality: From Antiquity to the Present*. New Brunswick, NJ, and London: Transaction Publishers, 2011.
Loux, Michael J. *Metaphysics: A Contemporary Introduction*. New York and London: Routledge, 2006.
Lowe, Edward Jonathan. *The Four-Category Ontology: A Metaphysical Foundation for Natural Science*. New York: Oxford University Press, 2006.
———. "A Neo-Aristotelian Substance Ontology: Neither Relational nor Constituent." In *Contemporary Aristotelian Metaphysics*, edited by Tuomas E. Tahko, 229–48. Cambridge: Cambridge University Press, 2011.
Lyon, Aidan. "Causality." In *British Journal for the Philosophy of Science* 18 (1967): 1–20.
Macdonald, Cynthia, and Graham Macdonald. "Emergence and Downward Causation." In *Emergence in Mind*, edited by Cynthia Macdonald and Graham Macdonald, 139–68. New York: Oxford University Press, 2010.
Machamer, Peter, Lindley Darden, and Carl F. Craver. "Thinking about Mechanisms." *Philosophy of Science* 67 (2000): 1–25.
Mackie, John Leslie. "Causes and Conditionals." In *American Philosophical Quarterly* 2/4 (1965): 245–64.

———. *The Cement of the Universe: A Study of Causation*. Oxford: Clarendon Press, 1974.

———. *Truth, Probability and Paradox: Studies in Philosophical Logic*. Clarendon Library of Logic and Philosophy. Oxford: Clarendon Press, 1973.

MacLeod, Miles, and Nancy J. Nersessian. "Modeling Systems-Level Dynamics: Understanding without Mechanistic Explanation in Integrative Systems Biology." *Studies in History and Philosophy of Science*, part C, *Studies in History and Philosophy of Biological and Biomedical Sciences* 49 (2015): 1–11.

Malebranche, Nicolas. *Malebranche: "The Search after Truth" / "Elucidation of the Search after Truth,"* translated by T. M. Lennon and P. J. Olscamp. Columbus: Ohio State University Press, 1980.

Manafu, Alexandru. "A Novel Approach to Emergence in Chemistry." In *Philosophy of Chemistry: Growth of a New Discipline*, edited by Eric Scerri and Lee McIntyre, 39–55. Boston Studies in the Philosophy and History of Science. Dordrecht: Springer Netherlands, 2015.

Marmodoro, Anna. "Aristotle's Hylomorphism without Reconditioning." *Philosophical Inquiry* 37 (2013): 5–22.

———, ed. *The Metaphysics of Powers: Their Grounding and Their Manifestations*. New York: Routledge, 2010.

———. "Power Mereology: Structural Powers versus Substantial Powers." In Paoletti and Orilia, *Philosophical and Scientific Perspectives*, 110–27.

Martin, Charles Burton. "Final Replies to Place and Armstrong." In *Dispositions: A Debate*, by D. M. Armstrong et al., edited by Tim Crane, 163–92. London: Routledge, 1996.

———. *The Mind in Nature*. Oxford and New York: Clarendon Press, 2008.

Mayr, Erasmus. "Powers and Downward Causation." In Paoletti and Orilia, *Philosophical and Scientific Perspectives*, 76–91.

Mayr, Ernst. "Teleological and Teleonomic: A New Analysis." In *Evolution and the Diversity of Life: Selected Essays*, 383–404. Cambridge, MA, and London: The Belknap Press of Harvard University Press, 1976.

McClure, Matthew Thompson. *The Early Philosophers of Greece*. New York: Appleton-Century-Crofts, 1935.

McGivern, Patrick, and Alexander Rueger. "Emergence in Physics." In *Emergence in Science and Philosophy*, edited by Antonella Corradini and Timothy O'Connor, 213–32. New York: Routledge, 2010.

McKitrick, Jennifer. "Manifestations as Effects." In Marmodoro, *Metaphysics of Powers*, 73–83.

McLaughlin, Brian P. "Emergence and Supervenience." *Intellectica* 25 (1997): 25–43.

———. "The Rise and Fall of British Emergentism." In *Emergence or Reduction? Essays on the Prospects of Nonreductive Physicalism*, edited by Ansgar

Beckermann, Hans Flohr, and Jaegwon Kim, 49–93. Berlin and New York: Walter de Gruyter, 1992.

Mellor, David Hugh. "Counting Corners Correctly." *Analysis* 42 (1982): 96–97.

———. "In Defense of Dispositions." *Philosophical Perspectives* 12 (1974): 283–312.

Menzies, Peter. "Probabilistic Causation and Causal Processes: A Critique of Lewis." *Philosophy of Science* 56 (1989): 642–63.

Menzies, Peter, and Christian List. "The Causal Autonomy of Special Sciences." In *Emergence in Mind*, edited by Graham Macdonald and Cynthia Macdonald, 108–28. Oxford and New York: Oxford University Press, 2010.

———. "Nonreductive Physicalism and the Limits of the Exclusion Principle." *Journal of Philosophy* 106 (2009): 475–502.

Menzies, Peter, and Huw Price. "Causation as a Secondary Quality." *The British Journal for the Philosophy of Science* 44 (1993): 187–203.

Mill, John Stuart. *A System of Logic*. New York, 1846.

Molnar, George. *Powers: A Study in Metaphysics*. Edited by Stephen Mumford. New York: Oxford University Press, 2003.

Moreno, Alvaro, and Jon Umerez. "Downward Causation at the Core of Living Organization." In *Downward Causation: Mind, Bodies and Matter*, edited by Peter Bøgh Andersen, Claus Emmeche, Niels O. Finnemann, and Peder Voetmann Christiansen, 99–116. Aarhus and Oxford: Aarhus University Press, 2000.

Morgan, Conwy Lloyd. *Emergent Evolution*. London: Williams & Norgate, 1923.

Mumford, Stephen Dean. "Causal Powers and Capacities." In Beebee, Hitchcock, and Menzies, *Oxford Handbook of Causation*, 265–78.

———. *David Armstrong*. New York: Routledge, 2007.

———. *Dispositions*. New York: Oxford University Press, 1998.

———. *Laws in Nature*. New York: Routledge, 2004.

———. "The Power of Power." In *Powers and Capacities in Philosophy: The New Aristotelianism*, edited by Ruth Groff and John Greco, 9–24. New York: Routledge, 2013.

Mumford, Stephen, and Rani Lill Anjum. "Causal Dispositionalism." In *Properties, Powers, and Structures: Issues in the Metaphysics of Realism*, edited by Alexander Bird, B. D. Ellis, and Howard Sankey, 101–18. New York: Routledge, 2012.

———. *Getting Causes from Powers*. Oxford and New York: Oxford University Press, 2011.

———. "A Powerful Theory of Causation." In Marmodoro, *Metaphysics of Powers*, 143–59.

Murphy, Nancey. "Emergence and Mental Causation." In Clayton and Davies, *Re-Emergence of Emergence*, 227–43.

———. "Reductionism: How Did We Fall into It and Can We Emerge from It?" In *Evolution and Emergence: Systems, Organisms, Persons*, edited by Nancey Murphy and William R. Stoeger, 19–39. Oxford and New York: Oxford University Press, 2007.

Nagel, Ernest. *The Structure of Science: Problems in the Logic of Scientific Explanation*. New York: Harcourt, Brace & World, 1961.

Nahm, Milton C. *Selections from Early Greek Philosophy*. Englewood Cliffs, NJ: Prentice-Hall, 1964.

Needham, Paul. "The Discovery That Water Is H_2O." *International Studies in the Philosophy of Science* 16 (2002): 205–26.

———. "Water and the Development of the Concept of Chemical Substance." In *A History of Water*, series 2, vol. 1, *Ideas of Water from Antiquity to Modern Times*, edited by Terje Tvedt and Terje Oestigaard, 86–123. London: Storbritannien, 2010.

———. "What Is Water?" *Analysis* 60 (2000): 13–21.

O'Connor, Timothy, and Hong Yu Wong. "Emergent Properties." In *Stanford Encyclopedia of Philosophy*. Summer 2015 ed. Edited by Edward N. Zalta. https://plato.stanford.edu/archives/sum2015/entries/properties-emergent/.

———. "The Metaphysics of Emergence." *Noûs* 39 (2005): 658–78.

Oderberg, David S. *Real Essentialism*. New York: Routledge, 2007.

Orilia, Francesco, and Michele Paolini Paoletti. "Three Grades of Downward Causation." In Paoletti and Orilia, *Philosophical and Scientific Perspectives*, 25–41.

O'Rourke, Fran. "Aristotle and the Metaphysics of Evolution." *The Review of Metaphysics* 58 (2004): 3–59.

Paoletti, Michele Paolini, and Francesco Orilia. "Downward Causation: An Opinionated Introduction." In Paoletti and Orilia, *Philosophical and Scientific Perspectives*, 1–21.

———. *Philosophical and Scientific Perspectives on Downward Causation*. New York: Routledge, 2017.

Paul, Laurie Ann. "Counterfactual Theories." In Beebee, Hitchcock, and Menzies, *Oxford Handbook of Causation*, 158–84.

Peacocke, Arthur. "God's Interaction with the World: The Implications of Deterministic 'Chaos' and of Interconnected and Independent Complexity." In *Chaos and Complexity: Scientific Perspectives on Divine Action*, edited by Robert J. Russell, Nancey Murphy, and Arthur Peacocke, 263–87. Vatican City State: Vatican Observatory; Berkeley, CA: Center for Theology and the Natural Sciences, 1995.

Pearl, Judea. *Causality: Models, Reasoning, and Inference*. New York: Cambridge University Press, 2000.

Perlman, Mark. "The Modern Philosophical Resurrection of Teleology." In *Philosophy of Biology: An Anthology*, edited by Alex Rosenberg and Robert Arp, 149–63. Oxford: Blackwell, 2010.

Peterson, Gregory R. "Species of Emergence." *Zygon* 41 (2006): 689–712.
Place, Ullin T. "Dispositions as Intentional States." In *Dispositions: A Debate*, by David Malet Armstrong, Charles Burton Martin, and Ullin T. Place, 19–32. London: Routledge, 1996.
Plato. *Phaedo*. In *Plato: Vol. I*, translated by Harold North Fowler, 200–403. Cambridge, MA: Harvard University Press, 1914.
———. *Timaeus*. In *Plato: Vol. VII*, translated by R. G. Bury, 16–253. Cambridge, MA: Harvard University Press, 1929.
Polanyi, Michael. *Knowing and Being*. Chicago: University of Chicago Press, 1969.
Popper, Karl R. *Conjectures and Refutations: The Growth of Scientific Knowledge*. London: Routledge & K. Paul, 1963.
———. *The Logic of Scientific Discovery*. New York: Basic Books, 1959.
———. *Objective Knowledge: An Evolutionary Approach*. Oxford: Clarendon Press, 1979.
———. *The Open Society and Its Enemies*. Vol. 2, *The High Tide of Prophecy*. New York: Harper & Row, 1962.
———. *Unended Quest: An Intellectual Autobiography*. London and New York: Routledge, 2002.
Popper, Karl R., and John Carew Eccles. *The Self and Its Brain*. Berlin: Springer, 1977.
Powell, Alexander, and John Dupré. "From Molecules to Systems: The Importance of Looking Both Ways." *Studies in History and Philosophy of Biological and Biomedical Sciences* 40 (2009): 54–64.
Psillos, Stathis. "Regularity Theories." In Beebee, Hitchcock, and Menzies, *Oxford Handbook of Causation*, 131–57.
Putnam, Hilary. "From Quantum Mechanics to Ethics and Back Again." In *Philosophy in an Age of Science: Physics, Mathematics, and Skepticism*, 51–71. Cambridge, MA: Harvard University Press, 2012.
———. "The Meaning of Meaning." In *Mind, Language and Reality: Philosophical Papers*, 2:215–71. Cambridge: Cambridge University Press, 1975.
———. "The Nature of Mental States." In *Mind, Language, and Reality: Philosophical Papers*, 2:429–40. Cambridge: Cambridge University Press, 1975.
———. *The Threefold Cord: Mind, Body, and World*. New York: Columbia University Press, 1999.
Quine, Willard V. O. *Ontological Relativity, and Other Essays*. New York: Columbia University Press, 1969.
———. *Quiddities*. Cambridge, MA: Harvard University Press, 1987.
———. *The Roots of Reference*. La Salle, IL: Open Court, 1974.
———. "Two Dogmas of Empiricism." *The Philosophical Review* 60 (1951): 20–43.
———. *Word and Object*. Cambridge, MA: MIT Press, 1960.
Raatikainen, Panu. "Causation, Exclusion, and the Special Sciences." *Erkenntnis* 73 (2010): 349–63.

Rea, Michael C. "Hylomorphism Reconditioned." *Philosophical Perspectives* 25 (2011): 341–58.

Robb, David. "The Properties of Mental Causation." *The Philosophical Quarterly* (1950–) 47 (1997): 178–94.

Robinson, H. M. "Prime Matter in Aristotle." *Phronesis* 19 (1974): 168–88.

Rosenberg, Alex. *The Atheist's Guide to Reality: Enjoying Life without Illusions.* New York: W. W. Norton, 2011.

Sachs, Joe. *Aristotle's Physics: A Guided Study.* 1st ed. New Brunswick, NJ: Rutgers University Press, 1995.

Salmon, Wesley C. "Causality: Production and Propagation." In *Causation*, edited by Ernest Sosa and Michael Tooley, 154–71. New York: Oxford University Press, 1993. First published in *PSA: Proceedings of the Biennial Meeting of the Philosophy of Science Association* (1980), vol. 2, edited by Peter D. Asquith and R. Giere (East Lansing, MI: Philosophy of Science Association, 1981).

———. *Causality and Explanation.* Oxford: Oxford University Press, 1998.

———. "Causality and Explanation: A Reply to Two Critiques." *Philosophy of Science* 64 (1997): 461–77.

———. *Scientific Explanation and the Causal Structure of the World.* Princeton: Princeton University Press, 1984.

Scaltsas, Theodore. *Substances and Universals in Aristotle's Metaphysics.* New York: Cornell University Press, 1994.

Schaffer, Jonathan. "The Metaphysics of Causation," *Stanford Encyclopedia of Philosophy.* Fall 2016 ed. Edited by Edward N. Zalta. https://plato.stanford.edu/archives/fall2016/entries/causation-metaphysics/.

———. "Overlappings: Probability-Raising without Causation." *Australasian Journal of Philosophy* 78 (2000): 40–46.

Schlick, Moritz. *Philosophy of Nature.* Translated by A. von Zeppelin. New York: Philosophical Library, 1949.

Scott, Alwyn C. *The Nonlinear Universe.* Berlin: Springer, 2007.

Shapiro, Larry. "Lessons from Causal Exclusion." *Philosophy and Phenomenological Research* 81 (2010): 594–604.

Sherman, Jeremy, and Terrence W. Deacon. "Teleology for the Perplexed: How Matter Began to Matter." *Zygon* 42 (2007): 873–901.

Shields, Christopher. "Causal Processes: An Aristotelian Alternative to an Unwarranted Humean Hegemony." Paper presented at Neo-Aristotelianism: A Conference on Neo-Aristotelian Metaphysics, Ethics, and Politics, Chicago, September 30–October 1, 2016.

Shoemaker, Sydney. "Causal and Metaphysical Necessity." *Pacific Philosophical Quarterly* 79 (1998): 59–77.

———. "Causality and Properties." In *Time and Cause: Essays Presented to Richard Taylor*, edited by Richard Taylor and Peter van Inwagen, 109–35. Dordrecht and Boston: Reidel, 1980.

———. "Kim on Emergence." *Philosophical Studies* 108 (2002): 53–63.
Sider, Theodore. "What's So Bad about Overdetermination?" *Philosophy and Phenomenological Research* 67 (2003): 719–26.
Silberstein, Michael. "In Defence of Ontological Emergence and Mental Causation." In Clayton and Davies, *Re-Emergence of Emergence*, 203–26.
Smart, John Jamieson Carswell. "Physicalism and Emergence." *Neuroscience* 6 (1981): 109–13.
Sober, Elliott. "Two Concepts of Cause." In *PSA: Proceedings of the Biennial Meeting of the Philosophy of Science Association (1984)*, vol. 2, edited by Peter D. Asquith and Philip Kitcher, 405–24. East Lansing, MI: Philosophy of Science Association, 1985.
Solmsen, Friedrich. "Aristotle and Prime Matter: A Reply to Hugh R. King." *Journal of the History of Ideas* 19 (1958): 243–52.
Sosa, Ernest, and Michael Tooley, eds. *Causation*. Oxford and New York: Oxford University Press, 1993.
Sperry, Roger Wolcott. "Discussion: Macro- versus Micro-Determinism." *Philosophy of Science* 53 (1986): 265–70.
———. "A Modified Concept of Consciousness." *Psychological Review* 76 (1969): 532–36.
Stephan, Achim. "Emergence—a Systematic View on Its Historical Facets." In *Emergence or Reduction? Essays on the Prospects of Nonreductive Physicalism*, edited by Ansgar Beckermann, Hans Flohr, and Jaegwon Kim, 25–48. Berlin and New York: Walter de Gruyter, 1992.
Stöltzner, Michael. "The Logical Empiricists." In Beebee, Hitchcock, and Menzies, *Oxford Handbook of Causation*, 108–27.
Storck, Michael Hector. "Parts, Wholes, and Presence by Power: A Response to Gordon P. Barnes." *The Review of Metaphysics* 62 (2008): 45–59.
Suárez, Francis. *On the Formal Cause of Substance: Metaphysical Disputation* 15. Translated by John Kronen and Jeremiah Reedy. Milwaukee: Marquette University Press, 2000.
Suppes, Patrick. *A Probabilistic Theory of Causality*. Amsterdam: North-Holland Publishing, 1970.
Swartz, Norman. "A Neo-Humean Perspective: Laws as Regularities." In *Laws of Nature: Essays on the Philosophical, Scientific and Historical Dimensions*, edited by Friedel Weinert, 67–91. Berlin and New York: Walter de Gruyter, 1995.
Tabaczek, Mariusz. "The Metaphysics of Downward Causation: Rediscovering the Formal Cause." *Zygon* 48 (2013): 380–404.
Tooley, Michael. *Causation: A Realist Approach*. Oxford: Oxford University Press, 1987.
Unger, Peter. *Ignorance: A Case for Scepticism*. Oxford: Oxford University Press, 1975.

van Gulick, Robert. "Reduction, Emergence, and the Mind/Body Problem." In *Evolution and Emergence: Systems, Organisms, Persons*, edited by Nancey Murphy and William R. Stoeger, 40–73. Oxford and New York: Oxford University Press, 2007.

———. "Who's in Charge Here? And Who's Doing All the Work?" In *Evolution and Emergence: Systems, Organisms, Persons*, edited by Nancey Murphy and William R. Stoeger, 74–87. Oxford and New York: Oxford University Press, 2007.

Varzi, Achille. "Mereology." *Stanford Encyclopedia of Philosophy*. Winter 2016 ed. Edited by Edward N. Zalta. https://plato.stanford.edu/archives/win2016/entries/mereology/.

von Wright, Georg Henrik. *Explanation and Understanding*. London: Routledge, 1971.

———. "On the Logic and Epistemology of the Causal Relation." In *Causation*, edited by Ernest Sosa and Michael Tooley, 105–24. New York: Oxford University Press, 1993.

Waismann, Friedrich. "The Decline and Fall of Causality." In *Turning Points in Physics*, edited by A. A. Crombie, 84–154. Amsterdam: North-Holland Publishing, 1959.

Wallace, William A. *Causality and Scientific Explanation*. 2 vols. Ann Arbor: University of Michigan Press, 1972–74.

———. *The Modeling of Nature: Philosophy of Science and Philosophy of Nature in Synthesis*. Washington, DC: Catholic University of America Press, 1996.

———. "A Place for Form in Science: The Modeling of Nature." *Proceedings of the American Catholic Philosophical Association* 69 (1995): 35–46.

Walsh, Denis. "Teleology." In *The Oxford Handbook of Philosophy of Biology*, edited by Michael Ruse, 113–37. Oxford and New York: Oxford University Press, 2008.

Williams, Neil E. "Puzzling Powers: The Problem of Fit." In Marmodoro, *Metaphysics of Powers*, 84–105.

Williamson, Jon. "Probabilistic Theories." In Beebee, Hitchcock, and Menzies, *Oxford Handbook of Causation*, 185–212.

Wilson, Jessica. "Metaphysical Emergence: Weak and Strong." In *Metaphysics in Contemporary Physics*, edited by Tomasz Bigaj and Christian Wüthrich. Leiden and Boston: Brill, 2015.

Wimsatt, William C. "Aggregativity: Reductive Heuristics for Finding Emergence." *Philosophy of Science* 64 (1997): S372–84.

———. "Emergence as Non-Aggregativity and the Biases of Reductionisms." In *Re-Engineering Philosophy for Limited Beings: Piecewise Approximations to Reality*, 274–312. Cambridge, MA: Harvard University Press, 2007.

Witt, Charlotte. *Ways of Being: Potentiality and Actuality in Aristotle's Metaphysics*. Ithaca, NY: Cornell University Press, 2003.

Wittgenstein, Ludwig. *Philosophical Investigations*. New York: Macmillan, 1968.
Wong, Hong Yu. "The Secret Lives of Emergents." In *Emergence in Science and Philosophy*, edited by Antonella Corradini and Timothy O'Connor, 7–24. New York: Routledge, 2010.
Woodward, James F. "Agency and Interventionist Theories." In Beebee, Hitchcock, and Menzies, *Oxford Handbook of Causation*, 234–62.
———. "Causation and Manipulability." *Stanford Encyclopedia of Philosophy*. Winter 2016 ed. Edited by Edward N. Zalta. https://plato.stanford.edu/archives/win2016/entries/causation-mani/.
———. "Interventionism and Causal Exclusion." *Philosophy and Phenomenological Research* 91 (2015): 303–47.
———. *Making Things Happen: A Theory of Causal Explanation*. New York: Oxford University Press, 2003.
———. "Mental Causation and Neural Mechanisms." In *Being Reduced: New Essays on Reduction, Explanation, and Causation*, edited by Jakob Hohwy and Jesper Kallestrup, 218–62. Oxford: Oxford University Press, 2008.
Worrall, John. "Structural Realism: The Best of Both Worlds?" *Dialectica* 43 (1989): 99–124.
Zimmermann, Rainer E. *Metaphysics of Emergence, Part I: On the Foundations of Systems*. Berlin: MoMo Berlin, KonTexte, Philosophische Schriftenreihe, 2015.

INDEX

Abrahamsen, Adele, 296n.91
absences as causes
  vs. absences as enabling conditions, 120, 193
  Deacon on, 35–36, 38, 99, 100, 103–4, 116, 119, 120–25, 128, 129, 132, 256, 258, 265–67, 268, 311n.21, 324n.76, 324n.86
  David Lewis on, 150–51
  in vector diagrams, 197
action-related view of causation (ArelVC), 162–63, 168, 177, 272
actuality. *See* potentiality and actuality
Aetius Doxographus, 286n.6, 288n.14
agency view of causation (AVC), 158, 159–63, 192, 271–72, 329n.44
  anthropocentricity of, 162, 168, 177, 192
  vs. ArelVC, 162–63
  vs. DVC, 192, 248
  vs. IntVC, 164, 166
  and possible worlds modality, 162
  vs. ProbVC, 160
aitia, 2, 285n.2
Alexander, Samuel
  on EM, 46, 48, 50, 52, 55, 305n.25
  on emergent laws, 52
  on emergent properties, 55, 305n.25
  on levels of complexity, 50, 52, 68
  on mental states and neural process, 58

  Space, Time and Deity, 50
Alexander of Aphrodisias, on *archē* vs. *aitia*, 285n.2
Allen, Sophie, 327n.21, 333n.15
analytic metaphysics, 46, 89–90, 135–36, 250, 308n.4, 326n.1 (chap. 4), 337n.7
  and Aristotelian metaphysics, 217, 218–20, 240–41, 249
  bare substratum theories in, 66, 208
  constituent ontology, 66, 88, 208, 238, 239, 250
  and dispositional/powers metaphysics, 36, 37, 187–88
  efficient causation in, 91
  graphic models of causal dependencies, 193–99
  hylomorphism in, 217, 218–41
  ontology of dispositions/powers, 136–37
  relational ontology, 66
  See also agency view of causation (AVC); counterfactual view of causation (CVC); dispositional/powers metaphysics; inferability view of causation (InfVC); manipulation view of causation (MVC); probability view of causation (ProbVC); process view of causation (ProcVC)

Anaxagoras, on cosmic intelligence (*nous*), 3, 287n.10
Anaximander, on the infinite (*apeiron*), 3, 286n.7
Anaximenes, on air, 2–3, 286nn.6–7
Anjum, Rani
  on causal dispositionalism, 189–90
  causal modeling theory of, 194–99
  on causal processes, 189–90
  on compositional vs. resultant powers, 195–98
  on demergence, 90, 315n.56
  on dispositions, 183, 189–90, 332n.11
  on emergent powers, 90
  on necessity of causal dependencies, 335n.37
  on strong EM and DC, 89, 90
Anscombe, Elizabeth
  on causation, 155
  on hydrogen bonding, 31
anthropocentricity, 162, 168, 177, 192
Aquinas, Thomas
  on accidental form, 18, 224
  Commentary on Aristotle's Metaphysics, 217–18
  De ente et essentia, 322n.63, 343n.85
  on efficient causation, 18–19, 254, 294n.70, 347n.6
  on *esse* and *essentia*, 18
  essentialism of, 18, 19, 294n.65
  on final causation, 19, 254, 347n.6
  on formal causation, 18, 217–18, 254, 339n.41, 343n.85, 347n.6
  on four elements, 230, 341nn.73–74
  on God, 18, 294n.70
  on inanimate beings, 347n.6
  In II Sent., 294n.64
  on interrelatedness of causes, 19, 254
  on material causation, 18, 254, 347n.6
  on matter and individuation, 235, 322n.63
  on matter as *materia signata*, 235, 322n.63, 343n.85
  on participation in God, 18, 294n.68
  on potentiality, 18, 295n.73
  on primary matter, 18, 230, 294n.64
  on primary vs. secondary efficient causes, 12, 18–19
  on principal vs. instrumental efficient causes, 19
  on principle of individuation, 235, 322n.63, 343n.85
  on proximate/secondary matter, 223, 233
  on substantial change, 223, 230, 233
  on substantial form, 18, 19, 122–23, 217, 223, 233, 342n.77
  on substantial unity by alteration of components (UAC/unity by *alteratio componentium*), 218, 219, 220, 221, 224, 227, 228, 229, 233, 236
  on substantial unity by arrangement of parts (USO/unity *secundum ordinem*), 217, 218, 219, 220, 221, 224, 227, 228, 236
  on substantial unity by contact and bond (USCC/unity *secundum contactum et colligationem*), 218, 219, 220, 221, 224, 227, 228, 236
  Summa theologiae, 294n.64, 294n.68
  on univocal vs. equivocal efficient causes, 19
  on virtual (*virtute*) presence of elements in mixed substances, 19–20, 218, 220, 227, 229–34, 295n.73, 341n.74, 342n.76
archē, 2, 12, 285n.2
Aristotle
  on accidental change, 9–10, 115, 233, 240, 241, 250–51

on accidental form, 9–10, 16, 20, 65–67, 115, 116, 205, 208, 224, 250–51, 308n.5
on chance, 16–17, 19, 154, 254–55, 293nn.60–61, 335n.42
on change vs. permanence, 201–2, 204
Deacon on, 36, 97, 99–100, 113, 118–19, 127–29, 256, 272, 299n.114, 311n.21, 321nn.56–57, 324n.76
De anima, 210
De caelo, 325n.95
De generatione et corruptione, 7, 9–10, 19–20, 230, 292n.43, 292n.45, 295n.73, 325n.95, 341n.74
and dispositional views of causation (DVC), 37, 39, 199
on efficient causation, 10, 11–12, 14, 15, 16, 93, 115–18, 119, 120, 127, 129, 130, 199, 208, 241, 249, 251–52, 253–54, 255, 257, 258, 263, 293n.53, 294n.70, 323n.70, 346n.106
on elements, 19–20
EM and metaphysics of, 38, 45, 63, 65, 66–67, 78, 87, 90–91, 93–94, 99–100, 300n.1 (chap. 1)
essentialism of, 6, 8, 9, 122, 123, 124, 128, 207–8, 214, 239, 240, 241, 249, 255, 257, 258, 290n.32, 290n.34, 290n.36, 291n.39, 294n.65, 297n.98, 300n.1 (chap. 1), 327n.20, 344n.99
on final causation, 9, 10, 12–14, 16, 19, 35, 36, 37, 78, 93, 94–95, 99, 117–18, 119, 121, 127, 128, 129, 180, 199, 206, 208, 216, 241, 247, 253–54, 255, 263, 265, 269, 272, 273, 291n.41, 292nn.43–45, 292n.51, 293n.53, 315n.57, 324n.74, 324n.76, 337n.17
on formal causation, 5–6, 8–10, 11, 14, 15, 16, 35, 36, 78, 87, 93, 94–95, 96–97, 115–18, 119, 121, 122, 124, 127, 129, 130, 180, 199, 208, 217–18, 227–28, 247, 249–50, 253–54, 255, 258, 265, 269, 272, 273, 290nn.35–36, 291nn.38–39, 293n.53, 297n.98, 300n.1 (chap. 1), 315n.57, 323n.74
on four elements, 7, 229–30, 289nn.28–29, 325n.95, 341n.74
on interrelatedness of causes, 14–15, 19, 254, 255
on material causation, 5–8, 15, 93, 94, 113–15, 127, 129, 199, 208, 253–54, 288n.22, 289n.29, 293n.53, 297n.98, 300n.1 (chap. 1), 324n.94
on mathematics, 4, 5
on matter and individuation, 235, 322n.63, 343n.85
on Megarian school of philosophy, 203–5, 213
Metaphysics, 5, 6, 8, 97, 202–3, 218, 271, 286n.5, 288–89n.23, 290n.32, 290n.36, 291n.38, 291n.41, 293n.53, 321n.57, 325n.95, 336n.2, 337n.10, 340n.61
Meteorology, 12
on necessity, 17, 19, 255
On the Generation of Animals, 11
On the Parts of Animals, 14, 292n.43, 292n.45, 297n.98
on *per accidens*/incidental causes, 16–17, 120–21, 155, 255–56, 293n.60
on *per se* causes, 16–17, 120–21, 155, 255–56, 293n.60, 335n.42

Aristotle (*continued*)
  Physics, 1, 5, 6, 8, 10–11, 12, 13, 14, 16, 97, 254, 290n.32, 290n.36, 291n.41, 294n.64, 321n.57, 346n.106
  on Plato, 5, 8
  on possession vs. exercise of knowledge, 210
  on primary matter, 6–8, 9, 15, 36, 114, 115, 116, 118, 123, 125–26, 128, 205, 207, 212, 219, 220, 223–24, 227, 228–29, 230, 231–34, 236–37, 239, 240, 245, 249–50, 251, 265, 283, 288–89n.23, 289n.28, 289n.30–31, 291n.39, 300n.1 (chap. 1), 324n.94, 325n.95, 341n.72, 342n.80, 345n.101
  on primary vs. secondary causes, 11–12
  on principle of individuation, 235, 322n.63, 343n.85
  on privation and absence, 120–21, 324n.86
  on properties, 65–66
  on proximate/secondary matter, 6, 15, 229, 233, 240, 288n.23
  on Pythagoreans, 4, 288n.14
  on substantial change, 7–8, 9–10, 115, 223, 228–29, 230, 233, 240, 241, 250–51, 288n.23, 289n.28, 297n.98
  on substantial form, 7–8, 9–10, 11, 13, 15, 16, 20, 36, 65–67, 97, 115, 116, 120, 122–23, 124, 125–28, 205, 207–8, 212, 218–19, 223–24, 225–26, 227–29, 230, 231–34, 236–37, 239, 240, 245, 250–51, 269–70, 284, 289n.28, 290n.33, 299n.114, 300n.1 (chap. 1), 305n.35, 308n.5, 325n.95, 342n.80, 345n.101
  on Thales, 286n.5
  on universals, 225–26, 290n.34
  on unmoved mover, 11, 276, 337n.13
  on Wisdom, 271
  See also Aquinas, Thomas; potentiality and actuality
Aristotle's hylomorphism, 66, 95, 126, 284, 289n.30, 299n.114, 322n.63, 323n.68, 341n.72, 343n.85, 345n.101, 346n.105
  characteristics of, 6, 15, 116, 207–8, 212
  and DC, 249–51, 253
  Deacon on, 36, 116, 118, 128
  and dispositional/powers metaphysics, 37, 214, 217, 234–41, 244, 245, 250, 253, 265
  and essentialism, 6, 37, 214, 237–41, 245, 250, 253, 255
  vs. incomplete entities version of hylomorphism, 223–25
  vs. mereological and structural version of hylomorphism, 218–23
  vs. mixed version of hylomorphism, 225–28
  vs. re-identification of the parts version of hylomorphism, 228–29
  role of virtual presence in, 229–34, 342n.80
Armstrong, David, 135–36, 174
  on categorical reductionism, 187
  on dispositions, 182, 210, 214
  on intentionality, 214
Aronson, Jerrold, 172
atomism, 3, 4, 5, 230, 287n.13
attractors
  Deacon on, 106, 107, 130, 257
  strange attractors, 47
autopoiesis, 102, 108, 111, 129
Averroës, 341n.74
Avicenna, 341n.74, 343n.85

Bacigalupi, Augustus, 116
Bacon, Francis, 21, 112
Bacon, Roger, 17
Bain, Alexander, 45
Barnes, Gordon
　on emergent properties, 341n.68
　hylomorphism of, 225–28, 233
　on substance-independent vs.
　　substance-dependent matter,
　　226–27
　on substantial forms, 225–28,
　　341n.69
Bechtel, William
　on DC, 33–34
　Discovering Complexity,
　　296n.91
　on interlevel vs. intralevel causation, 32, 33–34
　on mechanistically mediated effects (phenomena), 32, 33–34
　on strong EM, 34
Bedau, Mark
　on strong EM, 307n.50, 312n.30
　on teleology as value-centered, 13
　on weak vs. strong EM, 307n.50
Bénard cells, 106, 131
Berkeley, George, 22, 23
Bickhard, Mark, 125, 280
　on Aristotelian substance ontology, 95, 125, 324n.94
Big Bang, 26, 191
biological mechanisms, 27–34, 35, 274, 299n.112
　constitutive levels of, 32–33
　and downward causation, 32–34
　levels of complexity, 31
　and nonreductive physicalism, 29, 298n.100
　and process view of causation, 30
　as producing, underlying, or maintaining phenomena, 29–31
biological vitalism, 118, 299n.114

biology, 30, 45, 160, 236–37, 308n.1, 309n.8
　biochemistry, 29, 274
　environment shaping biological development, 56
　evo-devo field, 299n.112
　evolutionary biology, 71–72, 91, 271, 292n.51, 299n.112, 321n.53
　examples of DC in, 56, 57, 58
　vs. metaphysics, 236–37
　molecular biology, 29, 32, 46, 57, 274
　symbiosis in biological systems, 56
　systems biology, 26, 27, 34, 37, 46, 56, 76, 119, 255, 272, 274, 299n.112
　See also biological mechanisms; evolution
Bird, Alexander, 187, 304n.22
　on definition of powers, 182, 237, 332n.7
　on dispositions vs. powers, 182, 188
　on timing indeterminacy of radioactive decay, 215–16
Bohr, Niels, on principle of complementarity, 25
Bokulich, Peter, 121, 267
Bonaventure, St., 294n.63
Born, Max, on chance, 25
Bostock, David, 12, 292nn.43–44
Boyd, Richard, on natural kinds as homeostatic property clusters, 238, 250
Boyle's law, 143, 158, 177
Brentano, Franz, on intentionality of the mental, 214–15
Bridgman, Percy, operationalism of, 25
Broad, Charlie Dunbar, 46
　on emergent laws, 52–53
　on emergent properties, 48–49
　on synchronic vs. diachronic features of EM, 48–49, 312n.33

bundle theories of substance, 66, 88, 208, 238, 239, 250
Bunge, Mario, 46

Campbell, Donald Thomas
  on Aristotelian substance ontology, 95
  on DC, 56, 71–72, 74, 75, 76
  on evolution, 56–57
Campbell, Richard
  on complexity vs. organization, 304n.16
  on EM and quantum physics, 280–83
  on evolution, 79
  on generic processes, 282, 348n.6
  on laws of nature, 304n.22
  on mereological vs. global and Humean SUP, 303n.15
  Metaphysics of Emergence, 279, 304n.22
  on process metaphysics, 279–84
  on synchronic vs. diachronic EM, 79, 312n.34
Cartwright, Nancy, 201, 209, 238
Cashman, Tyrone
  on absences as causes, 122–25, 258
  on autogenesis, 108–12, 270, 320n.42
  on teleology, 100–102, 318n.7, 320n.48
causal closure of physics/causal inherence principle, 35, 86, 340n.54
  Hendry on, 313n.42
  Putnam on, 317n.70
  relationship to DC, 76–78, 83, 85
causal determinism, 17, 19, 21, 22, 25, 160, 164, 255, 293n.62
causal modeling, 175, 194–99, 213, 262, 266–67
causal preemption
  and CVC, 149, 176, 327n.23
  early and late preemption, 31, 149, 176, 192, 194
  simultaneous preemption/trumping, 149, 154, 162, 176
causal reductionism, 21–25, 54, 174, 180, 181
  and efficient causation, 21, 26, 35, 76, 100, 126, 168, 179, 248–49, 251, 269, 271–72, 338n.37
causal relevance, 140, 141, 143, 163
causation by disconnection, 165, 167, 178
causation by double prevention, 30, 31, 150, 165, 166, 167, 176, 178
causation by omission, 30, 31, 150–51, 165, 167, 173, 176, 178
causation by prevention, 30, 31, 173, 178
cell signaling networks, 26
Chakravartty, Anjan
  on powers metaphysics, 243–44
  on scientific realism, 242–44
Chalmers, David, on SUP, 303n.15
chance
  Aristotle on, 16–17, 19, 154, 254–55, 293nn.60–61, 335n.42
  Born on, 25
  and causal modeling, 199
  and emergent entities, 154–56
chaos, 26
Charlton, William, 9, 290n.32
chemistry, 46–47, 56, 115, 127, 308n.1, 309n.8
  biochemistry, 29, 274
  examples of DC in, 57, 76, 306n.41
  vs. metaphysics, 236–37
  natural kinds in, 238
  vs. physics, 145, 236–37, 313n.42
circularity
  and AVC, 160–61, 177
  and CVC, 147, 148, 176
  and IntVC, 168

Kim on DC and, 79–80, 81, 83, 84
  and mechanisms, 30–31
Clark, Stewart J., on condensed matter physics, 306n.41
Clayton, Philip
  on mind as emergent, 301n.5
  on Neoplatonism and EM, 300n.1 (chap. 1)
  on strong EM, 60, 301n.5
  on weak EM, 60
Collier, John, 304n.16
Collingwood, Robin, 159
computer science, 174
Comte, Auguste, 24
constituent ontology
  bundle theories of substance, 66, 88, 208, 238, 239, 250
  substratum theories of substance, 66, 208
Cornford, Francis M., 9, 290n.32
counterfactual view of causation (CVC), 146–51, 158, 271–72, 328n.24
  and causal preemption, 149, 176, 327n.23
  vs. DVC, 190, 191, 192–93, 248
  vs. IntVC, 166, 329n.56
  and overdetermination, 31, 149–50, 192
  and possible worlds modality, 147–48, 149, 166, 168, 176, 192–93, 329n.56
  vs. RVC, 147, 148, 149, 150
  See also Hume, David
Craver, Carl
  on DC, 33–34
  on interlevel vs. intralevel causation, 32, 33–34
  on mechanisms and causation, 30–31, 298n.105
  on mechanistically mediated effects (phenomena), 32, 33–34
  on strong EM, 34
  on teleology, 338n.37
  on weak EM, 34
CVC. *See* counterfactual view of causation

Darden, Lindley, 298n.105
Davidson, Donald
  on anomalism of the mental, 69, 247, 310n.11
  on anomalous monism, 69, 310n.11
  on SVC, 156
  on token-identity thesis, 310n.11
Davies, Paul, 72, 74, 75, 92
Dawkins, Richard, on evolution as Blind Watchmaker, 112
DC. *See* downward causation
Deacon, Terrence
  on absences as causes, 35–36, 38, 99, 100, 103–4, 116, 119, 120–25, 128, 129, 132, 256, 258, 265–67, 268, 311n.21, 314n.46, 324n.76, 324n.86
  on Aristotelian categories of causality, 35, 36, 97, 99–100, 113, 118–19, 127–29, 256, 272, 299n.114, 311n.21, 321nn.56–57, 324n.76
  on attractors, 106, 107, 130, 257
  on autocatalysis, 102, 106, 109, 130–31, 257, 260, 261, 262, 266, 268, 270
  on autogenesis, 108–12, 122, 129–30, 257, 261, 262, 263, 266, 270, 319n.38, 320nn.42–43, 320n.47
  on biological organisms, 100–101
  and causal nonreductionism, 112–29, 272, 321n.53
  "Complexity and Dynamical Depth," 319n.23

Deacon, Terrence (*continued*)
    on constraints, 35–36, 38, 74, 99, 100, 101, 103–4, 105, 106, 107, 108, 109, 116–17, 120–25, 128, 129, 130, 131, 132, 256–60, 265–67, 268, 311n.21, 314n.46, 320n.42, 320n.44, 322n.66, 323n.70
    on DC, 75, 103, 132–33, 258–59, 267–69, 273, 311n.21, 322n.66
    definition of EM, 102–3, 301n.4
    on dynamical depth, 99, 100, 102–12, 114, 130–33, 137, 248, 256–70, 272–73, 311n.21, 321n.50, 325n.103
    on efficient causation, 115–16, 119–20, 122, 123, 127, 128, 130, 257, 258, 263, 265, 266, 323n.68, 323n.70
    on EM, 35–36, 38, 39, 72, 73, 91, 97, 99–133, 245, 247–48, 256–70, 272–73, 301n.4, 318n.16, 325n.103
    "Emergence: The Hole at the Wheel's Hub," 102, 127, 265, 318n.16
    on emergent properties, 103, 104–5, 114
    on ententionality, 99, 113, 118, 119, 120, 128, 311n.21, 317n.1
    on event ontologies, 126–27, 259, 269–70
    on evolution, 106, 108, 110, 318n.7, 320n.45
    on examples of EM, 104–5, 106, 131–32
    on final causation, 35, 36, 127, 247–48, 320n.48, 324n.76
    on formal causation, 35, 36, 103, 115–18, 122–23, 124–25, 127, 128, 130, 155–56, 247–48, 257, 258, 263–65, 322n.66, 323n.68, 323n.74
    on homeodynamics/first-order EM, 104–5, 107–8, 111–12, 127, 131–32, 133, 257, 259, 260–61, 268, 319n.23, 321n.50, 323n.70
    on homuncular explanations, 36, 118, 120, 123, 130, 257, 299n.114
    Incomplete Nature, 35, 114, 118, 123–24, 126–28, 318n.16, 320n.45
    on intrinsic vs. extrinsic constraints, 103–4, 105, 106, 108, 131
    on living vs. nonliving structures and processes, 97, 99, 101–2, 107, 108, 109, 118, 129–30, 252, 262, 321n.61
    on material causation, 113–15, 127
    on morphodynamics/second-order EM, 72, 105–6, 107–8, 109, 111–12, 127, 130, 131, 133, 257, 258, 259, 260–61, 266, 268, 321n.50, 323n.70
    on orthograde vs. contragrade changes, 104–6, 108, 116, 117, 122, 128, 130, 256–57, 260–62, 266, 323n.70
    on patterns of organization, 72, 74
    on probability space and formal causation, 115–16, 117, 130, 263–64, 323n.74
    on quantum mechanics, 114, 126, 259, 321nn.60–61
    and reductionist/eliminative materialism, 36, 112–13, 117, 128, 257, 265, 267, 270, 325n.100
    on self-assembly process, 109, 110, 130–31, 260, 261, 262, 266, 268, 270
    on selfhood, 100–101, 106–8, 109, 111, 132–33, 252, 262, 263, 320n.43, 320n.48
    on self-organizing processes, 100, 101–2, 103, 131–32, 260, 319n.38

on self-reparation and self-reproduction, 106–7, 109, 111, 130, 131, 270
on *semeota* vs. *morphota*, 111
on spontaneous change, 101, 102–3, 104, 105–6, 115–16, 117, 130, 132, 133, 257, 259, 260–62, 263, 270
on substance metaphysics, 100, 114, 115, 125–29, 259, 269–70
on teleodynamics/third-order EM, 73, 106–8, 109, 111–12, 119, 124, 127, 130–31, 132, 257, 258–59, 260–61, 263, 266, 268, 270, 320n.43, 320n.48, 321n.50, 323n.70, 324n.76
on teleology, 99, 100–102, 111, 112, 118–20, 130, 252, 257, 258, 262–63, 264–65, 273, 318n.7, 320n.48, 321n.53, 321n.61, 325n.100, 325n.103
on thermodynamics, 101–2, 104, 105, 108, 109, 112, 127, 131–32, 133, 257, 259, 260, 261–62, 268, 323n.74, 325n.100
on topological influences, 103
Decaen, Christopher, 20, 295n.73, 342n.76
Democritus, atomism of, 3, 4, 230, 287n.13
Descartes, René
on Aristotle, 21
corpuscular metaphysics of, 279
on efficient causation, 21, 295n.77
on final causation, 21, 112
on God, 21
substantial dualism of, 188, 212–13
Dionysius, on atomism, 287n.13
dispositional/powers metaphysics
and analytic metaphysics, 36, 37, 187–88
and Aristotelian metaphysics, 37, 136, 137, 180, 188, 208, 209–12, 213–16, 234–45, 249, 259–61, 263, 265, 268–69, 270, 272–73, 315n.57
and bundle theory of substance, 238
categorical reductionism, 185, 186, 187, 188, 213
compositional vs. resultant powers, 195–98
and DC, 36–38, 137, 179–80, 248–56, 272
and Deacon's dynamical depth model of EM, 248, 260–70
dispositional reductionism, 185, 186, 187, 213
dispositional vs. categorical properties, 184–88, 212
dispositions and manifestations, 188–89, 192, 193, 207, 209, 210, 214, 221, 237, 241, 244–45, 249, 253, 260, 263, 265, 269, 270, 272, 332n.7, 332n.11, 334n.27, 334n.33, 335n.39, 335n.42, 343n.84
dispositions vs. conditionals, 183, 332n.11
dispositions vs. powers, 182–83
and EM, 36–38, 137, 179–80, 248, 255, 260–70, 272
and essentialism, 37, 181, 182, 191, 237–41, 245, 249–50, 252, 347n.1 (chap. 7)
and event ontologies, 241–42
fragility as disposition, 183–84, 189
identity theory regarding dispositional and categorical properties, 185, 186, 187, 188, 213
and laws of nature, 244–45
neutral monism (dual aspect monism), 185, 186, 187, 188, 213
ontological dualism of dispositional and categorical properties, 184–86, 188, 212–13, 237

dispositional/powers metaphysics (*continued*)
   ontological monism of dispositional and categorical properties, 185, 186, 187, 188
   pan-categoricalism (categorical eliminativism), 185, 186, 187, 188, 213
   pan-dispositionalism (dispositional eliminativism), 185, 186, 187, 188, 189, 213, 234–37, 344n.92
   powers as universals, 183–84
   pure necessity and pure contingency avoided by, 183, 191, 213
   relationship to hylomorphism, 217–18, 234–37, 238, 239–41, 249–51, 347n.1 (chap. 7)
   relationship to potentiality and actuality, 182, 183, 207, 209–13, 237, 239, 245
   relationship to teleology, 213–16, 241
   and scientific realism, 242–44, 245
   and scientific research, 136–37, 181
   solubility as disposition, 183, 184, 189, 211–12
   and spontaneous change, 260–61, 263
   and subjunctive conditionals, 189, 332n.11, 334n.27
   and teleology, 213–16, 241, 252–53, 264–65, 338n.37
   transitive vs. intransitive powers, 184
dispositional view of causation (DVC), 26, 37–38, 137, 139, 179–80, 181, 189–99, 256, 259–60
   vs. Aristotelian views of causation, 37, 39, 199
   vs. AVC, 192, 248
   and causal modeling theory, 194–99, 213
   causal necessity as suppositional in, 191, 192
   vs. CVC, 190, 191, 192–93, 248
   and Deacon's dynamical depth model of EM, 261–62
   vs. INUSConVC, 190–91, 248
   and Kim's causal exclusion argument against DC, 88–91, 139
   and laws of nature, 193
   vs. MVC, 248, 335n.35
   and pleiotropic causation, 38, 192, 193, 213, 261
   and polygenic causation, 38, 191–92, 193, 213, 261
   vs. ProbVC, 192, 248, 335n.35
   vs. ProcVC, 248, 335n.35
   vs. RVC, 190–91, 192, 248
   vs. SVC, 192, 213, 248, 335n.35
dissipative structures, 100, 129
DNA, 111, 318n.7, 320n.44
Dodds, Michael J., 8, 286n.7
Dougherty, Jude, on hylomorphism, 323n.68
Dowe, Phil, on ProcVC, 172–73, 174, 331n.71
downward causation (DC)
   Donald Campbell on, 56, 71–72, 74, 75, 76
   causal factors in, 71–74, 75, 77, 78
   and circularity, 79–80, 81, 83, 84
   Craver and Bechtel on, 33–34
   Deacon on, 75, 103, 132–33, 258–59, 267–69, 273, 311n.21, 322n.66
   and essentialism, 245, 249–50, 252, 253
   examples of, 56–58, 76, 78–79, 252, 306n.39, 306n.41
   Hulswit's definition of, 71
   and hylomorphism, 249–51, 253
   and interdependency of causes, 253–54

and irreducibility of emergents, 27, 54–55, 57–58, 61, 63, 68–69, 76, 137, 249, 252, 253, 273
Kim's causal exclusion/systematic overdetermination argument against DC, 78–91, 126, 139, 248–49, 273, 312n.32, 313n.35, 314n.43, 314n.45, 315n.55, 330n.59
and mechanisms, 32–34, 35, 59, 60
medium downward causation (MDC), 92–93, 96, 316n.64
nature of, 71, 76–78
objects of causation in, 71, 74–76, 77, 78
and ontology of levels/levels of complexity, 50–52, 53–54, 58, 60, 63, 68, 92
and probability, 254–56
relationship to causal closure of physics, 76–78, 83, 85
relationship to efficient causation, 76–77, 91–93, 248–49, 251–52, 253–54, 267–68, 316n.59, 316n.64
relationship to EM, 27, 34, 35, 38, 55–59, 63, 71, 78, 89–91, 103, 129, 132–33, 245, 248–56, 258–59, 267–69, 272–73, 300n.1, 301n.5, 316n.64
relationship to final causation, 93, 96, 252–54, 272
relationship to formal causation, 91–92, 93, 96, 249–51, 253–54, 272
relationship to material causation, 93, 253–54
strong downward causation (SDC), 92, 93, 316n.59, 316n.64
and supervenience (SUP), 49–50, 60, 81, 83, 85–86, 88, 96

weak downward causation (WDC), 93, 96
Driesch, Hans, biological vitalism of, 118, 299n.114
Ducasse, Curt J., on single events and causation, 155–57
Dudley, John, 293n.62
Dupré, John
 on DC, 57
 on polygeny of events, 192
DVC. *See* dispositional view of causation

eddies, 78–79, 106, 131, 250, 306n.39
efficient causation, 64, 83, 247
 Aquinas on, 18–19, 254, 294n.70, 347n.6
 Aristotle on, 10, 11–12, 14, 15, 16, 93, 115–18, 119, 120, 127, 129, 130, 199, 208, 241, 249, 251–52, 253–54, 255, 257, 258, 263, 293n.53, 294n.70, 323n.70, 346n.106
 and causal reductionism, 21, 26, 35, 76, 100, 126, 168, 179, 248–49, 251, 269, 271–72, 338n.37
 Deacon on, 115–16, 119–20, 122, 123, 127, 128, 130, 257, 258, 263, 265, 266, 323n.68, 323n.70
 Descartes on, 21, 295n.77
 Emmeche, Køppe, and Stjernfelt on, 92, 93, 316n.59, 316n.64
 Presocratics on, 2
 relationship to DC, 76–77, 91–93, 248–49, 251–52, 253, 267–68, 316n.59, 316n.64
 relationship to EM, 91
 Shields on, 241, 346n.106
Einstein's special theory of relativity, 298n.107
electrons, 90, 171, 230, 231, 342n.78

virtual presence of, 224–25, 232, 340n.60
El-Hani, Charbel Niño, 93, 272
Ellis, Brian, 173, 239
  on causal powers, 135, 189, 333n.17
  on natural kinds, 238
  on natural laws, 135
  on ontological dualism of dispositional and categorical properties, 185–86
  on quantum physics and powers, 333n.17
Ellis, George, 72, 74, 75, 92
  on causally effective goals, 94
emergence (EM)
  characteristics of, 46–60, 301nn.4–5
  as diachronic, 48, 52, 79, 81–82, 83, 303n.15, 306n.41, 309n.8, 312nn.33–34
  direct vs. mediated, 86–87
  and dispositional/powers metaphysics, 36–38, 137, 179–80, 248, 255, 260–70, 272
  dynamical version of, 35–36, 38, 39
  examples of, 48, 56–58, 67, 78–79, 104–5, 106, 131–32, 250, 252, 306n.41, 308n.5, 319n.23
  and fusion, 59, 115, 307n.46, 313n.41, 315n.55, 322n.64
  and global organization, 68, 101–2
  historical development of, 45–46
  irreducibility of emergents, 27, 47, 53, 54–55, 57–58, 61, 63, 68–69, 76, 111–12, 137, 249, 252, 253, 273, 279, 301n.5, 306n.41
  and levels of complexity, 50–52, 53–54, 58, 60, 63, 68, 301n.5
  and mechanisms, 27–34, 35, 59, 60
  mereological top-down versions of, 35, 36, 38, 39
  and mind-brain problem, 46
  negative definition of, 64–65, 89
  nonadditivity of causes in, 46–48, 63, 301n.5
  nondeducibility of emergents, 53, 54, 63, 68–69, 301n.5
  nonpredictability of emergents, 53–54, 63, 68–69, 301n.5
  and nonreductionist physicalism, 68–71
  novelty of complex processes, entities, and properties in, 27, 48–50, 63, 250–51, 280–81, 301n.5, 306n.41
  process metaphysics of, 279–84
  and quantum physics, 280–83, 303n.15, 342n.76
  relationship to boundaries/constraining conditions, 35–36, 72, 74, 92–93
  relationship to DC, 27, 34, 35, 38, 55–59, 63, 71, 78, 89–91, 103, 129, 132–33, 245, 248–56, 258–59, 267–69, 272–73, 300n.1 (chap. 1), 301n.5, 316n.64
  strong EM, 27, 32, 34, 54, 59–60, 71, 89–90, 137, 247, 301n.5, 306n.41, 307n.50, 309n.8, 312n.30, 313n.42, 316n.64
  and supervenience (SUP), 31, 49–50, 60, 81, 88, 103, 129, 247, 303n.15, 309n.8, 318n.16
  as synchronic, 48, 52, 53, 79, 306n.41, 309n.8, 312nn.33–34
  as transformational, 59, 307n.46
  weak/epistemic EM, 34, 54, 59–60, 307n.50, 309n.8
  See also Deacon, Terrence; downward causation (DC); emergent dynamical systems; emergent entities/substances; emergent laws; emergent properties

emergent dynamical systems, 52, 57–58, 63, 64, 93, 245, 250–51, 253, 254. *See also* Deacon, Terrence
emergent entities/substances
  and chance, 254–56
  and efficient causation, 251–52
  emergent entities/substances vs. emergent properties, 48, 65, 66–68, 250–51, 301n.5, 308n.5
  and interdependency of causes, 253–54
  Kim on, 301n.5
  Lewes on emergent effects, 45, 47–48, 64, 65
  Peterson on, 301n.5
  and teleology, 252–53
  water as emergent substance, 48, 67, 250, 308n.5
  See also emergent dynamical systems; emergent laws; emergent properties
emergent laws
  Alexander on, 52
  Broad on, 52–53
  vs. emergent properties, 65, 67–68
  Kim on, 54
  Stephan on, 67–68
  Wong on, 50
emergent properties, 37, 46, 303n.15, 314n.45
  in Aristotelian metaphysics, 65–66
  causal powers of, 55, 73–74, 77–78
  and constituent ontology, 66
  Deacon on, 103, 104–5, 114
  vs. emergent entities/substances, 48, 65, 66–68, 68, 250–51, 301n.5, 308n.5
  vs. emergent laws, 65, 67–68
  as epiphenomenal, 55, 58, 85, 249
  escape reaction to pain, 79, 80, 83–86
  fusion of, 59, 115, 307n.46, 313n.41, 315n.55, 322n.64

  Humphreys on, 313n.41, 315n.55
  as irreducible, 27, 47, 53, 54–55, 57–58, 61, 63, 68–69, 76, 111–12, 137, 249, 252, 253, 273, 279, 301n.5, 306n.41
  Kim on functionalization of, 54–55
  mental property of pain, 59, 80, 83–86, 91
  as nondeducible, 53, 54, 63, 68–69, 301n.5
  as nonpredictable, 34, 53–54, 63, 68–69, 301n.5
  novelty of, 27, 48–50, 63, 250–51, 280–81, 301n.5, 306n.41
  and relational ontology, 66
  vs. resultant properties, 50
  Stephan on, 65–68
Emmeche, Claus, 78, 272
  on causal factors in DC, 73, 74, 91–93, 95
  on constraining conditions, 92
  on efficient causation, 92, 93, 316n.59, 316n.64
  on final causation, 292n.51
  on formal causation, 92, 93, 322n.66
  Hulswit on, 95–97
  on inclusivity of levels, 50–51, 68
  on material causation, 113, 288n.22
  on medium downward causation (MDC), 92–93, 96, 316n.64
  on object of causation in DC, 75
  on strong downward causation (SDC), 92, 93, 316n.59, 316n.64
  on weak downward causation (WDC), 93, 96
Empedocles, on Love and Strife, 3, 287n.12
empiricism, 23, 24, 25–26
Engelhard, Kristina, 184–85

epiphenomena, 135, 283, 284
  emergent properties as, 55, 58, 85, 249
  holistic epiphenomenalism, 223
  and INUSConVC, 142
essentialism
  of Aquinas, 18, 19, 294n.65
  of Aristotle, 6, 8, 9, 122, 123, 124, 128, 207–8, 214, 239, 240, 241, 245, 249, 255, 257, 258, 290n.32, 290n.34, 290n.36, 291n.39, 297n.98, 300n.1 (chap. 1), 327n.20, 344n.99
  and DC, 245, 249–50, 252, 253
  and dispositional/powers metaphysics, 37, 181, 182, 191, 237–41, 245, 249–50, 252, 347n.1 (chap. 7)
  and hylomorphism, 6, 37, 214, 237–41, 245, 250, 253, 255
  Kripke on, 344n.99, 345n.100
  Oderberg on, 240, 345n.103
  Popper on, 344n.93
  Putnam on, 345n.100
  Quine on, 344n.93
  Wittgenstein on, 344n.93
Euclid of Megara, 203
event ontologies, 241–42, 259, 269–70
evolution
  Donald Thomas Campbell on, 56–57
  Richard Campbell on, 312n.34
  Dawkins on, 112
  Deacon on, 106, 108, 110, 318n.7, 320n.45
  evo-devo field, 299n.112
  evolutionary biology, 71–72, 91, 271, 292n.51, 299n.112, 321n.53
  and natural selection, 111, 112, 130, 136, 318n.7, 320n.45, 344n.93

Fair, D., 172
Farrell, John, on Deacon, 118

Farrer, Austin, 73, 74, 75
Feser, Edward
  on consciousness and intentionality, 338n.36
  on possible worlds modality, 147, 327n.20
  on potency and act, 337n.19
  Scholastic Metaphysics, 275
  on teleology and panpsychism, 216
final causation
  Aquinas on, 19, 254, 347n.6
  Aristotle on, 9, 10, 12–14, 16, 19, 35, 36, 37, 78, 93, 94–95, 99, 117–18, 119, 121, 127, 128, 129, 180, 199, 206, 208, 216, 241, 247, 253–54, 255, 263, 265, 269, 272, 273, 291n.41, 292nn.43–45, 292n.51, 293n.53, 315n.57, 323n.74, 324n.76, 337n.17
  Deacon on, 35, 36, 127, 247–48, 320n.48, 324n.76
  Emmeche, Køppe, and Stjernfelt on, 292n.51
  Hulswit on, 95
  relationship to DC, 93, 96, 252–54, 272
  See also teleology
Fine, Kit, hylomorphism of, 220–21, 222, 233
formal causation
  Aquinas on, 18, 217–18, 254, 339n.41, 343n.85, 347n.6
  Aristotle on, 5–6, 8–10, 11, 14, 15, 16, 35, 36, 78, 87, 93, 94–95, 96–97, 115–18, 119, 121, 122, 124, 127, 129, 180, 199, 208, 217–18, 227–28, 247, 249–50, 253–54, 255, 258, 265, 269, 272, 273, 290nn.35–36, 291nn.38–39, 293n.53, 297n.98, 300n.1 (chap. 1), 315n.57, 323n.74

Deacon on, 35, 36, 103, 115–18, 122–23, 124–25, 127, 128, 130, 155–56, 247–48, 257, 258, 263–65, 322n.66, 323n.68, 323n.74
Emmeche, Køppe, and Stjernfelt on, 92, 93, 322n.66
Hulswit on, 95, 96–97
Leibniz on, 22
modern rejection of, 20–21, 126
vs. natural laws, 96–97
Plato on Ideas/Forms, 4–5, 8, 18, 117–18, 207, 344n.99
relationship to DC, 91–92, 93, 96, 249–51, 253–54, 272
fragility as disposition, 183–84, 189
Frank, Philipp
on causal laws, 142–44
Philosophy of Science, 142–44

Galileo
on book of nature, 295n.75
law of falling bodies, 148
Gánti, Tibor, chemoton of, 110
Garson, Justin, 338n.37
Gasking, Douglas, 159
Gaye, R. K., 8–9
genetics, 111, 192, 318n.7, 320n.44
Gestalt psychology, 58
Gilbert, William, 20
Gillett, Carl, 313n.35
on direct vs. mediated DC, 86–87
on Kim's causal exclusion argument against DC, 86–87
on machretic determination, 86–87
Gillies, Donald, on ArelVC, 162–63
Glennan, Stuart S., on causation and mechanisms, 30–31, 298n.105
God
Aquinas on, 18, 294n.70
Descartes on, 21
as efficient cause, 21
Leibniz on, 22

Malebranche on, 22
as pure act, 276, 337n.13, 345n.102
Grosseteste, Robert, 17

Hall, Ned, on double prevention, 150
Handfield, Toby, 189
Hardie, W. F. R., 8–9
Harvey, William, 20
Hegel, G. W. F., 45, 300n.1 (chap. 1)
Heil, John
on identity of dispositional and categorical properties, 187
on natural intentionality, 214–15
on powers and manifestations, 189
Heisenberg, Werner, 286n.7, 345n.103
Hempel, Carl, on deductive-nomological model of scientific explanation, 145
Hendry, Robin F., on causal closure of physics, 313n.42
Heraclitus on perpetual change, 3, 145, 202, 242, 270, 284, 287n.8
Higgs field, 230
Hippolytus, 286n.6, 287n.9
Hobbes, Thomas, 21, 22
homeostasis, 32
Hooker, Cliff, 304n.16
Hulswit, Menno
on DC, 71, 78
on efficient causation, 96
on Emmeche, Køppe, and Stjernfelt, 95–97
on final causation, 95
on formal causation, 95, 96–97
From Cause to Causation, 326n.5
on natural laws, 96–97
on process metaphysics, 95–96
on substance, 95, 96
Hume, David, 91, 179, 194, 209, 214, 248, 272, 303n.15
on connections between causal relata, 22, 23, 30, 168, 181

Hume, David (*continued*)
   counterfactual view of causation, 23, 26, 31, 35, 38, 139, 146, 190–91, 192, 194, 249, 298n.104
   on events as causal relata, 23, 35, 67, 136, 157, 169, 181, 189–90, 207, 213, 242, 245, 249, 261, 296n.82, 304n.22, 334n.33
   influence of, 23, 26, 31, 38, 139–40, 142, 146, 182, 190–91, 192, 242, 298n.104, 346n.111
   regularity view of causation, 23, 26, 35, 38, 136, 139–40, 142, 145, 147, 156, 190–91, 192, 193, 249, 296nn.82–83, 298n.104, 304n.22, 334n.33
   on substance, 208, 238, 239, 250
   on temporal priority of causes before effects, 23, 189–90, 213, 261, 334n.33
Humphreys, Paul
   on characteristics of EM, 301n.5
   on complexity of emergence, 309n.7
   on emergent properties, 313n.41, 315n.55
   on entangled states in quantum mechanics, 308n.1
   on fusion EM, 59, 115, 313n.41, 315n.55, 322n.64
   on generative atomism, 313n.41
   on heterogeneous vs. homogeneous realizability, 311n.12
   on inferential vs. conceptual EM, 59–60
   on Kemeny-Oppenheim reduction, 305n.30
   on Kim's functionalization of emergent properties, 55
   on levels of complexity, 50, 51–52, 313n.41
   on necessitation and supervenience (SUP), 49
   on physicalism and EM, 309n.8
   on physicalism and naturalism, 309n.8
   on supervenience (SUP), 49–50, 303n.15, 305n.30, 309n.8
   on synchronic EM, 79, 312n.34
   on transformational emergence, 307n.46
   on weak EM vs. strong EM, 59–60
hydrogen atoms, 224–25, 232, 236, 340n.60, 342n.78
hylomorphism
   dispositionalist/powers version of, 234–37, 239–41
   and downward causation (DC), 249–51, 253
   incomplete entities version of, 223–25, 226, 239
   mereological and structural versions of, 218–23, 225, 227, 229, 232–34, 236, 239, 250, 257, 264, 339n.44
   mixed version of, 225–28, 239
   re-identification of the parts version of, 228–29, 233–34, 239
   relationship to dispositionalism, 217–18, 234–37, 238, 239–41, 249–51, 347n.1 (chap. 7)
   relationship to essentialism, 347n.1 (chap. 7)
   versions in analytic metaphysics, 217, 218–41, 339n.39
   and virtual presence, 229–34, 342n.80
   See also Aristotle's hylomorphism

identity of indiscernibles, 66, 88, 208
indeterminism, 53, 88, 89, 114, 125, 154, 160, 164
inferability view of causation (InfVC), 142–44, 145–46, 151, 176, 271–72

vs. DVC, 248
vs. SVC, 156, 157
interventionist view of causation
(IntVC), 158, 163–67, 168, 178,
272, 329n.50
vs. AVC, 164, 166
vs. CVC, 166, 329n.56
vs. DVC, 248
vs. MVC, 166
Woodward on, 163, 165–66, 167,
329n.56, 330n.59
INUS-conditions view of causation
(INUSConVC), 141–42, 143,
145–46, 271–72
vs. DVC, 190–91, 248
and overdetermination, 142, 146,
149, 176, 192
Irwin, Terrence, 9, 205, 290n.34

Jaworski, William
on EM, 340n.52
on embodiment thesis, 340n.51
hylomorphism of, 221–22, 257,
342n.79
Joachim, Harold H., 290n.35
Johnston, Mark, hylomorphism of,
219–20, 221, 222, 227, 233,
339n.44
Juarrero, Alicia
on context-sensitive constraints, 72,
74, 92
on human intentionality, 311n.21
on object of causation in DC, 75

Kant, Immanuel
on the categories, 24
on causation, 24
on synthetic a priori judgments, 24
on things-in-themselves/noumena,
24, 242
on transcendental illusion, 24
Kauffman, Stuart, 108–9

Kepler's laws of planetary motion,
148
Kim, Jaegwon
on causal exclusion/systematic overdetermination argument against
DC, 78–91, 126, 139, 248–49,
273, 312n.32, 313n.35, 314n.43,
314n.45, 315n.55, 330n.59
on causal factor in DC, 73–74
on causal-power actuality principle,
79, 81–82
on causal powers of emergent properties, 55
on characteristics of EM, 301n.5
on circularity and DC, 79–80, 81,
83, 84
on conceptual interpretation of DC,
79
and dispositional views of causation, 88–91, 139
on emergent laws, 54
on emergent properties and supervenience, 49–50
on functionalization of emergent
properties, 54–55
on inductive vs. theoretical predictability, 52–53
on levels of complexity, 51
on mind-body causation, 79
on Nagel's bridge laws, 54
on nonreductionist physicalism, 69
on object of causation in DC, 75
on Putnam's multiple realizability,
70
on reduction, 54–55
on reflexive DC as diachronic vs.
synchronic, 79
on supervenience (SUP), 49–50, 70
on underdetermination, 150
Kirwan, Christopher, 9, 290n.32
Koonin, Eugene, on minimal cells,
110

Koons, Robert, 66, 332n.11
  hylomorphism of, 222–23, 233, 340n.54
  on laws of nature, 346n.111
  on sustaining instruments theory, 223
Koslicki, Kathrin, hylomorphism of, 218–19, 220, 221, 222, 227, 257, 342n.79
Koutroufinis, Spyridon
  on constraints, 103–4
  on dynamical depth, 99, 321n.50
  on selfhood, 106–8, 132–33
  on self-organization, 105
  on teleodynamics and efficient causality, 323n.70
Krebs cycle, 30
Kripke, Saul
  essentialism of, 136, 239, 344n.99, 345n.100
  on possible-worlds semantics, 136, 239, 344n.99, 345n.100
  on rigid designators, 344n.99
Køppe, Simo, 78, 272
  on causal factors in DC, 73, 74, 91–93, 95
  on constraining conditions, 92
  on efficient causation, 92, 93, 316n.59, 316n.64
  on final causation, 292n.51
  on formal causation, 92, 93, 322n.66
  Hulswit on, 95–97
  on inclusivity of levels, 50, 68
  on material causation, 113, 288n.22
  on medium downward causation (MDC), 92–93, 96, 316n.64
  on object of causation in DC, 75
  on strong downward causation (SDC), 92, 93, 316n.59, 316n.64
  on weak downward causation (WDC), 93, 96

laminar flow, 104, 132
Lancaster, Tom, on condensed matter physics, 306n.41
laws of nature, 24, 72, 169, 182
  Richard Campbell on, 304n.22
  as descriptive vs. prescriptive, 52, 68, 131, 159, 193, 257
  and DVC, 193
  vs. formal causation, 96–97
  Koons on, 346n.111
  relationship to natural kinds, 144–45, 244–45
  and RVC, 140, 144–45, 146, 326n.3
Leibniz, Gottfried Wilhelm, 145
  on God, 22
  Monadology, 22
  on preestablished harmony, 22
  on teleology and formal causation, 22
leptons, 230, 231
Leśniewski, Stanisław, 297n.98
Lewes, George Henry
  on emergent effects, 45, 47–48, 64, 65
  on nonadditivity of causes, 47–48, 64
Lewis, David, 135–36, 194, 303n.15, 343n.91, 346n.111
  on absences as causes, 150–51
  on counterfactuals, 146, 182, 327n.23
  on possible worlds modality, 147–48, 344n.99
  on unrestricted composition, 339n.39
Locke, John
  on bare substrata, 66, 208, 250
  on causal connection, 22–23
  on conditional analysis, 183
logical empiricism/positivism, 30, 46, 135
Losee, John
  on CVC, 148–49
  on INUSConVC, 141

on MVC, 158–59
on ProbVC, 152
on ProcVC, 171–72, 173–74
Lowe, Edward Jonathan
  on dispositional/occurrent distinction, 187
  on hydrogen atoms, 224–25, 342n.78
  hylomorphism of, 223–25, 227, 340n.57, 342n.79
  on ontological dualism of dispositional and categorical properties, 185–87
Lyon, Aidan, 146

Macdonald, Cynthia, and Graham Macdonald, on Kim's causal exclusion argument against DC, 85–86, 90
Mach, Ernst, phenomenalism of, 25
Machamer, Peter, 31, 298n.105
Mackie, John L., 146, 335n.41
Malebranche, Nicolas, on God as cause, 22
Manafu, Alexandru, 306n.41
manipulability-based views of causation (M-basedVsC), 26, 31, 139, 157–68, 177, 192, 248, 271–72. *See also* action-related view of causation (ArelVC); agency view of causation (AVC); inferability view of causation (InfVC); manipulation view of causation (MVC)
manipulation view of causation (MVC), 163, 166, 177, 271–72
  vs. DVC, 248, 335n.35
  Losee on, 158–59
Marmodoro, Anna
  hylomorphism of, 228–29, 233, 343n.91
  on physical vs. metaphysical unification, 90
  on power structuralism, 90, 343n.91
  on Rea, 235–36
  and strong EM and DC, 89–90
  on structural power vs. substantial power, 90
Martin, Charles Burton, 187
material causation
  Aquinas on, 18, 254, 347n.6
  Aristotle on, 5–8, 15, 93, 94, 113–15, 127, 129, 199, 208, 253–54, 288n.22, 289n.29, 293n.53, 297n.98, 300n.1 (chap. 1), 324n.94
  Emmeche, Køppe and Stjernfelt on, 113, 288n.22
  Presocratics on, 2–3
  relationship to DC, 93, 253–54
mathematics, 135, 253, 265, 288n.14
  differential equations, 25, 81, 92, 142–44
  Galileo on, 295n.75
  in modern science, 20, 64, 76, 243, 252, 271, 281–82, 295n.75
  and physicalism, 68–69, 76
  Plato on, 4–5, 17
  and probability, 255–56
  Pythagoras on, 4–5, 288n.14
  in quantum physics, 281–82
Mayr, Erasmus
  on exercise events/characteristic change processes (CCPs), 89
  on Kim's causal exclusion argument against DC, 80–81
Mayr, Ernst, 321n.53
McGivern, Patrick, on DC in physics, 306n.41
McKitrick, Jennifer, 189
McLaughlin, Brian P., on EM and nomological SUP, 303n.15
mechanisms, 296n.91, 325n.103, 338n.37
  and downward causation, 32–34, 35, 59, 60

mechanisms (*continued*)
   and EM, 27–34, 35, 59, 60
   levels of, 32–34
   and natural kinds, 238
   relationship to phenomena, 29–31, 32–33, 298n.104
   See also biological mechanisms
Megarian school of philosophy, 203–5, 213
Mellor, David Hugh, 187
Mendel, Gregor, laws of inheritance, 143
Menzies, Peter
   on AVC, 159, 160–62, 164
   on CVC, 153
   on free action and causation, 159, 160–61, 162
mereology, 126, 297n.98, 339n.39
   mereological and structural versions of hylomorphism, 218–23, 225, 227, 229, 232–34, 236, 239, 250, 257, 264, 339n.44
   mereological top-down versions of EM, 35, 36, 38, 39
Mertonians, 17
metabolic networks, 26
Milesians, 2–3
Mill, John Stuart
   on causation, 46–48, 64, 140, 156
   and EM, 45, 46–48, 64
   on heteropathic vs. homopathic effects/laws, 46–47, 52, 64
   on nonadditivity of causes, 46–48, 64
mind-body causation, 46, 56, 79–81, 136, 301n.5
modular substance theory, 66
molecular biology, 29, 32, 46, 57, 274
Molnar, George
   on conditional analysis of powers, 332n.11
   on dispositions and powers, 182, 237
   on ontological dualism of dispositional and categorical properties, 185–86
   on physical intentionality, 214–16
   on pleiotropy of causal powers, 192, 335n.41
   on polygeny of events, 192, 335n.41
   on properties as tropes, 332n.6
   on S-properties, 186
Moreno, Alvaro, 94, 272
Morgan, Conwy Lloyd, 46, 48
   Emergent Evolution, 58
multiple realizability, 70, 88, 247, 311n.12
Mumford, Stephen
   on categorical properties, 184–85, 187
   on causal dispositionalism, 189–90
   causal modeling theory of, 194–99
   on causal processes, 189–90
   on compositional vs. resultant powers, 195–98
   on demergence, 90, 315n.56
   Dispositions, 187, 334n.26
   on dispositions, 182–83, 184–85, 187, 189–90, 237, 252, 332n.11, 334n.26
   on emergent powers, 90
   on necessity of causal dependencies, 335n.37
   on strong EM and DC, 89, 90
Murphy, Nancey, 72, 73, 74, 75, 77–78, 92

Nagel, Ernest, on reduction, 54, 305n.30
natural kinds, 230, 267, 290n.34
   Boyd on, 238, 250
   and mechanisms, 238
   Quine on, 344n.93
   relationship to manifestations of powers, 189, 244–45, 269, 270

relationship to natural laws, 144–45, 244–45
natural selection, 111, 112, 130, 136, 318n.7, 320n.45, 344n.93
necessity
  Aristotle on, 17, 19, 255
  blind material necessity, 16, 17
  causal necessity as suppositional, 17, 145, 191, 192, 202, 203, 245, 255, 263, 335n.37, 335n.42
  conceptual necessity, 49
  Hobbes on, 21, 22
  Hume on necessary connection, 23, 181
  logical necessity, 49
  metaphysical necessity, 49
  necessary conditions and causation, 140–41, 142, 152, 154
  nomological necessity, 49
  Spinoza on, 22
Neoplatonism, 17, 45, 300n.1 (chap. 1)
neuron causal diagrams, 153, 194–95, 336n.47
neuroscience, 27, 136
  examples of DC in, 56, 58
Newton, Isaac
  on causality, 20–21
  constraints in Newtonian mechanics, 72
  laws of motion, 143, 321n.60, 322n.62
nominalism, 207, 238
nomological (law-based) sufficiency view of causation, 80
nonadditivity of causes, 46–48, 63, 301n.5
nonreductionist physicalism (NP), 78, 217, 221, 222
  and biological mechanisms, 29, 298n.100
  Davidson's anomalism of the mental, 69, 247, 310n.11
  defined, 43

and EM, 68–71
Kim on, 69
problems with, 68–71
Putnam's multiple realizability, 70, 247, 311n.12
Shoemaker's manifest and latent properties, 70–71, 247
versions of, 69–71, 247
See also downward causation; supervenience

occasionalism, 21, 22
O'Connor, Timothy
  on Kim's causal exclusion argument against DC, 83–85, 90, 313n.41
  "The Metaphysics of Emergence," 83–85
Oderberg, David
  on active and passive power, 239, 240, 345n.101
  on dispositions and essentialism, 240, 345n.103
  hylomorphism of, 239–40, 346n.105
  on laws of nature, 244
  on natural teleology, 216
  on qualitative vs. quantitative characteristics, 135
Ohm's law, 158, 177
Oppenheim, Paul, on deductive-nomological model of scientific explanation, 145
Orilia, Francesco
  on agent causation, 89
  on causal relata as events, 87–88, 315n.50
  on causal relata as tropes, 88
  on DC as noncausal, 314n.46
  on examples of DC, 57–58, 306n.41
  on IntVC, 330n.59
  on Kim's causal exclusion argument against DC, 87–89, 90, 330n.59
  on powers vs. exercise events, 88–89

O'Rourke, Fran, 9, 291n.39
overdetermination
   and compositional powers, 195
   and CVC, 31, 149–50, 192
   Kim's causal exclusion/systematic overdetermination argument against DC, 78–91, 126, 139, 248–49, 273, 312n.32, 313n.35, 314n.43
   neural causal diagram of, 194
   Putnam on, 317n.70
   and RVC/INUSConVC, 142, 146, 149, 176, 192
Oxford Handbook of Causation, The, 180
Oxford school, 17, 294n.63

panpsychism, 36, 127, 216
Paoletti, Michele Paolini
   on agent causation, 89
   on causal relata as events, 87–88
   on causal relata as tropes, 88
   on DC as noncausal, 314n.46
   on examples of DC, 57–58, 306n.41
   on exercise events, 88
   on IntVC, 330n.59
   on Kim's causal exclusion argument against DC, 87–89, 90, 330n.59
   on power events, 88–89
Parmenides
   on change as illusion, 3, 202, 284, 287n.9
   multiplicity denied by, 202, 270
Paul, Laurie Ann, on CVC, 147, 328n.24
Peacocke, Arthur, 72, 74, 75, 92
Pearl, Judea, on IntVC, 162–64, 167
Peirce, C. S., 95
Pemberton, John, 201, 209
per accidens/incidental causes, 16–17, 120–21, 155, 255–56, 293n.60
Pereira, Antonio Marcos, 93, 272

per se causes, 16–17, 120–21, 155, 255–56, 293n.60, 335n.42
Peterson, Gregory R., on characteristics of EM, 301n.5
phase space, 93, 117, 118, 257, 258
philosophy of mind, 46, 56
philosophy of science, 26, 91, 159, 259, 273
   anticausality position in, 175
   and Aristotle, 288n.22
   causal pluralism in, 175, 179
   limited scope position in, 175
   and scientific realism, 242
   See also analytic metaphysics
physical intentionality, 214–16, 267
physical monism, 69, 70, 76, 268
physics, 115, 127, 167, 174, 274
   vs. chemistry, 145, 236–37, 313n.42
   condensed matter physics, 306n.41
   elementary particles in, 224–25, 230–31
   examples of DC in, 57, 76, 306n.41
   and indeterminacy, 88
   vs. metaphysics, 236–37
   natural kinds in, 238
   relationship to other sciences, 45, 54, 56, 308n.1, 309n.8
   relativity theory, 136, 241–42, 298n.107
   See also quantum physics; thermodynamics
Pickavance, Timothy, 66, 332n.11, 346n.111
Pittendrigh, Colin, 321n.53
Place, Ullin
   on dispositions as intentional states, 214–15
   on physical intentionality (PI) vs. mental intentionality (MI), 214–15
Plato
   on Anaxagoras, 3
   Aristotle on, 5, 8

on causality, 4
on geometry and matter, 4–5
on Ideas/Forms, 4–5, 8, 18, 117–18, 207, 344n.99
on mathematics, 4–5, 17
on participation, 18
Phaedo, 3
on sensual experience, 4, 5
Timaeus, 4
pleiotropic causation, 38, 192, 193, 213, 261, 335n.41
Plotinus, 17, 45, 300n.1 (chap. 1)
Poincaré, Henri, conventionalism of, 25
Polanyi, Michael, on boundary conditions, 72, 74
polygenic causation, 38, 191–92, 193, 213, 261, 335n.41
Popper, Karl, 46, 187, 237
  on DC, 56
  on EM and nondeducibility, 53
  on EM and nonpredictibility, 53
  on essentialism, 344n.93
positivism, 24, 25–26
possible worlds modality, 173, 187
  abstractionism regarding, 148
  and AVC, 162
  concretism regarding, 147–48
  and CVC, 147–48, 149, 166, 168, 176, 192–93, 329n.56
  Feser on, 147, 327n.20
  Kripke's possible-worlds semantics, 136, 239, 344n.99, 345n.100
potentiality and actuality
  Aristotle on active vs. passive potencies, 202–4, 207, 211–12, 239, 242, 245, 337n.7
  Aristotle on actuality, 8, 9, 11, 126, 188, 201–3, 205–8, 209, 241, 273, 336n.2, 341n.74, 346n.106
  Aristotle on nonrational and rational active potencies, 203–4

Aristotle on potentiality, 6–7, 8, 9, 11, 15, 18, 114–15, 123, 126, 188, 201–8, 209–10, 211–12, 239, 241, 242, 245, 273, 283, 336n.2, 337n.7, 337n.10, 338n.23, 341n.72, 341n.74
  kinds of act, 277
  kinds of potency, 276
  logical vs. real potencies, 210, 276
  on possibility vs. potentiality, 204–5
  primary matter and potentiality, 6–7, 8, 9, 15, 18, 114–15, 123, 126, 205, 207, 212, 220, 224, 225, 228–29, 240, 265, 276, 283, 341n.72, 345n.103
  and quantum fields, 283–84
  relation between potentiality and actuality, 205–8
  relationship to causation, 9, 11, 120, 121–22, 154, 182, 183, 206–12, 241
  relationship to dispositional/categorical distinction, 188, 212–13, 237
  relationship to dispositions/powers, 182, 183, 207, 209–13, 237, 239, 245
  relationship to teleology, 208
  relationship to virtual presence, 19–20, 220, 227
  Scholastics on, 210, 275–77
  substantial form and actuality, 8, 15, 205, 207–8, 224, 225, 226, 227–28, 229, 231, 240, 241
  unmoved mover/God as pure actuality, 276, 337n.13, 345n.102
Powell, Alexander, on DC, 57
Price, Huw
  on AVC, 159, 160–62, 164
  on free action and causation, 159, 160–61, 162

primary vs. secondary causation, 11–12, 18–19
principle of individuation, 66, 88, 208, 235, 322n.63, 343n.85
probability and DC, 254–56
probability view of causation (ProbVC), 26, 142, 151–55, 177, 328n.30
　vs. AVC, 160
　vs. DVC, 192, 248, 335n.35
　vs. RVC, 152
process view of causation (ProcVC), 26, 139, 178, 272, 335n.35
　Dowe on, 172–73, 174, 331n.71
　vs. DVC, 248
　Salmon on, 30, 33, 168–74, 331n.68, 331n.70
Psillos, Stathis
　on RVC, 144, 156
　on SVC, 156
psychology, 27
　examples of DC in, 58
　Gestalt psychology, 58
Putnam, Hilary
　on causal closure of the physical, 317n.70
　on causality of the mental, 317n.70
　on essentialism, 345n.100
　on hidden structures of things, 239
　on multiple realizability/compositional plasticity, 70, 247, 311n.12
　on physical realization, 311n.12
Pythagoras, 4–5, 288n.14

qualitative vs. quantitative change, 32, 42, 70, 135, 253, 264, 297n.93, 323n.74
quantum physics, 104, 127, 151, 225, 313n.42, 330n.59
　Copenhagen interpretation, 25, 26, 114, 126, 196, 259
　Deacon on, 114, 126, 259, 321nn.60–61
　and elementary particles, 230–31, 280, 282, 283, 284, 325n.95, 333n.17, 342n.76
　and EM, 280–83, 303n.15, 342n.76
　entangled states, 57, 308n.1, 342n.76
　indeterminacy in, 88, 89, 114, 125, 345n.103
　and mathematics, 281–82
　metaphysical implications of, 279–81
　quantum fields, 114, 115, 125, 126, 136, 280–84, 321n.60
　and substances, 241–42, 269
quarks, 230, 231
Quine, Willard, 172, 221, 237
　on analytic and synthetic statements, 136
　on dispositions, 187
　on essentialism, 344n.93
　on natural kinds, 344n.93

Rea, Michael
　hylomorphism of, 234–37, 238, 322n.63, 343n.84, 344n.92
　on powers, 234–36, 238, 343n.84, 344n.92
　on principle of individuation, 235
　on science and metaphysics, 236–37
reductionist/eliminative materialism, 41–42, 69, 216–17, 221, 271–72, 274
　and Deacon, 36, 112–13, 117, 128, 257, 265, 267, 270, 325n.100
　See also causal reductionism; nonreductionist physicalism (NP)
referential opacity, 215, 216
regularity view of causation (RVC), 139–46, 158, 271–72, 326n.2 (chap. 4), 326n.4, 326n.5
　vs. CVC, 147, 148, 149, 150, 152
　vs. DVC, 190–91, 192, 248

and INUS conditions, 141–42, 143, 150
and laws of nature, 140, 144–45, 146, 326n.3
and necessary conditions, 140–41, 142
and overdetermination, 142, 146, 149, 176, 192
problem of similarity, 145, 157
Psillos on, 144, 156
and sufficient conditions, 141, 142
vs. SVC, 156
See also Hume, David; inferability view of causation (InfVC); INUS-conditions view of causation (INUSConVC)
Reichenbach's common cause principle, 169
relativity theory, 136, 241–42, 298n.107
RNA, 111
Ross, W. D., 8, 9
Rueger, Alexander, on DC in physics, 306n.41
Russell, Bertrand, on causal lines, 168
Russell, Robert, 322n.62

Sachs, Joe, 291n.38
Salmon, Wesley
"Causality: Production and Propagation," 170
on conjunctive forks, 169, 171
on interactive forks, 169, 170, 171
on mark transmission, 168–69, 170, 171, 172, 178
on perfect forks, 330n.64
on ProbVC, 328n.30
on ProcVC, 30, 33, 168–74, 331n.68, 331n.70
on simultaneity of cause and effect, 169, 170–71
Scaltsas, Theodore, 228
Schaffer, Jonathan, 326n.1 (chap. 4)

on causal overlapping, 154
Schlick, Moritz, 24–25
Scholastics
on potency and act, 210, 275–77, 337n.19
on virtual presence, 322n.64
scientific realism
Chakravartty on, 242–44
entity realism, 242–43, 244, 346n.107
relationship to laws of nature, 244–45, 346n.111
structural realism, 242, 243–44, 346n.107
Scott, Alwyn
on Aristotle's causes, 94, 113
on circular causal loops in DC, 81–83
on final causality, 292n.51
on Kim's causal exclusion argument against DC, 81–83, 90
on material causality, 113
The Nonlinear Universe, 94
Scotus, Duns, 205, 294n.63
Seibt, Johanna, 280
self-organizing processes, 100, 101–2, 103, 105, 131–32, 260, 319n.38
Shannon, Claude, on information, 101
Sherman, Jeremy, 131, 321n.61
Shields, Christopher, on efficient causation, 241, 346n.106
Shoemaker, Sydney, 187
on manifest vs. latent properties, 70–71, 247
on micro-manifest vs. micro-latent powers, 47
Silberstein, Michael, 94, 272
Simplicius
on Anaximander, 286n.7
on Anaximenes, 286n.6
on *archē* vs. *aitia*, 285n.2

simultaneity of cause and effect, 33, 192, 195, 262, 298n.107
  Salmon on, 169, 170–71
  vs. temporal priority of cause and effect, 189–90, 213, 261, 334n.33
singularity view of causation (SVC), 26, 155–57, 177, 271–72
  vs. DVC, 192, 213, 248, 335n.35
  vs. InfVC, 156, 157
  Psillos on, 156
  vs. RVC, 156, 157
Smart, John Jamieson Carswell, 46
Snell's law, 158–59, 177
Sober, Elliott, 328n.30
social science
  examples of DC in sociology, 58
  vs. natural science, 145, 167
  social neuroscience, 56
sodium cations, 232
sodium chloride, as soluble in water, 183, 189, 211–12
solubility
  as disposition, 183, 184, 189, 211–12
  of sodium chloride in water, 183, 189, 211–12
Sosa, Ernest, 153–54
Sperry, Roger Wolcott, 46, 78, 115
  on causal factor in DC, 73–74
  on DC, 56–57, 73, 75, 76, 306n.39
  on object of causation in DC, 75
Spinoza, Baruch
  on final causes, 112
  on necessity, 22
Srivastava, Alok, 116
statistics, 145, 167, 171, 175
Stephan, Achim
  on characteristics of EM, 301n.5
  on emergent events, 65, 67
  on emergent laws, 67–68
  on emergent properties, 65–68
Stjernfelt, Frederic, 78, 272

  on causal factors in DC, 73, 74, 91–93, 95
  on constraining conditions, 92
  on efficient causation, 92, 93, 316n.59, 316n.64
  on final causation, 292–93n.51
  on formal causation, 92, 93, 322n.66
  Hulswit on, 95–97
  on inclusivity of levels, 50, 68
  on material causation, 113, 288n.22
  on medium downward causation (MDC), 92–93, 96, 316n.64
  on object of causation in DC, 75
  on strong downward causation (SDC), 92, 93, 316n.59, 316n.64
  on weak downward causation (WDC), 93, 96
Storck, Michael, on substantial forms, 290n.33, 341n.69
Suárez, Francis, 205, 226
substance dualism, 188, 212–13, 222
substantial unity by alteration of components (UAC/unity by *alteratio componentium*), 218, 219, 220, 221, 224, 227, 228, 229, 233, 236
substantial unity by arrangement of parts (USO/unity *secundum ordinem*), 217, 218, 219, 220, 221, 224, 227, 228, 236
substantial unity by contact and bond (USCC/unity *secundum contactum et colligationem*), 218, 219, 220, 221, 224, 227, 228, 236
sufficient conditions and causation, 141, 142, 151, 154
supervenience (SUP), 69, 247, 303n.13
  and DC, 49–50, 60, 81, 83, 85–86, 88, 96
  and EM, 31, 49–50, 60, 81, 88, 103, 129, 247, 303n.15, 309n.8, 318n.16

Humphreys on, 49–50, 303n.15, 305n.30, 309n.8
Kim on, 49–50, 70
Suppes, Patrick, 151
surface tension, 104, 132, 250
symbiosis in biological systems, 56, 305n.35

Tabery, James
  on mechanisms and causation, 30–31
  on weak EM, 34
teleology, 3, 93, 136, 223, 247
  of Aristotle, 12–14, 19, 35, 208, 214, 241, 292nn.43–45, 292n.51, 293n.61
  Deacon on, 35, 99, 100–102, 111, 112, 118–20, 130, 252, 257, 258, 262–63, 264–65, 273, 318n.7, 320n.48, 321n.53, 321n.61, 325n.100, 325n.103
  and dispositional/powers metaphysics, 213–16, 241, 252–53, 264–65, 338n.37
  Leibniz on, 22
  modern rejection of, 20–21, 112, 126
  vs. reductive materialism, 216–17
  Silberstein on, 94
  as value-centered, 13
  See also final causation
Thales, on water, 2, 286n.5, 286n.7
theories of information, 100–101
thermodynamics
  Deacon on, 101–2, 104, 105, 108, 109, 112, 127, 131–32, 133, 257, 259, 260, 261–62, 268, 323n.74, 325n.100
  gradients of entropy, 101, 104, 131, 132, 133, 211, 257, 259, 260, 268, 325n.102
  increase in global entropy, 101, 104, 132, 133, 257, 259, 268

laws of thermodynamics as descriptive, 130, 257
maximum entropy production principle, 101, 109
Second Law of thermodynamics, 102, 108
thermodynamic equilibrium, 104, 105, 131–32, 133, 257, 259, 260, 262, 268
Tooley, Michael, 153–54
Tredennick, Hugh, 9, 290n.32

UAC. *See* substantial unity by alteration of components
Umerez, Jon, 94, 272
underdetermination, 149, 150, 176, 192
Unger, Peter, on mereological nihilism, 339n.39
unmanipulative causes, 161, 162–63, 166, 177, 178
USCC. *See* substantial unity by contact and bond
USO. *See* substantial unity by arrangement of parts

van Gulick, Robert
  on causal closure of physics, 77
  on object of causation in DC, 75
  on patterns of organization, 72–73, 74
van Inwagen, Peter, on restricted composition, 339n.39
Varzi, Achille, 297–98n.98
vector modeling of causes, 195–99, 213, 262, 266–67
Venn, John, 145
verificationism, 160, 183
Vienna Circle, 24
virtual presence, 240, 308n.5, 322n.64, 341n.69, 342n.76, 342n.79
  Aquinas on, 19–20, 218, 220, 227, 229–34, 295n.73, 341n.74, 342n.76

virtual presence (*continued*)
  of electrons, 224–25, 232, 340n.60
  and hylomorphism, 229–34, 342n.80
  relationship to potentiality and actuality, 19–20, 220, 227
  of sodium cations, 232
viscosity, 67, 104, 132
vitalism, 46, 118, 299n.114
von Wright, Georg Henrik, 158–59, 160, 329n.44

Waismann, Friedrich
  "The Decline and Fall of Causality," 25
Wallace, William, 5, 13–14
  on matter in modern science, 115, 322n.62
water
  eddies in, 78–79, 106, 131, 250, 306n.39
  as emergent entity, 48, 67, 250, 308n.5
  as having substantial form vs. accidental form, 67, 308n.5
  laminar flow of, 104, 132
  molecules of, 67, 104–5, 132, 172, 211, 308n.5
  sodium chloride's solubility in, 183, 189, 211–12
  surface tension of, 104, 132, 250
  Thales on, 2, 286n.5, 286n.7
  viscosity of, 67, 104, 132
Whitehead, Alfred North
  on mereology, 297n.98
  process metaphysics of, 36, 95, 100, 127, 259, 279
Wicksteed, Philip H., 9, 290n.32
Williams, Neil, on power holism, 335n.39
Wilson, Jessica, 47
Wimsatt, William, 34, 280
  on aggregativity, 297n.93
Witt, Charlotte, 204, 337n.7
Wittgenstein, Ludwig, 135, 174, 237, 344n.93
Wong, Hong Yu
  on causal closure and EM, 83, 85
  on emergent vs. resultant properties, 50
  on Kim's causal exclusion argument against DC, 83–85, 90, 313n.41
  "The Metaphysics of Emergence," 83–85
Woodward, James
  on fundamental physics and causation, 167
  on IntVC, 163, 165–66, 167, 329n.56, 330n.59

Mariusz Tabaczek is a researcher and lecturer at the Thomistic Institute in Warsaw, Poland.